American Warblers

DOUGLASS H. MORSE

American Warblers

AN ECOLOGICAL
AND
BEHAVIORAL
PERSPECTIVE

HARVARD UNIVERSITY PRESS

Cambridge, Massachusetts

and London, England · 1989

Typeset in Galliard and designed by Marianne Perlak.

Library of Congress Cataloging in Publication Data
Morse, Douglass H., 1938–
    American warblers: an ecological and behavioral perspective /
    Douglass H. Morse.
    p.   cm.
    Bibliography: p.
    Includes index.
    1. Wood warblers—North America.   I. Title.
QL696.P2618M67   1989   89-31622
598.8'72—dc20   CIP
ISBN 0-674-03035-4 (alk. paper)

Contents

Preface

Wood warblers (Parulinae) are in large part responsible for leading me to a career in biology and then shaping the directions I would later pursue. Like many bird-watchers who grew up in the northeastern United States, I have followed these birds avidly and taken their identification and study as a personal challenge. In this part of the country one might hope to see twenty-five or more species during a single day at the height of the spring migration. Many warblers remained to breed, enough so that they were among the most common of species in the countryside. I found warblers in the hedges that grew around the stone walls of the farm where I lived and in the nearby forests. In addition to familiar species such as Yellow Warblers and Yellowthroats, there were rare migrants, such as Mourning or Cape May warblers, that were easily missed. Getting those and other stealthy warblers was essential for a successful birding year, that is, one with a long list!

My early experiences were a natural background on which to rest my new-found scientific knowledge in later years. This book is an outgrowth of fieldwork that I commenced in the early 1960s. My first serious research on paruline warblers was partly the result of finding myself in the coastal forests of Maine with time to spare, and partly of reading Robert MacArthur's paper on warblers of the northeastern spruce forests, published in *Ecology* in 1958. MacArthur's study dealt with five species of *Dendroica* warblers that coexist during the breeding season. That now-famous paper was seminal in developing my interest in niche partitioning among species and in the mechanisms by which species fit into communities—in part, how communities take form. Although not primarily a theoretical paper, it was clearly of great importance to MacArthur's later work, which was among the most influential of that time and which still casts a long shadow. Aside from color, MacArthur's five warblers seemed as similar as peas in a pod. The result reported in his paper—that a precise pattern of niche partitioning took place—was considered so remarkable that verification seemed warranted, a procedure common to many sciences but one only too seldom pursued in ecology.

At the same time broader questions in biology—such as G. Evelyn Hutchinson's (1959) memorable "Why are there so many species?"—were being raised. In part Hutchinson's question was about how similar species coexist, which in turn limits how many can fit into a community. It is not surprising that MacArthur's work was of great interest to anyone intrigued by these issues. Although the basis for community structure remains a controversial and even contentious issue, the questions MacArthur addressed were then even more controversial than they are now. In the mid-1950s, the notion that niche partitioning could assume the myriad and sometimes subtle forms that we now take for granted was subject to challenge, although perusal of the literature from that time reveals that the crucial information already existed.

During the summer of 1962 I had the opportunity to work in a spruce forest near the ones studied by MacArthur. The habitat was similar to his, although the species composition of warblers differed somewhat. I quickly verified MacArthur's basic conclusion of niche differentiation among the species, but I realized that he had barely uncovered the possibilities that this system presented. Consequently I soon diversified my work to explore related topics, such as whether males or females of pairs partitioned their niches, the mechanisms by which partitioning among and within species was carried out, the singing behavior of the different species, and the behavior of adults and their recently fledged dependent young. I followed new fledglings and observed their participation in the mixed-species foraging flocks led by Black-capped Chickadees. In later years I studied the warblers of several small, spruce-clad islands near the principal study area, sites so small that each supported only a few of the mainland species. These islands made it possible for me to test the effect of missing species on those remaining in the depauperate habitats. Given the difficulty of experimentally manipulating these birds, such natural experiments were the best opportunities available for evaluating the role of interspecific interactions in resource-allocation patterns and species composition.

These projects, plus more limited investigations, gave me a chance to study a wide variety of warbler species; they are the basis on which I approached the task of writing this book. I set out to relate my results to current ecological, behavioral, and evolutionary theory and to incorporate other research on warblers, and different organisms as appropriate, into a synthesis. Although I focus on North America, I do so only because of the geographic distribution of wood warblers. My intent is much broader: to explore problems of general biological interest.

By now, the literature on warblers is a rich one, including studies of warbler species alone and ones that incorporate warblers into a broader treatise. It includes a plethora of research papers, as well as several major research monographs: on the Prairie Warbler (Nolan 1979), Kirtland's Warbler (Mayfield 1960; Walkinshaw 1983), and Golden-cheeked Warbler (Pulich 1976). I hope that by assessing this rich and varied literature in light of current theory I can throw new light on aspects of ecology and evolution. If this approach identifies new questions amenable to research on warblers or other comparable taxonomic groups, my effort will have been successful.

As befits such a beautiful and varied group, a number of general treatises have also been written about warblers (e.g., Chapman 1917; Griscom and Sprunt 1957). These books, encyclopedic in scope, covering the species one by one, or at least those that occur within a certain geographic area, contain relatively little detail about individual species. As a result, they offer little opportunity for synthesis. The same pattern generally holds in regional avifaunas; although a few pages may be devoted to a species, much of the same information is usually repeated in one source after another, each time in combination with the details of local distributions. Bent's "Life Histories of North American Warblers" (1953) brought together a vast amount of information published up to the time of its appearance, but it contains little in the way of synthesis, not surprising in light of its format of species accounts. In addition to these limitations, all the treatises are well over a quarter of a century old, and some are much older. In the last two decades an intense interest in organismal and community biology has grown and a wealth of new theories and ideas has arisen. It is my strong conviction that field biology is currently long on theory and short on testing of that theory. Clearly, the time is ripe for a synthesis of what we now have, so that we can set priorities for future advances. It is in that spirit that I have written *American Warblers*.

Birds, and warblers in particular, offer many advantages for the field ecologist. They are easily observed, for they are primarily visual and auditory species, as we are. Being active, diurnal vertebrates, warblers exhibit a wide range of behavior that can be readily studied and documented. Because of the great number of opportunities for observation, plus the profusion of closely related species, much interesting theory has been generated from studies of warblers (e.g., MacArthur 1972). Unfortunately, it is not easy to follow up these ideas with rigorous experimental studies, and thus many of the hypotheses have been subsequently tested on organisms suitable for manipulation.

Even though it is difficult to carry out manipulative experiments with birds, it is not impossible to do so. Birds provide examples of behavior-modified ecology that may not exist in many other animals, thereby warranting the extra effort of carrying out field manipulations with them. Many questions about warblers have been well mined already, and although further attention to them may fill in the life histories of individual species, their contribution to broader biological questions may be limited.

In some ways, writing this book places me in a rather awkward position. I advocate strongly the rigorous testing of hypotheses, but I also believe in airing ideas that have not been completely tested. The competition literature in general, and that involving bird studies in particular, has been severely criticized for what its detractors take to be a lack of rigor (e.g., Connor and Simberloff 1979). I applaud the recognition that alternative hypotheses must be tested, yet the lack of rigorous experimentation does not constitute evidence for the alternative. Several workers have argued that since so few experimental confirmations of competition exist from the field, one may conclude that it is not an important phenomenon (e.g., Connell 1975). This argument must also be challenged, as Schoener (1983) has done; further, Schoener pointed out that the number of experimental studies is rapidly growing. That does not mean that this subject is adequately explored for most major groups of organisms; competition may occur only intermittently (Wiens 1977; Dunham 1980). My own inclination, notwithstanding my past work, is that if researchers had made their work as rigorous as possible, this controversy would have dissipated by now. Still, it is easy to see how ecology's shift from a descriptive and comparative science to an experimental one caught many workers unprepared.

With this background in mind, I set out to interpret the literature in the most parsimonious way. Thus, I draw conclusions from nonexperimental studies, and ones that I might otherwise challenge, because they provide the best information available. Waiting for the best possible study to be done is often impractical if one's goal is to synthesize what is current in an active field. Therefore, wherever necessary I emphasize the conditional nature of my conclusions while using the results to the degree I found justified. A book of this sort offers a writer the luxury of some speculation and interpretation that might not always be appropriate in a research journal. By pointing out deficiencies I might goad someone into investigating questions in a different way!

I commenced this book during a sabbatical year I spent with the Animal Ecology Group at Uppsala University. There, basking in the unaccustomed luxury of time to think about some problems that had interested me for a long while, I found it possible to rough out the thrust of the book to follow.

First of all I thank Professor Staffan Ulfstrand at Uppsala for providing me with a stimulating and worry-free atmosphere in which to launch this project. I was able to air some of my early ideas with him and members of his group. I have since explored many ideas with my colleagues in the Ecology and Evolutionary Biology Group at Brown University, although they may not always have realized my ulterior motives. Martha Cheo's editorial pen proved most helpful when I was experiencing great difficulty in mastering the organization of subject matter. Many colleagues provided indispensable critiques of chapters of this book: Anthony W. Diamond, John T. Emlen, Jr., Millicent S. Ficken, Russell Greenberg, Thomas C. Grubb, Jr., Donald E. Kroodsma, Eugene S. Morton, Val Nolan, Jr., John R. Probst, C. John Ralph, Thomas W. Sherry, and Staffan Ulfstrand. I am especially indebted to Frank B. Gill for undertaking the huge task of critiquing the entire manuscript for me. Several others furnished useful information: William H. Buskirk, Frances C. James, Daniel T. Jennings, Nedra Klein, Kenneth Meyer, Helmut C. Mueller, David A. Spector, Robert W. Storer and Thompson Webb III. Howard Boyer and his colleagues at the Harvard University Press brought this project from raw manuscript to finished product. My field research on warblers was funded by the National Science Foundation.

Brian Regal did all of the artwork except Figure 3.8, which was drawn by Elizabeth Farnsworth. Raymond Paynter kindly made specimens available for illustrations from the bird collections of the Museum of Comparative Zoology at Harvard University.

I am indebted to the following publishers and individuals for permission to use materials in their care: Academic Press, American Institute of Biological Sciences *(BioScience)*, American Ornithologists' Union *(Auk, Ornithological Monographs)*, Ballière Tindall *(Animal Behaviour)*, Birkhäuser Verlag *(Experientia)*, Blackwell Scientific Publications *(Journal of*

Animal Ecology), Cooper Ornithological Society *(Condor)*, Cornell Laboratory of Ornithology *(Living Bird)*, Cranbrook Institute of Science, Ecological Society of America *(Ecology, Ecological Monographs)*, Macmillan *(Nature)*, Oxford University Press *(Journal of Heredity)*, *Scientific American*, Smithsonian Institution Press, Society for the Study of Evolution *(Evolution)*, Springer-Verlag *(Behavioral Ecology and Sociobiology, Oecologia)*, United States Forest Service, University of Notre Dame Press *(American Midland Naturalist)*, and Wilson Ornithological Society *(Wilson Bulletin)*.

American Warblers

1 Distribution and Evolutionary History

Traditionally, taxonomists place New World warblers within the great assemblage of nine-primaried oscines, allying them with New World honeycreepers, New World blackbirds, tanagers, and certain finches (Figure 1.1). Most New World warblers are small, insectivorous, and foliage-gleaning, and those characteristics are the group's distinguishing features, although a minority differ markedly from this description (Figure 1.2). Even in the temperate zone we find species that exhibit ecological (and often morphological) convergence toward the thrushes, nuthatches, and flycatchers. The thrushlike forms, notably the Ovenbird *(Seiurus aurocapillus)* and waterthrushes, have adopted a largely terrestrial existence: they walk over the ground, picking food from the substrate and hunting among dead leaves. The Black-and-white Warbler *(Mniotilta varia)* has become a trunk and large-limb forager, adeptly crawling about vertical surfaces in a manner suggestive of nuthatches. It has, though, retained the ability to glean adeptly in foliage, a trait of small nuthatches (Morse 1967c, 1970a). Several species habitually sally into the air column, capturing insects in the manner of true flycatchers. This adaptation is perhaps best exhibited by different species of redstarts *(Setophaga, Myioborus)*, unrelated species probably representing independent evolutionary convergences to this feeding style (Parkes 1961). Sherry (1979) and others have shown the importance of this foraging characteristic in the American Redstart *(Setophaga ruticilla)*, which has more ecological similarities to flycatchers than it does to some co-occurring species of warblers. Sherry includes the redstart within the flycatcher guild.

Clearly, what defines a warbler is not straightforward from an ecological viewpoint. Certain species are placed within the subfamily Parulinae by some taxonomists, in adjacent groups by others. Perhaps this state of affairs is not surprising, as the nine-primaried New World oscine group is one of the most explosively radiating avian groups at the present time (Sibley and Ahlquist 1983).

The very occurrence of such explosive evolution among parulines could

Figure 1.1 Representative nine-primaried oscines (Emberizidae) of subfamilies other than paruline warblers. *From upper left, clockwise:* Northern Cardinal (*Cardinalis cardinalis:* Cardinalinae), Common Grackle (*Quiscalus quiscula:* Icterinae), American Tree Sparrow (*Spizella arborea:* Emberizinae), Scarlet Tanager (*Piranga olivacea:* Thraupinae), Eastern Meadowlark (*Sturnella magna:* Icterinae), Bananaquit (*Coereba flaveola:* Coerebinae). (Illustrations by Brian Regal.)

Figure 1.2 Some examples of warbler (Parulinae) diversity. *From upper left, clockwise:* Black-and-white Warbler *(Mniotilta varia)*, American Redstart *(Setophaga ruticilla)*, Golden-winged Warbler *(Vermivora chrysoptera)*, Prairie Warbler *(Dendroica discolor)*, Ovenbird *(Seiurus aurocapillus)*, Wrenthrush *(Zeledonia coronata)*, Black-cheeked Warbler *(Basileuterus melanogenys)*, Common Yellowthroat *(Geothlypis trichas)*. *Center:* Yellow-breasted Chat *(Icteria virens)*. (Illustrations by Brian Regal.)

account for the difficulty in drawing tight lines around the group. In New World warblers we see a pattern that is not unusual among successful groups of organisms: a trend (here represented by small, brightly colored, foliage-gleaning species) accompanied by divergent pathways, which could be either ancestral or derived (usually the latter). The divergent characteristics of these "atypical" species or lines adapt them to different guilds (groups of species exploiting resources in the same way: Root 1967). My purpose in this book is not to provide an exhaustive review of paruline taxa, but to assess their ecological, behavioral, and evolutionary attributes in order to determine the impact of a monophyletic taxon, representing a certain life-style, on its environment.

Geographic Distribution

The New World warblers consist of about 130 species, depending on the authority cited. Although similar numbers of species of warblers breed in several segments of the Americas, number alone does not convey an adequate picture of their importance in different avifaunas. Many warblers that breed in eastern North America have wide ranges, the numbers of species of warblers in their communities are high, and local density is often high as well. Indeed, warblers can make up the single most abundant element of the avifauna, and in some places in eastern North America warblers' density exceeds that of all other birds combined. In some northeastern spruce-fir forests parulines, largely of the genus *Dendroica,* make up half or more of the species and three-fourths or more of the individuals in a local avifauna (Morse 1976b, 1977b). The impact is somewhat mediated by the birds' small size, yet it is impressive.

Although several species also nest in western North America, many are geographic replacements of each other, as in complexes (groups of closely related species) of the genera *Vermivora* and *Dendroica,* and the within- and between-habitat diversity found in eastern North America is not approached. Extensive geographic replacement of species by closely related ones does not occur in the eastern United States, perhaps because comparable opportunities for isolation do not exist (Mengel 1964). Western complexes have eastern counterparts, however. For example, the Black-throated Green Warbler of the *Dendroica virens* complex, a common, widespread species, is the eastern representative of that group, and three similar, largely allopatric species, separated either geographically or altitudinally—Hermit *(D. occidentalis),* Townsend's *(D. townsendi),* Black-throated Gray *(D. nigrescens)* warblers—are its western counterparts. The high diversity and extensive sympatry of the eastern warblers suggest that they are an ecologically more mature group than the western ones. Some eastern species may have invaded the west recently, as suggested by migration routes of Alaskan-breeding Blackpoll Warblers *(D. striata)* and western-breeding redstarts. Both species initially move eastward from these areas in migration, as if to retrace ancestral invasion routes.

Other warblers live solely in Central and South America, many of them confined to highland regions. The number of resident genera declines rapidly as one moves south through Mexico and Central America. South of Mexico, an increasing proportion of the resident species belong to the tropical genera *Basileuterus* and *Myioborus. Basileuterus* itself is composed of about 22 species, but because many of these are allopatric and similar to one another, there has been taxonomic disagreement about their status as species (AOU 1983). *Myioborus* (the tropical redstarts) has a similar distribution, and we can anticipate future shifts in the status of some of its species, also. In Central and South America, then, only one or two species of resident warblers are often found within a limited geographic area, where they are but a small component of a rich avifauna.

Considerable endemism occurs among resident West Indian parulines; at least four genera are confined to the area, sometimes to a single island. Warbler diversity at both regional and habitat levels is often intermediate between that seen in eastern North America and the tropical mainland. Warbler populations may be dense here, sometimes making up a large percentage of the avifauna (e.g., Kepler and Kepler 1970; Johnston 1975), probably as a result of island effects (lowered diversity, high density of certain species). Some workers (see Keast and Morton 1980) have argued that tropical residents, especially warblers, may be limited by the large numbers of wintering birds, especially warblers, that breed in North America. This as yet unresolved question is reviewed in Chapter 10.

During the nonbreeding season the geographic range of warblers declines considerably, as the temperate-zone breeders vacate their breeding sites: for the most part breeding and wintering zones of these species are exclusive of each other. These evacuations are the predictable consequence of the insectivorous habit. Those species occupying the highest latitudes during the winter have developed the ability to use other foods. The Yellow-rumped Warbler *(Dendroica coronata),* for instance, often subsists on Bayberry *(Myrica pennsylvanica)* and is confined in winter to where this plant grows. Even so, warblers evacuate Canada except for oceanic or

Great Lakes areas along its southern fringe; and most, if not all, species disappear from much of the northern part of the United States, with only low-diversity assemblages remaining in all but the southern fringe of the United States.

Evolutionary History

The relations of a few species usually placed with warblers are still open to question. Rather than dwell on these problems, I outline the most probable origin of the parulines and propose a likely evolutionary scenario for a major group of warblers, the eastern *Dendroica* species. I conclude with details of a probable microevolutionary shift in a warbler population. Such a shift may not result in speciation, but it can provide the raw variability that enhances the potential for eventual speciation.

Griscom (1957) has reviewed the early efforts of European taxonomists to classify the warbler specimens from the New World that first reached their collections in the 1700s and early 1800s. Before recognizing that these birds were distinct from Old World families familiar to them, these workers placed some of our commonest species in Palaearctic or Holarctic genera, such as *Motacilla* (wagtails), *Parus* (titmice), *Certhia* (creepers), *Turdus* (thrushes), and *Muscicapa* (Old World flycatchers). Alexander Wilson and John James Audubon began to refer them to the Old World warblers (Sylviidae), and 14 species were accordingly placed in the genus *Sylvia* before 1835. Recognition that these forms were not Old World warblers led, in 1838, to the proposal of a new family, Sylvicolidae, with species formerly referred to *Sylvia* being renamed *Sylvicola*. This new family did not receive immediate or enthusiastic approval. As recently as 1885 the great British ornithologist R. Bowdler Sharpe, in his *Catalogue of Birds of the British Museum*, accepted the family, but reluctantly, and predicted that it would be found an unnatural one. He suggested that *Vermivora* and the "brown warblers" were wrens, the Yellow-breasted Chat *(Icteria virens)* was a vireo, the American Redstart and other flycatching warblers were Old World flycatchers, and tropical forms were tanagers.

Although the core of the parulines has been generally accepted, regardless of its relation to other groups, the nomenclature has been confused, especially at the family level. Unfortunately, the genus *Sylvicola* was preempted, and these species were then referred to *Dendroica*, today the largest North American genus of paruline warblers. This change should have resulted in the family name shifting to Dendroicidae, but it was erroneously changed to Mniotiltidae, after the Black-and-white Warbler. The family name was subsequently shifted to the then-current name of the Northern Parula Warbler *(Parula americana)* and its allies, at that time Compsothlypidae; however, *Compsothlypis* was also preempted, so the genus became *Parula* and the family Parulidae. Although suffering an unfortunate past, the family (or subfamily) taxonomy is now somewhat more stabilized as a result of the 1961 International Code of Zoological Nomenclature (Stoll et al. 1961), which freezes the familial names in the instance of such generic changes in the future.

The shift to recognize the New World warblers as a subfamily is relatively new, being officially adopted by the sixth edition of the American Ornithologists' Union's *Check-list of North American Birds* (1983), the result of an ongoing effort to sort out the confusing mélange of nine-primaried oscines. The Check-list Committee's procedure was to merge the various New World nine-primaried oscines, formerly comprising several families, into one large, somewhat ungainly family, the Emberizidae *sensu lato,* and to recognize the former families as subfamilies. Given the uncertainty about relations among many species, it is unlikely that the current check-list gives a definitive designation for the warblers.

Boundaries of the Parulines

As with any group lacking a substantive fossil record, the evolutionary history of paruline warblers is unknown. Relations among the oscine birds (songbirds) are a matter of debate (Voous 1985). In recent years authorities have often separated the oscines, themselves identified by a distinctive pattern of syringeal musculature, into five groups: the larks, swallows, ten-primaried forms, nine-primaried forms, and corvines—a distinction that will suffice for our purposes. Although most appropriate taxa can be placed in one of these five groups, certain aberrant forms cloud the boundaries of these oscine groups and their principal divisions, especially the so-called nine- and ten-primaried oscine groups. The placement of certain taxa in either the nine-primaried or ten-primaried oscine group, and relations of certain taxa within these two great groups, pose some of the most difficult problems in avian taxonomy. Even the number of functional primaries may not always provide a reliable guide to relationship.

At least two controversies regarding paruline affinity have developed. The Olive Warbler *(Peucedramus taeniatus)* of the southwestern United

States and Mexico is similar in many ways to most insect-gleaning parulines; in fact, Webster (1958) found it to be so superficially similar to
Dendroica paruline warblers that he proposed merging it in that genus.
However, George (1962) has linked it with the parulines' ten-primaried
ecological equivalent, the Old World sylviines, on the basis of its hyoidean
musculature and several other traits. Another mystery species, the
Wrenthrush *(Zeledonia coronata)*, an enigmatic, skulking bird restricted to
the mountains of Costa Rica and western Panama, has usually been affili-
ated with ten-primaried groups because of its superifical similarities to the
thrushes (Turdinae), although it was considered distinct enough to warrant its own family (Zeledoniidae). On the basis of electrophoretic studies
(Sibley 1968, 1970) and a reassessment of its skeletal morphology (Hunt
1971), however, the Wrenthrush appears to be a paruline that has secondarily become adapted to a terrestrial existence in dense thickets.

Other problems of paruline higher taxonomy involve the assignment of
certain taxa within the subfamilies (or families) of the nine-primaried
oscines. Three different genera, the widely distributed Yellow-breasted
Chat, the tropical Rosy-breasted chats *(Granatellus* spp.), and the West
Indian White-winged Warbler *(Xenoligea montana)*, formerly placed in
Microligea with the Green-tailed Ground Warbler *(M. palustris)*, all are of
questionable affinity, but they are currently placed within the Parulinae by
the sixth edition of the AOU *Check-list* (1983).

Undoubtedly the most frequently questioned of these classifications is
the Yellow-breasted Chat, which has from time to time been proposed as
a member of totally different families, such as the New World blackbirds,
vireos, or honeycreepers. Even the superficial morphological and behavioral characteristics of this species are unusual. At about 18 centimeters (7.5
inches) and up to 25 grams, it is by far the largest of the warblers. Its nest,
eggs, down and molt patterns, mouth lining, song, courtship, and other
behavioral characteristics are also most unusual (Ficken and Ficken
1962b). Yet, in support of its inclusion within the Parulinae, it may be
linked by tropical species to more conventional paruline genera, such as
the yellowthroats *(Geothlypis)*. Furthermore, electrophoretic (Avise et al.
1980) and DNA-DNA hybridization studies (Sibley and Ahlquist 1982)
suggest that although most distinctive, Yellow-breasted Chats are more
closely related to warblers than to any other group. Both the Rosy-
breasted chats *(Granatellus)* and the White-winged Warbler *(Xenoligea
montana)* have possible affinities with the tanagers (McDonald 1987).

The precise relations within the warbler-honeycreeper-tanager axis are currently assumed to be especially close. New World honeycreepers had been treated as a separate family (Coerebidae) by some workers, but others merged them with warblers or tanagers. In the newest AOU *Check-list* (1983), the Bananaquit *(Coereba flaveola)* is placed in its own subfamily (Coerebinae), between the warblers and tanagers, and the remaining honeycreepers (four genera) with the tanagers.

The genus *Basileuterus,* a widespread tropical group, probably illustrates as clearly as any the difficulties of defining the taxonomic boundaries of the warblers. Some of these species (e.g., the Black-cheeked Warbler, *B. melanogenys*) are almost indistinguishable, morphologically, from the tanagers, such as the bush-tanagers of the genus *Chlorospingus* (see Moynihan 1962). *Basileuterus* cannot be treated as an unimportant sidetrack of paruline evolution, since it is divided by some workers into as many as 22 species (most are allopatric), ranging from Mexico to Chile and Argentina, making it one of the largest and most widespread of paruline genera. *Basileuterus,* if treated comparably to largely North American paruline genera, might well be divided into more than one genus. In fact, the AOU *Check-list* (1983) divides the *Basileuterus* within its range (Panama and northward) into two genera.

The paruline radiation has led to a profusion of genera, and Griscom (1957) has noted that the proportion of monotypic genera, nearly one-half in some classifications at the time he wrote, is higher than that for any other oscine family. Twenty-five genera were recognized then, and he introduced a scheme to reduce that figure to 18, but it has not gathered many adherents, nor has lumping in general. The AOU *Check-list* (1983) recognizes 27 genera, including the forms recognized by Griscom (some of questionable paruline ancestry) as well as the aberrant Wrenthrush. Given the current confusion over relations of these forms, no one is eager to revise the group radically in the absence of new information.

Area of Origin

Both Lönnberg (1927) and Mayr (1946) have postulated that the paruline warblers originated in the North American tropics, a prediction based largely on migration patterns (from temperate to tropical zones) and on the existence of a rather wide variety of resident genera in the tropics. When these birds likely evolved, the North American tropics were considerably larger than they are today. Although North America was separated from South America during most of the period between the Cretaceous-

Tertiary boundary and the Miocene, some nine million years ago, and its landmass did not extend south of Nicaragua, tropical or subtropical conditions—now confined to part of Mexico and Central America—prevailed for long periods over much of North America as far north as near the Canadian border (Mayr 1964). This distribution could account for the presence of most of the sedentary genera (no genera exist exclusively south of Nicaragua except for the Wrenthrush). It could also account for the migratory patterns of many of the current North American migratory species: these species are concentrated in Mexico and northern Central America, the number of species and individuals subsequently declining as one proceeds southward into southern Central America and South America (see Chapter 10).

The presence of two genera with a preponderance of species in South America, *Basileuterus* and *Myioborus,* and many fewer species of other genera there, suggests early colonizations of South America by warblers, or even prewarbler (Mayr 1964) stock, and a secondary invasion of Central America following reestablishment of a land bridge in the Miocene. The water gap between North and South America was not an absolute barrier; it may have had stepping-stone islands from time to time. The ancestors of *Basileuterus* and *Myioborus* may have crossed to South America in this way.

Possible Speciation Patterns

Although the North American landmass provided great expanses of tropical or subtropical habitat through much of the Tertiary, climatic patterns began to approach those of the present in the latter part of that epoch. The once-extensive broadleaf forests became restricted, to the extent that the southern Appalachian region likely was a refugium for them by the Pliocene. During the Pliocene, a period often associated with extensive speciation, a change to a less tropical climate probably divided the Appalachians into disjunct climatic areas at times.

Data on birds are equivocal, but Pliocene climatic changes could have been the driving force responsible for the high diversity of warblers in eastern North America today (Hubbard 1971). This explanation is consistent with Mengel's (1964) argument that many warbler lines evolved in the eastern North American broadleaf forests, which could account for the large numbers of both deciduous- and coniferous-nesting species found here today. Although Mengel (1964) maintained that Pleistocene glaciation played a major role in current warbler diversity, it seems necessary that much of this diversity arose in a pre-Pleistocene era. The southern Appalachians of the Pliocene may have provided the environment for the earlier speciation. If warblers were primarily adapted to deciduous forests, they would be able to radiate outward from these refugia into the invading coniferous woodlands. And, if this scenario were repeated at different times and in different places, there would have been more than enough isolated populations to explain today's unusual diversity. The greater length of the Pliocene period for speciation events to take place also enhances the possibility of extensive isolation. Closely related species-pairs often are allopatric, or exhibit interspecific territoriality if they are sympatric; the spruce-woods warblers show little sign of such constraints. Against that backdrop of several million years, the succeeding Pleistocene glaciation may have served as a filtering mechanism, possible enhancing both rapid evolution and the extinction of many lines, as previously isolated populations came together. Species such as the Kirtland's Warbler *(D. kirtlandii)* may represent analogous incipient casualties. In my mind there is no better explanation for the remarkable diversity of sympatric species among eastern parulines, whose diversity at a community level is matched by few other groups of temperate-zone birds.

A deciduous origin could also account for the ability of many coniferous-specializing breeders, the spruce-woods warblers, to prosper if forced to exist on broadleaf resources, as on their wintering grounds (Greenberg 1979, 1984b), and also to exploit outbreaks of defoliating insects on local deciduous vegetation in their breeding grounds (Morse 1970a, 1971b). It would also explain their ability to exploit sites that differ greatly from their summer coniferous haunts during migration (Morse 1980c). Some of these species readily occupy broadleaf habitats in the mountains of the southern Appalachians today (Brooks 1947).

Spruce-Woods Warblers: An Example of Speciation Patterns

Ornithologists have had more success in plotting evolutionary relationships at a lower taxonomic level, where the problems are those of speciation rather than formation of families or subfamilies. Mengel (1964, 1970) has proposed that several groups of North American warblers speciated in response to recurrent Pleistocene glaciation. The Black-throated Green Warbler and its allies (the *Dendroica virens* complex) (Stein 1962) nicely illustrate such a pattern. This group consists of several disjunct populations, most of which have reached probable species status but are still so similar that they collectively warrant the title of superspecies. Their differences are most easily explained by glacier-induced habitat changes that made intermediate areas periodically unfavorable and the divergent evolution that accompanied isolation (Figure 1.3). Mengel's (1964) model of warbler speciation proposed that the *virens* complex arose as a result of at least two episodes of Pleistocene glaciation; we can only guess how many populations were swamped during interglacials or exterminated during the glacials (Hubbard 1971). Most component species are largely or totally allopatric today, either geographically or altitudinally; where Her-

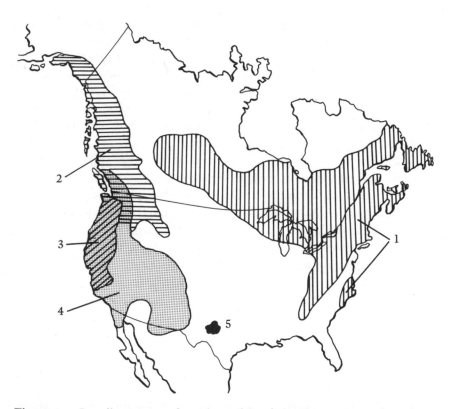

Figure 1.3 Breeding ranges of members of *Dendroica virens* superspecies group. 1, Black-throated Green Warbler, including disjunct southeastern coastal population (the *waynei* race); 2, Townsend's Warbler; 3, Hermit Warbler; 4, Black-throated Gray Warbler; 5, Golden-cheeked Warbler. (Modified from Mengel 1964, with more recent records added.)

mit and Townsend's warblers overlap in western Oregon, hybridization occurs (Burleigh 1944; Morrison and Hardy 1983).

In many areas the eastern representative of the complex, the Black-throated Green Warbler, lives largely or entirely in coniferous forests, but it occupies deciduous forests as well (Brooks 1947; Mengel 1964). Mengel argued that occupation of deciduous forests may be an ancestral tendency and that exploitation of coniferous forests may be a secondary trait. A small distinct race of the Black-throated Green Warbler *(D. v. waynei)* is confined to parts of the southeastern coastal plain. This distribution is consistent with Mengel's argument.

The Yellow-rumped Warbler exhibits another pattern that is probably a consequence of the glacial episodes. It is currently separated into eastern and western races, a pattern repeated by several other forest-dwelling species separated by the backbone of the Rocky Mountains. Where eastern and western populations come together at mountain passes, widespread introgression of the two races currently occurs, prompting Hubbard (1969) to lump the Myrtle (white-throated) and Audubon's (yellow-throated) warblers into a single species, the Yellow-rumped Warbler. Perhaps the more northerly extension of the Yellow-rumped Warbler's geographic range has prevented the isolation required for speciation into eastern and western species, as seen in the Black-throated Green Warbler complex. At present, the boreal coniferous forest provides a corridor around the north of the Great Plains. During Pleistocene glaciation, however, populations may have been separated by inclement conditions existing down the spine of the Rocky Mountains (discussed in Chapter 13).

Mengel has used the close similarities of the Yellow-rumped Warbler races to argue that the Townsend's and Hermit warblers, western disjuncts of the Black-throated Green Warbler group, branched off the main line prior to the most recent (Wisconsin) glaciation, as did the Golden-cheeked Warbler *(D. chrysoparia)* of the Edwards Plateau of Texas. Mengel believed the members of the Black-throated Green Warbler group to have differentiated further than the races of the Yellow-rumped Warbler. He also noted that other disjunct eastern and western populations of warbler species—Nashville *(Vermivora ruficapilla)* and Wilson's *(Wilsonia pusilla)* warblers—imply separation during the Wisconsin glaciation. These disjunctions also support a more remote separation of some members of the Black-throated Green Warbler complex since they have reached full species status since separation. He further suggested that the

Black-throated Gray Warbler represents an even earlier disjunction than the Hermit, Townsend's, and Golden-cheeked warblers, whose divergence he attributed to simultaneous or sequential disjunctions. This scenario implies that the divergence of *Dendroica* species above the superspecies level occurred before the Pleistocene. If this conclusion is valid, the rich diversity of species currently found in the eastern boreal forest may not be especially new, by passerine standards at least (see Olson 1985).

The Blackpoll Warbler, the most northerly breeding species of warbler distributed across this vast expanse of boreal coniferous forest, does not exhibit similar differentiation. Adaptations permitting it to exploit the most northerly areas may minimize the time that populations might become isolated into eastern and western segments. Although this explanation seems plausible, the closely related Bay-breasted Warbler *(D. castanea)* occupies a more southerly range in this forest but is not divided into more than one species or race, either. Hubbard (1971) suggested that these two species evolved in eastern (Appalachian) and western (Ozark–Northern Plains) glacial refugia, which expanded their ranges longitudinally, though today the species are usually contiguous rather than sympatric (Morse 1979).

Yellow-throated and Pine Warblers: A Possible Example of Rapid Evolutionary Change

Evolutionary changes in the paruline warblers may proceed far more rapidly than those noted above, but I have no evidence that speciation follows apace. The Yellow-throated Warbler *(Dendroica dominica)* usually breeds in deciduous forests of the southern and central United States; in some areas, such as the Delmarva Peninsula of Delaware, Maryland, and Virginia, east of the Chesapeake Bay, it occupies pine forests. Here, at the northeastern edge of its range, it occurs primarily in Loblolly Pines *(Pinus taeda),* though it sometimes strays into adjacent deciduous growth (Ficken et al. 1968). Beaks and skulls of these birds are some 10 percent longer than those of deciduous-adapted populations nearby. Further, birds collected 50 years earlier in the same area, when the Delmarva Peninsula was more heavily farmed and supported much less mature pine forest than at present, had beak dimensions similar to those of deciduous-breeding populations elsewhere. Ficken et al. (1968) noted that the only other population of this species with comparably long beaks is located on the panhandle of Florida, another locality where these birds inhabit pine forests. They also suggested the possibility of a similar trend commencing in North Carolina, in an area that is now primarily coniferous. Thus, the long-billed birds occupy pines and short-billed ones occupy deciduous forests; moreover, beak length in these populations has probably shifted rapidly. Even though one cannot eliminate the alternative that the Delmarva birds are an invading population, this explanation is unlikely, because the only forms with comparably long beaks inhabit the other end of their North American range, in western Florida. Since the species is widespread in collections, it is unlikely that the Delmarva birds represent a previously unknown race that moved into the now pine-clad area. If the rapid shift in beak length is real, one is prompted to ask what caused it.

The invasion of pinelands by Yellow-throated Warblers is of particular ecological interest: it is one of the few situations where more than one species of *Dendroica* warbler breeds in the southern Loblolly and Longleaf *(Pinus palustris)* pine forest. The unusual beak morphology of Delmarva Yellow-throated Warblers may facilitate coexistence with Pine Warblers *(D. pinus),* which are ubiquitous and abundant throughout most of the extensive range. When I observed recently returned Yellow-throated Warblers on cold, early-spring mornings it became apparent that the two species were segregated. It was too cold for insects to fly, so the Yellow-throated Warblers concentrated their activities on the old, open cones still retained by the trees, probing incessantly in them. When it became warm enough for insects to fly, the birds ceased their cone-probing and switched to flycatching (Figure 1.4). During these periods the Pine Warblers gleaned pine foliage in a way typical of most *Dendroica* warblers, but I never observed them probing cones. Thus, Yellow-throated Warblers seemed to be exploiting a refugium in the cones and perhaps coexisting with the Pine Warblers because of it.

With a study skin in hand, one can easily demonstrate that Pine Warbler skulls are too broad and their beaks too short to reach to the middle of the cones (Figure 1.5). In the same way one can verify that Delmarva Yellow-throated Warblers have beaks and skulls long and narrow enough to probe inside the open cones of loblolly pines, a prime hiding place for arthropod prey.

This apparent rapid shift of bill length by Delmarva Yellow-throated Warblers may be a consequence of character displacement (Brown and Wilson 1956) and has been cited as an example of it (Futuyma, 1979), yet Grant (1972) rightly warned that in order to admit it as a bona fide example, one needs to demonstrate that the long beaks of the Delmarva birds

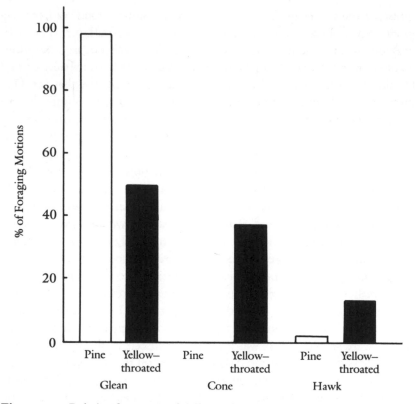

Figure 1.4 Relative frequency of different foraging movements (gleaning foliage, probing in cones, hawking for insects) by Pine Warblers and Yellow-throated Warblers in Loblolly Pine forests on the Delmarva Peninsula, Maryland. (Modified from Ficken et al. 1968.)

are more than part of a natural north-south cline in bill length for pine-dwelling Yellow-throated Warblers. Unfortunately, we cannot determine beyond doubt that the Yellow-throated Warblers on the Delmarva Peninsula in the 1960s were the offspring of the population present around 1900. Given the rarity of pine-dwelling races of Yellow-throated Warblers in this part of their range, however, it is likely that the present birds descended from those that occupied the area in 1900.

Individuals of the other primarily pine-forest population of Yellow-throated Warblers, in the panhandle of western Florida, also have extremely long beaks, but they are slightly shorter than those of the Delmarva birds (Ficken et al. 1968). Even so, one cannot easily invoke Grant's interpretation as an explanation, since the pattern is based on only two points. Further, the north-south cline need not eliminate the possibility of character displacement. Otherwise, one faces a problem in accounting for coniferous and deciduous races of Yellow-throated Warblers' existing side by side, exhibiting coinciding north-south clines in beak length—the longest-beaked population of deciduous-frequenting individuals, near the northern limit of the species' range, has a beak length shorter than that of the pine-adapted population near the southern limit of the range in western Florida (Sutton 1951; Ficken et al. 1968).

Why don't the pine-dwelling Yellow-throated Warblers glean foliage more heavily during inclement conditions, since their congener, the Pine Warbler, gleans in a fashion characteristic of most *Dendroica* warblers? One explanation is that Pine Warblers prevent Yellow-throated Warblers from doing so, thereby forcing them into the refugium of pine cones at critical times. Pine Warblers, extremely aggressive birds (Morse 1967c, 1970a), prevail over Yellow-throated Warblers in encounters between the two species (Ficken et al. 1968). Further, Pine Warblers are considerably

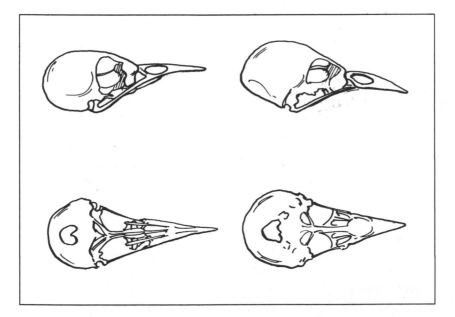

Figure 1.5 Skulls of Yellow-throated (*left*) and Pine (*right*) warblers, illustrating differences in length and breadth. (Modified from Ficken et al. 1968.)

more abundant in these areas than are Yellow-throated Warblers. Levels of aggression often are especially high at times of potential crisis (Morse 1980a), and aggression increases immediately following return to the breeding grounds, when territoriality is being established. Yellow-throated Warblers have only recently returned at such times to establish their territories. They are therefore highly mobile and have regular contact with conspecifics and Pine Warblers alike—the result: high levels of interaction (also see Morse 1976a). Because Pine Warblers reach high population densities in these areas, there is a strong potential for interspecific contact.

Why have Yellow-throated Warblers established distinct pine-dwelling populations in the Delmarva and west Florida sites and not in other places? First, both populations may be somewhat isolated: that certainly is true for the Delmarva population. Explaining the population in western Florida is difficult because we lack careful field studies. This area is on the edge of the species' range, however, and it may even be a curious disjunct. Being on the Gulf of Mexico, it is isolated to the south. Also, between these birds and their nearest conspecifics to the west (members of the central white-lored *albilora* race) is a gap as great as 300 kilometers (Burleigh 1944; Sutton 1951). The connection of the west Florida birds to more easterly Yellow-throated Warblers is not known (Sutton 1951).

The Dendroica petechia *Group: Clinal Variation*

The Yellow, Mangrove, and Golden warblers (currently all treated as members of *D. petechia*) exhibit the largest breeding range of any paruline species (Alaska to Galápagos, Labrador to Martinique), including both migratory and sedentary populations as well as congruent and isolated ones (Figure 1.6). These birds usually occupy marshy habitats and, to a lesser extent, low scrubby vegetation. Many North American Yellow Warblers nest about human habitations, such as gardens and orchards (Bent 1953), others in dry thickets of hawthorn *(Crataegus),* or even spruce forests on islands (Hebard 1961; Morse 1973). The populations in North America are nevertheless associated mostly with fresh water. The tropical species are largely coastal, however, and are usually associated with mangroves, although on some small islands they may extend into other habitats, especially arid scrub. Thus, the tropical populations are much more nearly linear in distribution (around the edges in a narrow fringe), and

Figure 1.6 Breeding ranges of Yellow Warblers: "typical" Yellow Warblers (area within heavy solid line), "Golden" Warblers (areas within dashed lines, plus the coastal area in western South America, traced with a dashed line), and "Mangrove" Warblers (coastal areas traced with cross-hatched line).

opportunities for isolation would consequently seem much greater. So, too, would islands facilitate isolation, as in the West Indies.

Variation, most visibly in the amount of chestnut coloration about the head (Figure 1.7), is for the most part clinal. North American nesting populations (the true Yellow Warblers, formerly *D. aestiva*) have no dark coloration on the head. Members of the so-called Golden Warbler group (sometimes designated as *D. petechia, sensu stricto*), most of which have a chestnut-colored cap (pileum), reside along most of the chain of the Greater and Lesser Antilles (and, recently, Florida Bay and the Keys) and hence westward along the South American coast to Venezuela. To the west and along both coasts of Central America and Mexico, resident *D. petechia* are characterized by a chestnut-colored hood and are called Mangrove Warblers. They have in the past been recognized as a separate species, *D. erithachorides*. Yet a strong cline between hooded and capped forms occurs in the Paraguaná Peninsula of Venezuela, where *D. petechia*'s range is discontinuously confined to mouths of rivers that reach the sea in an otherwise desertlike area.

Two populations do not conform to these clines, implying that the pattern is not as simple as I have suggested. One of these populations occupies the island of Martinique, in the midst of the Lesser Antilles. It is a hooded form in the midst of capped populations: the nearest hooded forms are well over a thousand kilometers away, to the west-southwest. An invasion of these sedentary birds from northwestern South America or from the Central American mainland seems highly unlikely. Although the Martinique birds might be the last remnants of an earlier radiation of hooded forms, the overall distribution of the Mangrove and Golden war-

Figure 1.7 Head coloration of male Yellow Warblers from different groups: "typical" Yellow Warbler *(left)*, "Golden" Warbler *(right)*, and "Mangrove" Warbler *(center)*. Dark areas are chestnut-brown, light areas are bright yellow. For breeding ranges, see Figure 1.6. (Illustrations by Brian Regal.)

blers renders this explanation unlikely. It seems more likely that the trait for feather color about the head is controlled by only one or a few genes and that the Martinique birds are an independent development of a "mangrove" form (see Remsen 1984). Initial analyses of mitochondrial DNA by Nedra Klein (pers. comm., 1988) suggest that the Martinique birds are closely related to the chestnut-capped birds of adjacent islands but less closely related to geographically distant *D. petechia* populations. The existence of Golden Warbler populations in the Greater Antilles with almost no chestnut cap feathers further suggests the lability of this character. The data support the present procedure of treating all members of this group as a single species; the different forms are then considered much more akin to color morphs.

The second population that differs from the color clines of the complex inhabits the Galápagos and adjacent South American mainland. These birds, on the basis of coloration, are Golden Warblers, although the closest populations, on the Pacific coast of Panama and northern Colombia, are Mangrove Warblers. Similarly, the Cocos Island population is composed of Golden Warblers, although the nearest mainland birds, along the Pacific Coast of Central America, are Mangrove Warblers. Perhaps these birds have shifted from the mangrove to the golden type in a manner analogous to, but in the opposite direction from, the Martinique birds.

Warblers as a Model for Studies in Evolutionary Ecology

The high local and regional diversity of warblers in the eastern United States and Canada has long interested zoogeographers and ecologists. Pleistocene glaciation probably played an important role in producing this diversity, but even prior to that the southern Appalachians may have been a key area of differentiation, presenting alternating deciduous and coniferous habitats as a consequence of climatic fluctuation. Speciation patterns in the southern Appalachians could be responsible for the development of coniferous-nesting species from deciduous-dwelling ones. More recent events associated with episodes of Pleistocene glaciation probably account for superspecies complexes, whose component species and populations are largely or totally allopatric and quite similar to each other. The superspecies complexes and more strongly differentiated, but still similar, cohabiting species suggest that the assemblages are not young on an evolutionary time scale and that high diversity may have lasted a long time. The rationale behind this conclusion is that the superspecies complexes have resulted from episodes of glaciation and that the more strongly differentiated forms must predate them.

If we can accept this putative calibration of relative species age, the period of distinctness and coexistence of the rich eastern contingent is likely to be relatively long, perhaps a few million years old. This conclusion is noteworthy because it suggests that this diverse assemblage is not new but has been relatively stable over long periods of time. Considerable environmental change has happened over this time, and species ranges probably moved alternatively south and north in response to the waxing and waning of glaciers.

This pattern is of theoretical importance. It enhances our understanding of community diversity over time and our perspective on the diversity of members in a closely related group. Traditional thinking (Odum 1969) has supposed that diversity generates stability, yet subsequent arguments have made prediction and causation much less clear (e.g., May 1973). The coexistence of numerous congeners, especially among large and active vertebrates, is unusual; its rarity could argue that such relations are unstable. Evidence from warbler populations suggests the possibility that such relations need not be unstable. If so, it raises the question of why such assemblages have not arisen in the temperate zone more frequently, and what special conditions have favored them in the parulines of eastern North America. For example, this relation differs from that of the parulines' European ecological equivalents, the sylviine warblers (Cody 1985; Chapter 15).

Nevertheless, under certain conditions evolutionary changes can take place among paruline warblers much more rapidly than the patterns I described, although it is not clear that they are of significance for speciation. The Yellow-throated Warblers of the Delmarva Peninsula suggest that rapid directional selection contributes to distinctness. Changes in beak length in Delmarva warblers are large in relation to the level of morphological differences in long-separated species, which partition more by space than by feeding mechanisms; as a result, they may have opened a new resource to the Yellow-throated Warblers: insects in Loblolly cones. Although one cannot assume that this change represents a significant step toward speciation, it indicates that the *Dendroica* genome is capable of sudden changes. This should not be surprising, from what we know about patterns of change in the evolutionary record. These patterns often imply that most change in a lineage comes in short bursts, with longer periods of stasis interspersed, in a manner resembling the punctuated-equilibrium scenario of Eldredge and Gould (1972). But they need not require the invocation of any special evolutionary processes; perhaps they reflect what happens when new ecological niches open (Wright 1982). For our purposes, this example helps to explain how other episodes of ecological change, such as those of the Pleistocene and earlier, could have generated the diversity seen today. Note that the deciduous-coniferous shift in the habitat of the Delmarva Yellow-throated Warblers is analogous to the one proposed for warblers in the southern Appalachians (Hubbard 1971). The pattern I have proposed for the Yellow-throated Warbler could have occurred again and again in the latter region. Further, such episodes may have required only a short period for evolution to take place.

The tendency of Yellow-throated Warblers in other pine areas to develop long bills, a trait found in other coniferous-dwelling races of *Dendroica* as well, suggests that for this trait the group responds to similar selective pressures in a predictable way. This morphological character can be easily tied to ecological variables. By contrast, the Yellow Warbler group exhibits a patchier pattern of color variation, conforming to broad geographic trends; some populations exhibit head colors characteristic of populations hundreds or thousands of kilometers away. These differences may reflect drift acting on a few genes in small island populations, or color variation may be under little selective pressure where the birds are in depauperate avifaunas such as the Galápagos.

These examples illustrate some of the opportunities that paruline warblers provide for testing important ideas in evolution and ecology. Warblers' wide range of diversity, abundance, and population size; their extensive geographic variation; the substantial span of phylogenetic distances between recent geographic isolates and long-established yet ecologically similar species; and the apparent differences in rates of evolutionary change all provide an unusually rich opportunity for analysis. Indeed, warblers may be a model group for many such studies.

2   The Breeding Season

Warblers that nest in the temperate zone are highly migratory birds. Their year can be separated into distinct periods: breeding (summer), migratory (fall and spring), and wintering. The biology of warblers is most often studied in the breeding season, because the birds are nesting in North America, where the majority of New World ornithologists work, and most researchers are free for fieldwork in the summer. Warblers' vocal and visual displays at this time add to their conspicuousness and facilitate study.

Unfortunately some workers have interpreted the birds' life histories largely or entirely on the basis of summer study. Important as this period may be, most migratory warblers spend no more than one-third to one-quarter of their time on the breeding ground (e.g., Schwartz 1980), and they are also likely to face serious problems during migration and on the wintering grounds. Several authors (e.g., Lack 1968; Fretwell 1972) have argued that the winter season plays the greatest role in limiting populations of birds that spend this period at higher latitudes. Is it possible that the arduous exercise of migration has liberated warblers from the rigors of winter? This is unlikely, but regardless of the answer, the question deserves attention. If these birds are regulated mainly on their wintering grounds or migration lanes, or if all three of the warblers' habitats have an impact on population dynamics, interpretations of breeding biology might take a different view from the current one (Morse 1980c). Only the most recent work (e.g., Keast and Morton 1980) is beginning to illuminate the lives of migrants at other seasons, yet there are enough studies to demonstrate that the predominant attention paid to the breeding season may result in a seriously oversimplified view of warbler biology.

Despite these reservations, I find it reasonable to commence this study of New World warblers with the breeding season because it is the start of a bird's life. And since most of our information comes from this period, we can begin to identify questions we will want to keep in mind about the other seasons.

Even the quality and quantity of information from different parts of the breeding period vary. Although foraging and other activities have been intensively studied during the period before the young fledge, we have very little information on the same individuals during the molt that immediately follows. This state of affairs is not surprising: during most of the breeding season warblers are highly active and easily observed; during the molting period they tend to be inactive, silent, and hard to observe. These problems do not diminish the importance of data from the "quieter" moments of a bird's life, but merely target them as topics for future emphasis. In spite of obvious gaps we have a great deal of evidence from warblers' life histories with which to test several basic ecological questions. Warbler studies provide some of the most detailed information on niche partitioning and species packing (MacArthur 1958, 1968), subdivision of the niche by males and females (Morse 1968), and the social attributes of guild members (Morse 1971a, 1976a).

Probably more anecdotal information exists on warbler reproduction than for any other aspect of their biology. A myriad of papers and notes document where nests are built, how long incubation takes, and the like. Much of this literature is compiled in Bent (1953) and Verner and Willson's (1969) bibliography. Yet it is striking how little quantitative information describes basic aspects of warbler reproductive biology: whether males contribute to nest building or incubation, whether warblers rear more than one clutch, or whether they engage in polygamy. Further, some of the "information" is almost certainly erroneous, partly as a consequence of the similarity in the sexes of some species or of unwarranted assumptions. Some observations are simply reported without further comment. A quote from Williams (in Bent 1953) about Hooded Warblers *(Wilsonia citrina)* is illuminating in this regard: "As to whether the male ever assists the female in incubating the eggs I am unable to say. Early in my acquaintance with the Hooded Warbler I thought I saw a male in the act of incubating the eggs in the nest, and I so recorded it. As I gained experience and familiarity with the species, I noted that some females had much more black on the head than others, and I am not sure that the incubating birds may not have been one of these well-marked females." It is difficult to eliminate dubious reports, although as new data accumulate unusual or unlikely observations can be more critically evaluated. Fortunately, the exhaustive analysis by Nolan (1978) of the Prairie Warbler *(Dendroica discolor)* and, for some aspects of the breeding cycle, those by Mayfield (1960) and Walkinshaw (1983) of the Kirtland's Warbler provide yardsticks upon which to base future comparisons. This state of affairs accounts for what may seem an unduly strong reliance upon these important works.

Arrival at the Breeding Grounds

Northern breeders vary in arrival times on their nesting grounds: earliest and latest species differ by well over a month (Figure 2.1). This difference is a function of the species' habits and their migration distances. Moreover, the two variables may not be independent of each other. For instance, Yellow-rumped Warblers are able to subsist in northern climes

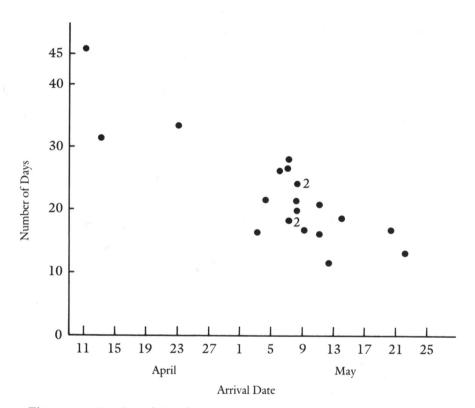

Figure 2.1 Number of days from arrival date to nesting date for several species of Maine warblers. The number 2 is used to indicate two points located at the same spot. The one-tailed Spearman Rank Correlation (r_s = 0.70) between these two variables is highly significant (p = 0.001).

over the winter on berries and other noninsect material, which allows them to remain near the breeding grounds; they are among the earliest of spring migrants. By contrast, some conspecifics winter as far south as Central America, which may account for the species' continued presence south of the breeding grounds in the spring migration.

At the opposite extreme is the Blackpoll Warbler, probably the most southerly wintering species of warbler that nests in North America. The peak of its migration generally passes through eastern North America after movement of most other species has nearly ceased. This species has the most northerly breeding range of any strictly forest-dwelling *Dendroica* warbler, in the boreal forests to the tree line, with the result that it is unable to claim its breeding areas until relatively late in the spring. The factors of distance and time are also clearly correlated for these birds. Their movement over the last part of their northward journey is rapid; although they move a conventional 50 kilometers per day through most of the United States, those that nest in Alaska often average 325 kilometers per day across Canada (Bent 1953).

By arriving first on their breeding grounds, migrants maximize the probability of finding high-quality territories or of reclaiming favorable sites (Nolan 1978). Balanced against this advantage, a conservative strategy of arriving later on the breeding grounds minimizes dangers of severe early-season weather or other threats. This selective pressure is well illustrated by occasional massive die-offs of early-arriving insectivores as a result of late snowstorms, unusual cold, or other inclement conditions (e.g., Griscom 1941). But selection for arrival time may act more strongly on the few earliest individuals in normal years than on the majority of a population at risk of meeting an occasional catastrophe. The actual arrival time is presumably a compromise between these two conflicting selective pressures.

Arrival dates of a given species vary considerably, even within a population, and are mostly associated with sex and age. Males arrive on their breeding grounds before females, and older birds before first-year birds (reviewed by Gauthreaux 1982). Initially, female arrival dates may not appear to be under selective pressure comparable to those of males, since the males claim the territories. Yet females obtain other advantages by arriving early, some of which are shared with males. Since females probably select the site upon which they nest, and only incidentally the male that goes along with it (Nolan 1978), they stand to gain from having the choice of the lot, in the same way as the males do. Both males and females exhibit site tenacity (a tendency to occupy the same area in successive years), but females do not have as high a return rate as males. In the few studies with a large enough banded population to discern this difference (Nolan 1978), females that did not reclaim the same site often could not do so because another female had arrived first and had formed a pair bond with the male territory holder, who was frequently the latecomer's mate of the preceding year. If the site was of high quality, or if the territory holder was a successful male, the female might incur a loss in fitness by not arriving sooner. In addition, early-arriving females would share with males the advantage of maximum time for renesting in case of failure, or even second broods (Nolan 1978), and the possibility of synchronizing nesting with early flushes of arthropod prey. These shared advantages suggest that the females' later arrival is not a mere consequence of selective pressure on males to arrive as early as possible.

Female warblers are enough smaller than males (Ridgway 1902) that later arrival times may be a selective advantage for them; unfavorable area-volume ratios will exacerbate heat loss by small individuals. Differences in arrival time by males and females may also be due to differences in the distances between their breeding and wintering grounds. In some species it is now clear that males and females winter in different geographical areas. For instance, male Dark-eyed Juncos *(Junco hyemalis)* winter significantly farther north, on average, than females (Ketterson and Nolan 1976, 1985), and this pattern could be widespread. I know of no relevant data on warblers.

I am also unaware of systematic efforts to compare fat conditions and other important correlates of bodily conditions among male and female warblers that have just arrived on the breeding grounds, although Nolan and Mumford (1965) found that male and female Prairie Warblers that had just crossed the Gulf of Mexico in spring did not differ in the amount of fat they carried. It is well known that some nonpasserines breeding in the Arctic bring most of the resources necessary to lay a clutch of eggs rapidly, before the local conditions would permit the acquisition of such resources (e.g., Ross' Goose, *Chen rossii:* Ryder 1970). Bringing one's own resources is essential for reproductive success in those birds, and similar, if not absolute, advantages apply for warblers if they can push forward their nesting dates.

Shorebirds in which females are larger than males and establish the territory migrate first (Oring in Gauthreaux 1982). Although it did not separate the roles of size and behavior in the timing of arrival, this study suggests that special egg-laying needs do not hold back migration in these species. Sexual role reversal does not occur among warblers, however, so we cannot test whether male warblers arrive at the breeding ground before females because of their greater size.

Age-related differences in arrival times of the American Redstart are readily observed because second-year and older males are unusual in having different coloration. Adults have black and orange plumage, and younger birds have brown and yellow plumage that resembles female plumage. In New York, the first resident second-year males arrived about a week later than the last of the older males (Ficken 1962; Ficken and Ficken 1967). Although younger birds often do breed (Morse 1973), they are believed to have less chance of establishing an adequate site, and subsequently reproducing successfully, than do older birds. A conservative strategy of arriving on the breeding grounds later, thereby minimizing early-season dangers, might be favored if the chance of obtaining the best sites is low anyway. In one study (Morse 1973) of the success of old and young birds, the younger males occupied low-quality sites significantly more often than did old birds, but their nesting success in high- or low-quality sites did not differ significantly from that of old males in those same sites; thus, the main difference between young and old birds lay in the sites they selected. In that older birds often evicted young ones, the result appears to be attributable to adult dominance rather than to a low level of habitat discrimination by the young. Maybe first-year birds arrive on breeding grounds later because many territories are normally occupied by returning males, which have an ability to retain them in competition with a second-year bird or an older one—a pattern that accords with the site tenacity of many species. Although differences in arrival times are conspicuous in old and young redstarts, second-year males of other warbler species also appear to arrive later than old birds (Nolan 1978; Walkinshaw 1983; Francis and Cooke 1986). This pattern of adult-young relations may thus be widespread.

Nest Sites

Most temperate-zone warblers build small open nests, either in trees or on the ground. Arboreal foragers often nest lower than might be expected on the basis of their activity patterns, and some of them actually nest on the ground. For instance, members of the genus *Oporornis*, the Kentucky *(O. formosus)*, Mourning *(O. philadelphia)*, Connecticut *(O. agilis)*, and Mac-

Gillivray's *(O. tolmiei)* warblers, regularly nest on the ground, although they forage mostly above the ground, usually in dense low cover (Bent 1953). In numerous species of *Dendroica,* females forage lower than do the males. Nests are usually located in the parts concentrated on by the females, often toward the lower parts of it (Morse 1968). Only two members of this large genus, the Kirtland's and Palm *(Dendroica palmarum)* warblers, habitually nest on the ground (Mayfield 1960).

Among the few warblers with unusual nesting requirements is the Prothonotary Warbler *(Protonotaria citrea),* a conspicuous, large warbler that typically breeds in and adjacent to swamps of the southern and central United States. Prothonotary Warblers are hole nesters, making them one of only two warbler species exhibiting this trait. Given their size, similar to common hole nesters such as Tree Swallows *(Tachycineta bicolor),* House Wrens *(Troglodytes aedon),* and Carolina Wrens *(Thryothorus ludovicianus),* and their inability to excavate their own nest sites, they compete with these species for holes. This factor could account for the remarkable attention these birds give to searching and inspecting, which can occupy much of their time for several days after their return (Morse, unpublished). Since they are migrants and arrive in their breeding areas after other species have claimed many of the sites, the Prothonotary Warbler's breeding population may be limited by these other native species, not to mention permanent residents such as Common Starlings *(Sturnus vulgaris)* and House Sparrows *(Passer domesticus)*. Perhaps as a consequence, this species is strongly territorial. Walkinshaw even reported observing Prothonotary Warblers driving House Wrens from the warblers' sites. Yet interspecific competition for nest sites could account for Prothonotary Warblers being largely confined to swamplike habitat. Most nests are located over the water or near it (Walkinshaw 1938, 1941), at least when claimed in the spring. Many sites that are dry later in the season are in areas of standing water at the time of nest selection (Brewster 1878). In that Prothonotary Warblers readily accept artificial nest boxes (Walkinshaw 1941; Petit et al. 1987), it should be possible to determine experimentally whether they are limited by available nest sites.

The only other warbler known to use natural holes is the Lucy's Warbler *(Vermivora luciae),* a native of the southwestern United States and adjacent Mexico. In common with the Prothonotary Warbler, it frequently chooses cavities other than tree holes, including large overhanging pieces of bark and even the remarkable large, globular nests of the Verdin *(Auriparus flaviceps)* (Bent 1953). These sites provide insight into the origin of hole nesting, a trait which subjects its practitioners to remarkably different selective pressures.

Northern Parula Warblers habitually place their nests in either Old-man's Beard lichen (*Usnea* ssp.) or in the superficially similar Spanish Moss (*Tillandsia usneoides*), a bromeliad (Morse 1968). The nest is otherwise conventional: an open cup. The literature routinely suggests that Northern Parula Warblers are limited by the presence of *Usnea* or Spanish Moss (e.g., Chapman 1917; Bent 1953), but the habitats in which this species usually nests contain more than adequate amounts of these materials for supporting nests, and Bent (1953) describes several nests having no *Usnea* or Spanish Moss. Northern Parula Warblers may be limited by the presence of other warblers and other small insectivorous birds (Morse 1968) rather than by nesting materials.

Except for hole nesters, we have no evidence that nest sites limit the populations of paruline warblers. This does not mean that the problem is unique to hole nesters, but nest-site limitation seems negligible in other species. Species that nest in crotches between branches appear to be selective in their choices, but it is highly unlikely that the number of sites within a territory limits them.

Time of Nesting

As is typical of most migrants, warblers in the temperate zone usually breed soon after arrival from their wintering areas in spring. Late-arriving species begin to breed more rapidly than do those that arrived earlier. The Yellow-rumped Warbler, the only member of the spruce-forest group that winters north of the tropics, arrives on its breeding grounds two weeks or more before associated species, such as the Magnolia (*Dendroica magnolia*), Cape May (*D. tigrina*), Black-throated Green, Blackburnian (*D. fusca*), and Bay-breasted warblers. In Maine, though, its nesting cycle averages only a day to a week earlier than that of the others, overlapping their cycles broadly (Figure 2.1). This difference suggests that arrival date is influenced by distance and nesting date by resource conditions. The insect crops that spruce-woods warblers eat are most abundant when demands on parents, at least females, are heaviest (Figure 2.2): when birds are incubating or have small young, their foraging time is severely curtailed (Morse 1968, 1976b). If Yellow-rumped Warblers nested earlier, clutches would be produced before the food supply peaked, but a comparable wait by other species would not improve their opportunities. Given

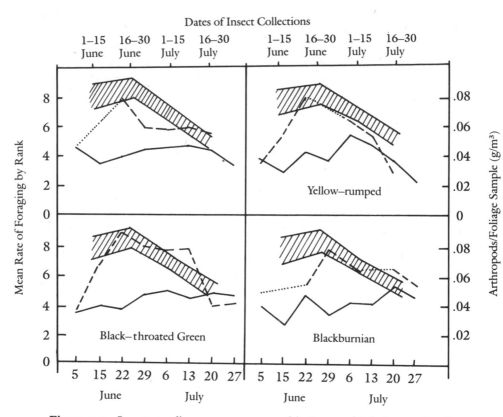

Figure 2.2 Insect standing crops as measured in June and July in a study of Maine warblers (cross-hatched area indicates the mean weight per sample ± 1 standard error) peaked when estimated demand for food was highest. Demand was estimated on the basis of mean foraging rates for males (solid lines) and females (dashed lines). A foraging rate of 1 = 1–2 motions per minute, a rate of 8 = 21–25 motions per minute. (Modified from Morse 1968, 1977b.)

the high foraging speeds of female warblers during the nesting period (Morse 1968, 1981), they are hard pressed to gather enough food, even at the peak of availability. Earlier nesting attempts might well increase the probability of starvation in the young.

Just as there are differences among species in the time between arrival on the breeding grounds and commencement of breeding, one could predict this pattern within a species over a latitudinal gradient. Certainly passerine breeding commences in the south and proceeds northward in wavelike form; Yellow-rumped Warblers in Maine nest earlier than those in northern Canada (e.g., Todd 1963) and Yellow Warblers in the western United States before those in Canada (Phillips 1951). Birds at the northerly sites should breed with more dispatch, however, than the more southerly nesters. Prairie Warblers exhibit this pattern in the southern and midwestern United States (Nolan 1978).

Clutches: Size, Number, and Success

Most clutches of temperature-zone parulines consist of 3 to 5 eggs, with 4 being the most frequent number (Figure 2.3). Clutch sizes within a species are likely to increase with latitude (Lack 1954), and the data in Figure 2.3 presumably reflect that pattern.

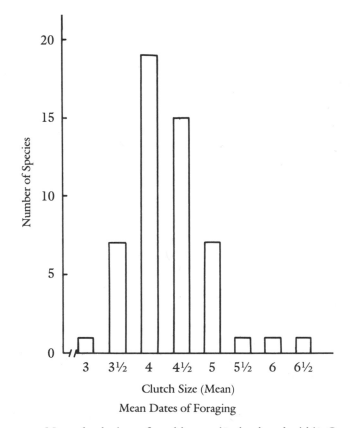

Figure 2.3 Mean clutch sizes of warbler species that breed within Canada and the United States. (Data from Bent 1953.)

Species associated with outbreaks of Spruce Budworm *(Choristoneura fumiferana)* in the boreal coniferous forests of eastern North America are an interesting exception to the typical pattern of clutch sizes; the Cape May, Bay-breasted, and Tennessee *(Vermivora peregrina)* warblers sometimes lay 6 to 8 eggs—more than other warbler species. The Bay-breasted Warbler, the only species for which adequate data exist, lays more eggs during an outbreak than at other times (5.8 ± 0.7 vs. 5.1 ± 0.7 eggs) (MacArthur 1958). Clutch sizes for Cape May and Tennessee warblers probably exhibit a similar pattern because they have a wide range of extremes in clutch size (Bent 1953; MacArthur 1958).

Although the expanded clutch size of these three species probably results from their encountering a temporarily abundant food source, clutches of co-occurring congeners (Yellow-rumped, Black-throated Green, and Blackburnian warblers) do not increase similarly at these times (MacArthur 1958); further, the population density of these species may even decline (Kendeigh 1947; Morris et al. 1958). This decrease in density could be caused by decreased availability of the birds' favored prey—other lepidopteran larvae in this season (Morse 1976b)—in the presence of the budworms, or it could be the result of direct interaction among the species themselves.

At the other end of the spectrum, virtually all species with low clutch sizes (3.0 or 3.5 in Figure 2.3) have distinctly southern or southwestern breeding ranges—examples are the Swainson's *(Limnothlypsis swainsonii)*, Bachman's *(Vermivora bachmanii)*, Tropical Parula *(Parula pitiayumi)*, Olive, Grace's *(Dendroica graciae)*, Red-faced *(Cardellina rubrifrons)*, and Hooded warblers, and the Painted Redstart *(Myioborus pictus)*—commencing a pattern continued among tropical species (Chapter 14). Only one species with a mean clutch size larger than 3.5, the Kentucky Warbler ($\bar{x} = 4.5$), has as southerly a breeding range as the species with small clutches, but its range largely overlaps only one of the latter species, the Hooded Warbler. Although southern populations could make up the differences in clutch size by producing more clutches than their northern counterparts, we lack enough evidence to say that southern warblers are more likely to rear two clutches than the more northerly ranging species (Bent 1953).

Temperate-zone parulines are primarily single-brooded. Many apparent exceptions occur when a first nesting fails or when a single bird does not find a mate until late in the season and thus fledges its young at times that would be consistent with second broods. Many, if not all, warblers will renest quickly upon failure of a first clutch, and as many as nine such efforts have been made by Prairie Warblers (Nolan 1978).

Unambiguous examples of late-forming warbler pairs are usually hard to establish, but in studies of warblers on small islands, previously established males paired as late as 28 June (Morse 1971a). Mates acquired by the island birds might have participated in earlier unsuccessful nesting attempts elsewhere, but this could not be ascertained. Female Prairie Warblers with brood patches have been observed to arrive at study areas (Nolan, pers. comm., 1987), indicating that females too move in mid-season.

Second clutches (as opposed to replacements) are produced by a small minority of individuals. Bent (1953) reported authentic double broods only for the Yellow-throated Warbler, on the basis of a single report from South Carolina, but his treatise preceded the detailed monographs that have since set the standard. Nolan (1978) found that 18 of 86 pairs of Prairie Warblers successfully completing their first broods commenced a second. Most individuals that attempted a second brood were the first of their population to fledge their clutch (Figure 2.4). Extrapolation from Nolan's results leads to the conclusion that double-brooding is most often attempted by the individuals that nest earliest. Nolan found authentic second broods of Prairie Warblers in 10 of the 11 years that he searched for them; this implies that though only a minority of the population produces second clutches, double-brooding is a regular and normal phenomenon. Since Prairie Warblers suffer high egg and nestling mortality (Nolan 1978), about 80 percent of the population will initially fail to fledge a clutch. The time constraint imposed by such losses effectively removes many pairs from the pool of those with the time to attempt a second brood (versus renesting). Thus, only about 4 percent of Prairie Warbler pairs attempt genuine second broods.

Kirtland's Warbler is similar in this respect; Walkinshaw (1983) reported 11 instances of second nestings following a successful first brood (a nearly 50 percent effort), and these also involved the earliest nesters. In previous studies of the same species Mayfield (1960) reported two such instances, but stated that it probably occurred more often than his data revealed. In less extensive studies of Golden-cheeked Warblers and Ovenbirds, neither Pulich (1976) nor Hann (1937) reported attempts of successful pairs to undertake second broods.

The maximum yearly reproduction of nearly all warblers is modest in comparison with many residents or short-distance migrants, such as chick-

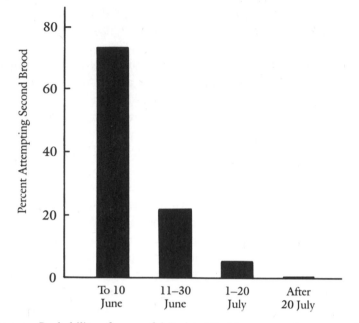

Figure 2.4 Probability of successful Prairie Warblers attempting a second brood as a function of when their first brood leaves the nest. $N = 86$. (Modified from Nolan 1978; used by permission of the American Ornithologists' Union.)

adees, Golden-crowned Kinglets *(Regulus satrapa)*, and nuthatches, having similar life-styles and breeding in the same areas as the warblers. It also falls far below that of many routinely multibrooded species, such as Robins *(Turdus migratorius)* or Song Sparrows *(Melospiza melodia)*. (See Figure 2.5.) This point is often ignored by workers who emphasize only the number of eggs in a clutch, rather than total numbers of eggs laid in a year (see Whitcomb et al. 1981).

Phases of the Nesting Cycle

In this section I present a thumbnail sketch of the activities that characterize warblers at different stages of their tenure on the breeding grounds; some activities are described further in subsequent chapters. Since Nolan's (1978) work on Prairie Warblers has treated several aspects of the subject in much greater detail than any other study, certain parts of this section take the form of a synthesis of parts of his monograph.

After arrival at the breeding ground. Behavior of male Prairie Warblers returning to their territories differed, depending on whether other males were present in the territory or in adjacent areas. The first birds to return foraged rapidly and moved steadily, often over a much larger area than their preceding year's territory, singing infrequently. Subsequent arrivals reached full song more quickly than the early birds: singing appears to be associated with the territorial activities of both the singer and his neighbors. Nolan noted three birds that apparently arrived before they were ready to stake their claims. They foraged quietly for one or more days on or near their territory of the preceding year before assuming territorial behavior.

Since males precede females to the breeding grounds by several days, direct sexual interactions do not occur immediately. Establishing and defending territories take up much of the early period. Nolan found that in southern Indiana the single largest number of territories was claimed by Prairie Warblers on the first day that males arrived, followed by comparable numbers of claims by birds arriving the rest of the first week, declining to single claims on the sixteenth and seventeenth days; these figures are taken from data summed over eight years (Figure 2.6). The initial dates of

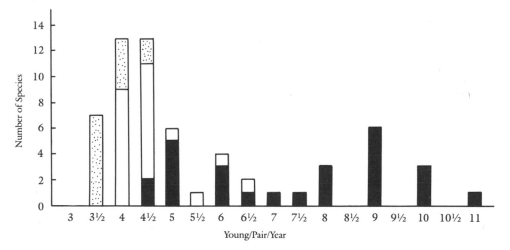

Figure 2.5 Average number of young per pair per year of small oscine species in Maine. Open bars = Neotropical migrant warblers; dotted bars = other Neotropical species; solid bars = residents and short-distance migrants. Most large numbers result from multiple broods in a year. (Data from Palmer 1949.)

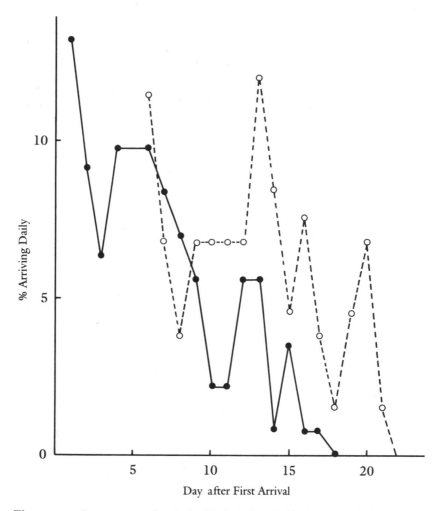

Figure 2.6 Percentages of male (solid circles) and female (open circles) Prairie Warblers arriving on successive days after the arrival of the first male (Day 1). Percentages shown are mean percentages of samples taken over eight years for males (*N* = 143) and seven years for females (*N* = 132). (Modified from Nolan 1978; used by permission of the American Ornithologists' Union.)

arrival over this eight-year period varied by 11 days (April 11–22), which Nolan attributed to differences in the weather. Variance in weather probably also accounted for differences in the length of time required to repopulate the site, which ranged from 9 to 17 days. Repopulation sometimes took place rapidly in years with late springs, but in other years no such relation was found; thus, one can make no simple comparisons, and these data are unique so far as I can tell.

With one type of exception, contests for territories were always won by the bird already in possession. Territory holders flew quickly and directly at intruding males and usually chased them away with ease. But if the territorial owner of the preceding year returned in spring and found another male already on its territory (or if they both arrived on the same morning), the former owner always prevailed in the end. In this situation the earlier-arriving male initially chased the newly arrived former owner around and around in a ritualized circular pursuit. These bouts sometimes lasted as long as two hours, sometimes only a few minutes, but they always ended in favor of the original owner. Returning territory holders engaged in fewer fights than new claimants. Nolan attributed this difference to individual recognition of song; if so, it indicates long-term memory of vocalizations. Shorter-term song recognition has been demonstrated in Ovenbirds (Weeden and Falls 1959). Alternatively, if both former owners and new claimants behaved like returned territory holders, response to this behavior might produce the same result (few fights)—long-term memory notwithstanding.

Female Prairie Warblers arrived on their breeding area about five days after the males (extremes in Nolan's study were 1 to 9 days). Although nearly as many females arrived on their first day as on any other day, their arrivals were somewhat more spread out than the arrivals of the males, but the extreme between first and last arrival dates (16 days) is similar to the males' (Figure 2.6).

Upon arrival, female Prairie Warblers moved about slowly, deliberately, and silently. They frequently approached males that were singing, and over two-thirds of them (26 out of 37) established pair bonds within a few hours, several of them temporary. Those requiring more than one day to establish a pair bond included some of the earliest arrivals.

Nolan believes that some returning males reached their territories during the night, as three such birds sang on their territories for the first time as it grew light in the early morning. Another 16 birds commenced to sing on their territories within 1–2 hours of daylight, and 10 more between then and 10:30. No birds appeared to arrive on their sites in the afternoon. This pattern, observed in birds banded in previous years, suggests that they reach the approximate area of their territories during the night and that those not scoring a "direct hit" reach their target quickly early in the morning. Several times Nolan observed Prairie Warblers moving southward shortly after dawn; these birds may have been making the hypothesized fine-orientation flights.

All these observations, which involved previous territory holders, suggest that most of the sites were settled virtually immediately. Other workers have proposed that recolonization does not happen this fast, that during the morning one often sees exploratory behavior by individuals; this behavior is taken to be a search for satisfactory territory (Ickes and Ficken 1970). It would be difficult to know whether birds exhibited true habitat selection, which must take place when they are nesting in an area for the first time, or whether they are searching for last year's territory, as described by Nolan, unless one worked with known individuals.

About three-quarters of the 75 banded returning males studied by Nolan reoccupied their previous year's site, and those that did so almost always occupied at least 80 percent of that territory. The remaining quarter moved to territories as far away as 3,400 meters ($\bar{x} = 710$ meters), the limits to this analysis being the size of the area surveyed by Nolan. Contrary to prediction, site fidelity did not appear to increase with the number of times that a bird returned; however, this result should be adaptive for birds of early-successional vegetation, whose quality of territory could change over a lifetime. The fidelity of Prairie Warblers may not be as strong as that of some Kirtland's Warblers, populations of which fluctuate in density within a local area over a few years and sometimes eventually dwindle to a few males returning to past sites but unable to attract females (Mayfield 1960).

Many fewer females returned to their former territories than did males; only 14 of 144 females were known to reoccupy their last year's territory. This 10 percent reoccupancy rate compares with a 47 percent annual reoccupancy rate for the males. In another 23 instances females returned and nested in the study area, but in different sites (16 percent)—the figure was 26 percent for males. In two such instances females joined males in territories that overlapped the territories of their mates of the preceding year; the former mates were present, and one of them had not yet paired. In a majority of instances in which a female returned to the study area but subsequently moved, a male was present in her former territory but already had a mate; in all instances in which the female nested in her former site a male was present and was not yet mated. Therefore, females do not assure themselves of their site simply by appearing at it: the male's concern is whether another male is present in their former territory; the female's is whether a male is present *and* without a mate. Whether females pay a price for their tardiness in lowered reproductive success, in a way similar to seabirds exhibiting strong site tenacity (Morse 1980a), is unknown, although the cost should be lower in warblers because they do not have the high long-term remating frequencies that characterize seabirds.

Courtship. Weather can affect the timing of the nesting cycle. Prairie Warblers initiate pair formation and nest building with the onset of warm weather; either cold or rainy weather inhibits the initiation of nesting. Once the nesting cycle has begun, extremes of sunlight and rain prompt greater incubation and brooding. These patterns are widespread among songbirds in the temperate zone (Grubb 1985).

Male Prairie Warblers' response to females that approached them was to attack and chase them. Females usually flew low to the ground and for shorter distances than males; if similarly attacked, intruding males flew at a greater height and often somewhat upward, a response that would take them off a territory. Ficken (1963) reported a similar pattern in American Redstarts. The result of the differences in male and female behavior at such times is that intruder males leave the territories, but females do not. This difference in itself may signify the bird's sex to the defending male. In about one-fifth of the first meetings male Prairie Warblers initially responded to females sexually, performing a slow display flight characteristic of subsequent courtship displays. In sexual chases males "captured" the females about two-thirds of the time, either driving them to the ground or onto a perch. This contact typically resulted in a struggle, in which the male usually repeatedly pulled the female's tail. Most females did not tolerate this for more than a few seconds. Males that were able to capture the female had a higher percentage of eventual pairings than males that were not successful. Males performed exaggerated display flights ("butterfly flights," "moth flights," gliding) much more frequently than sexual chases, an activity that also declined as pairing progressed (Figure 2.7). As time elapsed, females were less likely to retreat from males and became more aggressive.

Pair-formation displays subsequently became less frequent, and the males' singing declined (songs changed in type as well). During this period the male often followed the female very closely. The female spent most of her time during the courtship phase searching for a nest site or foraging. In about 10 percent of Nolan's observations of nest-site selection, males also visited sites actively, and sometimes females joined them at sites. At times females drove the males out of certain sites, some of which they eventually chose. Such behavior seems highly unusual in warblers,

Figure 2.7   Some displays of male warblers (from Ficken and Ficken 1962a). *From upper left, clockwise:* "fluff" of Chestnut-sided Warbler; "wings out" of Chestnut-sided Warbler; courtship display of American Redstart; distraction display of American Redstart. (Redrawn by Brian Regal from painting by Dilger in Ficken and Ficken 1962a.)

although Ficken (1963) observed similar behavior in American Redstarts. Approaches by other males are met by resident males chasing them from the area and immediately returning to the female. Courtship feeding, important in some passerine groups (e.g., Royama 1966), occurs in a few species of warblers (Mayfield 1960), but not the Prairie Warbler (Nolan 1958).

Nest building. Nest building follows courtship and is typically interspersed with bouts of feeding and maintenance activity. Thus, an observer assumes he is seeing a bird feeding, only to note on closer inspection that the bird seldom pecks at the vegetation inspected. The true behavior is clearly revealed when a bird finds a satisfactory piece of nesting material and flies off with it. Males and females are often close together at this time, but it is not certain whether males of most species participate directly in either the procurement of nesting material or in nest building proper, although male Prothonotary Warblers build "dummy" nests in early spring (Bent 1953). Bent (1953) listed reports of male Magnolia, Black-throated Blue *(Dendroica caerulescens),* Black-throated Green, and Yellow-throated warblers helping to build nests, and Verner and Willson (1969) reported several other such examples. Given the similar coloration of males and females of most species, reports of this behavior, as well as of other primarily female behaviors performed by males, must be considered with extreme caution. I conclude that, with the exception of Prothonotary Warblers, nest building by males is, at most, infrequent and requires careful study. I do not consider it accidental that the most detailed studies (those of Kirtland's and Prairie warblers) have not recorded this behavior. Males of several species occasionally enter the nest (references in Nolan 1978), which could be a basis for the reports. Nolan proposed that this behavior is directed to the female rather than to the nest itself.

Male Prairie Warblers spend most of their time at this season watching females at a moderate distance from the nest (20–30 meters), although sometimes they follow them. They are more apt to follow females when the females move slowly or are away from the nest for longer than usual. Such behavior by the male minimizes the possibility of copulation by other males, and for this reason we might expect those that nest in denser vegetation to follow their females more often than do Prairie Warblers. Singing is frequent during this period, and sometimes males swoop on females.

The search for nesting material can take the hunter into parts of the habitat that it would seldom if ever visit to forage for food; for example, foliage-gleaning *Dendroica* may venture down to the ground, where they gather bits of grasses or mosses. Prairie Warblers picked up 75 percent of their material within 25–30 meters of the nest, although special items (lining materials, gathered late in building) sometimes required longer flights.

Nest building takes several days, the exact length depending in part on weather conditions. Some female Prairie Warblers are not able, after migration, to build rapidly. Second and subsequent nests are produced more rapidly than those for the first brood (Figure 2.8), perhaps because the latter are larger and more insulated. Young and old birds do not differ in nest-building times.

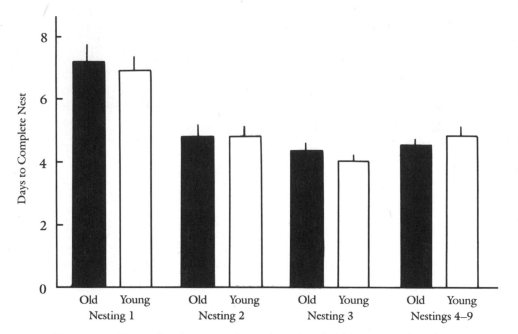

Figure 2.8 Days taken by young (second-year) and old (past second-year) female Prairie Warblers ($\bar{x} \pm 2$ s.e.) to build first and subsequent nests. Vertical lines extending above the bars represent standard errors. (Modified from Nolan 1978; used by permission of the American Ornithologists' Union.)

Egg laying and incubation. Prairie Warblers normally lay one egg each day, and a clutch is typically completed in not over a day more than the number of eggs in the clutch. Prairie Warblers almost invariably lay in the early morning, as is typical for other species as well. Females of several species sometimes commence incubation the day before the last egg is laid, and female Prairie Warblers usually spend the night on the nest before laying the last egg. Female Prairie Warblers are on the nest about 20 percent of the time on the day the penultimate egg is laid, triple the frequency at the beginning of egg laying.

Males and females of many warbler species frequently forage close together during most of the egg-laying period (Bent 1953; Verner and Willson 1969). This probably prevents the female from copulating with other males. A male Yellow Warbler foraging with its mate during this period immediately chased another male that had silently come within 15 meters of the female, even though the resident male foraged and sang prior to finding the encroaching male (Morse 1966b). The priority allotted to aggressive behavior is a sign of the importance accorded such intruders.

Males incubate little, if at all, even though this time may be a critical one for females. Both Bent (1953) and Verner and Willson (1969) listed records of male incubation, but neither Nolan (1978), nor Mayfield (1960), nor Walkinshaw (1983) observed this behavior in their more intensive studies.

Because the males do not contribute to incubation, females spend as much as 80 percent or more of the daylight hours on the nest once their clutch is complete (Mayfield 1960; Morse 1968; Nolan 1978). This high level of attentiveness could be a consequence of thermoregulatory considerations or it could be a means of minimizing nest predation. Under some circumstances the thermoregulatory explanation seems unlikely; for instance, Prairie Warblers are also very attentive on warm days, when thermoregulatory constraints should be less significant. Yet Nolan found that attentiveness was lower at cooler temperatures, perhaps because the females were experiencing difficulty in balancing their energy budget.

This type of evidence makes the second explanation attractive. Nests are less vulnerable to predation if they are constantly covered—that is, if the eggs are hidden. The incubating bird must, of course, leave on occasion to replenish itself or be fed by its mate at the nest—the less frequently the better, for activity about the nest brings attention to it. The female bird sitting on her nest is usually inconspicuous as one looks down on her, and her activities about the nest are extremely furtive. Merely being at the nest to defend it would not suffice against jays, crows, and squirrels, the principal nest predators in my spruce-forest study areas, or opossums and snakes elsewhere. Parents can defend their nests against a nest parasite, the Brown-headed Cowbird *(Molothrus ater),* so their presence may be an advantage in interactions with this species. Since cowbirds usually lay their egg when the host is away from the nest (Nolan 1978), however, the main advantage of nest sitting here, too, may lie in concealing the nest. Cowbirds may actually observe the nest behavior of hosts (Thompson and Gotfried 1976).

I have one indirect test of the relative advantages of thermoregulation and predator avoidance in determining incubation strategy. Female warblers on small islands not supporting their principal nest predators, Blue Jays *(Cyanocitta cristata)* and Red Squirrels *(Tamiasciurus hudsonicus),* foraged more slowly than they did in nearby forests containing many of these predators (Morse 1981). This pattern persisted even though standing crops of insects did not differ in the two areas. If the slower foraging was accompanied by greater time off the nest, which I could not measure, it would suggest that the mainland behavior is primarily related to avoiding the depredations of nest predators.

The failure of males to contribute to incubation may be related to the problem of nest predation (frequent exchanges of individuals can increase the danger of nests being found) and to the demands on the males to defend their territory against the incursions of both conspecifics and other species. The foraging speeds attained by females at certain times indicate a high premium on having a foraging area that is not depleted in an unpredictable way by other individuals, including mates (Morse 1968; Holmes 1986). Although males are much more likely to confront other males than females in territorial conflicts (Morse 1976a), elimination of the other males alone should diminish the presence of other females and enhance foraging conditions for their mates.

Males of several species regularly feed their females at the nest. Yellow Warblers performed this behavior frequently (Morse 1966b), but it was largely or completely absent in the spruce-woods species (Morse 1976b, 1968). Feeding the female decreases the time that she has to spend foraging for herself and thus is a direct input by the male into the costs of incubation. I have interpreted these differences in contributions by male warblers in the context of energy demands; for instance, at high population densities more of a male's energy is devoted to territorial defense, leaving little or no time for feeding the female (Morse 1968). Alternatively, they can be considered in terms of danger from predation. Hence the differences in contributions by male Yellow Warblers in swampy alder habitats and males of the spruce-woods species reflect the lesser threat of nest predation in swamps. At the least, squirrel predation will be lower in swamps since they do not typically inhabit them. Verner and Willson (1969) found several citations of male spruce-woods warblers feeding their mate on the nest; it would be interesting to compare these feeding conditions with those of the birds in my studies.

Regardless of whether thermoregulatory, predatory, or other factors determine the frequency that males bring food to their mates, the separation of male and female activities creates the possibility for bigamous relations; the male has the opportunity to intrude on neighboring territories but risks cuckoldry (Alatalo et al. 1981) and having his territory foraged by others if he does so. Observations of marked birds on territories reveal that paired males regularly leave their territories (Ford 1983). Warblers provide two of the three data sets that permit analysis of the timing of such visits. The vast majority of intrusions by territorial male Yellow Warblers (80.7 percent) in other territories takes place during nest building and egg laying; at these times resident females are fertilizable (Ford 1983), assuming synchrony of the nesting pairs in an area. Ford figured from Nolan's data that about 55 percent of Prairie Warbler incursions in other territories were made during this time. Nolan remarked how frequently Prairie Warblers made incursions while females were incubating, and he noted Kendeigh's (1945) comment that male Chestnut-sided Warblers *(Dendroica pensylvanica)* often ventured beyond their territorial boundaries at this time.

Males' movements could serve several purposes, but one likely explanation is that the wandering individuals are at least ready to copulate with other females. These individuals might also establish second territories and mates. It is becoming clear that several species of small passerine birds are facultatively polygynous, and accumulating evidence from marked birds (e.g., Ford 1983; Gowaty 1985) suggests that this trait may eventually be recorded in a majority of species. For example, Eliason (1986) found that 8–30 percent of a Blackpoll Warbler population were bigamous over a three-year period, and Nolan observed that about 10 percent of his Prairie Warblers were so paired. This percentage increased as the summer went on, probably because of a shortage of unpaired territory holders.

Polygyny can be simultaneous or, more often, in sequence. If it is sequential, females are often designated primary and secondary, according to the order in which the male forms pair bonds with them. The secondary female often lags the primary female by several days in the nesting cycle. When the first female has begun to incubate, the second may be at the pre-egg-laying or early-egg-laying stages, when copulation and insemination occurs. What seems clear is that polygyny is not an aberrant condition, as Verner and Willson (1969) suggested in referring to reports of polygyny by warblers and other species as "abnormal."

Gowaty (1985) has noted that polyandry also may be an important, though overlooked, force in the evolution of behavior in female birds. Ironically, polyandry might be the consequence of males leaving the territory or not attending females at critical times to seek matings with other females! Presumably the tendency to search is balanced against the tendency to guard, but little is known about this matter, or of how receptive the females are to other males (see Ford 1983).

Alatalo and colleagues (1981) have reported a pattern of sequential bigamy in the European Pied Flycatcher *(Ficedula hypoleuca)* that parallels covert copulation; such flycatcher males have two distinct territories, separated by at least one territory of another bird. Nolan (pers. comm., 1987) noted that one bigamous male had territories 1.3 kilometers apart, which the bird visited daily. This pattern has not been recorded for any other paruline warbler.

Bigamous males are often unable to feed their females as frequently as can monogamous birds, which may impose a cost if the frequency of feeding is normally high, as in the Yellow Warbler, through forcing the female to be away from the nest for more time. One might test the importance of the male's incubation feeding by comparing nests in which males are removed with unmanipulated controls. No such experiment has been run on warblers, but in nests from which male Prairie Warblers naturally disappeared, the females increased their activities. Lyon and Montgomerie (1985) demonstrated that female Snow Buntings (*Plectrophenax nivalis*) without the male's food supplement were less successful in rearing their clutch because of nest predation during the times they spent foraging. This difference may be considered a cost of bigamy and a factor that should mediate male tendencies toward this behavior. Snow Buntings live in a more rigorous climate than most nesting warblers, and thus the failure of males to feed females may exact from them a larger cost than from warblers; nevertheless, one would expect that species performing extensive incubation feeding, such as Yellow Warblers, would suffer a substantial cost. Secondary female Wilson's Warblers spent less time brooding and made more feeding trips than did either monogamous females or primary females (Stewart et al. 1977). In this species, a population with low nest predation exhibited bigamy, but another population with higher nest predation was strictly monogamous (Stewart 1973). If the cost is primarily to the second mate, the male's tendency toward bigamy should be enhanced.

The origin of male food bringing has been a subject of considerable interest. Nolan (1958) pointed out that males bring food to their nests regardless of the female's presence and that, even when she is there, the male often does not direct the food specifically to her. Often the male even ate the food or left with it. Nolan proposed that this behavior facilitates the male's ability to commence feeding the young as soon as they hatch. He further suggested that incubation feeding may have developed from "anticipatory" food bringing, and that courtship feeding merely represents an expansion of this behavior to an earlier point in the breeding cycle. Nolan provided no direct evidence that would permit testing of this hypothesis, but information on the characteristics of the food brought to the nest would be enlightening, since food appropriate for females and for newly hatched young would differ considerably. Nolan indicated that the food brought early in the cycle is slightly larger than is appropriate for very young nestlings and that the male must learn what size is appropriate, but one could also propose that the food is brought for the female rather than for the young. In this view the male could equally well learn when the young hatched. Further, the male's ability to provide satisfactory food for the young in subsequent visits might provide insight into the origin for this behavior, which enhances the male's contribution to the overall nesting endeavor. The frequency of this behavior also differs markedly among species—a maximum of three times per day in the Prairie Warbler—and tends to increase during the incubation period, which is consistent with Nolan's hypothesis that food bringing is associated with the male's learning of the hatching of the young. But with species that exhibit this behavior frequently, such as the Yellow Warbler (see Nolan 1958), feeding probably enhances the female's condition markedly: it was coupled with an unusually high (91.9 percent) total time spent incubating by the female (Kendeigh 1952).

Nestlings. Both males and females feed the nestlings. Initially the female broods the newly hatched young about as much as she brooded the eggs. This activity can limit the number of feeding trips she makes for the young, but the yolk sac of the young may obviate pressure for more frequent departures. After the third day the frequency of brooding decreases, however, falling to a much lower level after the young have begun to develop thermoregulatory abilities, about five days after hatching (Figure 2.9). Early in the nestling period males of some species feed females on the nest, and part of this food is often fed to the young by the female. Male and female contributions to the young are variable. Male Kirtland's Warblers make major contributions early in this period, but the female eventually performs half or more of the feedings, although nests vary greatly. No clear pattern of age-related male and female contributions emerges in the Prairie Warbler (Nolan 1978), which also vary from nest to nest. What is clear is that males often make as large a contribution to nesting activities at this time as their mates. The frequency of male visits by both Kirtland's and Prairie warblers is relatively low in comparison to most other small passerines, including Yellow Warblers and Common Yellowthroats *(Geothlypis trichas)* (Nice 1943), but this might be due to the large amounts of food carried by some warblers (Lawrence 1953), a vari-

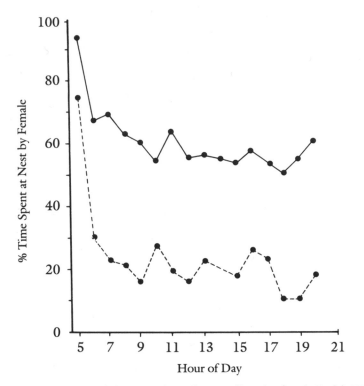

Figure 2.9 Percentage of day spent brooding nestlings by female Prairie War-blers. Solid lines = days 1–5, data from eight day-long watches; dashed line = days 6–9, data from 11 day-long watches. (Modified from Nolan 1978; used by permission of the American Ornithologists' Union.)

able that also changes over time. Number of trips is also likely to vary with clutch size (Perrins 1979); unfortunately, differences in clutch size were too small to allow evaluation of this factor in either Mayfield's or Nolan's studies.

Female Prairie Warblers typically brood their young the first six or seven nights, but thereafter brooding declines rapidly, perhaps being the exception on the last one or two nights the young are in the nest. Male warblers typically do not brood or shade young, although Walkinshaw (1959) observed one male Prairie Warbler doing both, and Mayfield (1960) noted a single example for the Kirtland's Warbler.

Foraging rates of male spruce-woods warblers increase as their young grow, but those of females decrease (Morse 1968), which suggests that the incubation period is the most critical one for females. The high variability of male and female contributions is partly related to how much the male participates—in the Prairie Warbler, at least. The female picks up much of the slack if the male ceases to feed his young. This trait exposes her to exploitation by bigamous males, as it should not be in the male's interest to contribute extra efforts under these circumstances if it could otherwise help another brood of its own fathering. The effect may be severe on such females, however. The lightest female Nolan ever weighed had been abandoned by her mate, and it would be surprising if such experiences did not threaten her survival, as well as that of her young.

By the time the young are ready to fledge, nests have become quite conspicuous because of begging calls from the young and the activity of parents feeding them; this increases the danger of nest predation. Nolan found more nest failure at this time, although his sample was small; and Perrins (1979) noted that predation is higher on large (and hungry) broods of Great Tits *(Parus major)* than on smaller broods, possibly because their Weasel *(Mustela nivalis)* predators are attracted to noisy sites. Comparable forces probably act on most small passerines, and the short fledging time is probably an adaptation to such escalating dangers. Young of many species can barely fly when leaving the nest (e.g., Kirtland's Warbler: Mayfield 1960). This trait is a natural phenomenon, not merely a response to human disturbance. Nestling departure ages are usually shorter for ground-nesting warblers and other birds (Mayfield 1960) than for tree nesters, but 9–12 days is the normal range for warblers. Both Blue Jays and Red Squirrels regularly take nestlings in my coniferous forest study area. Several times I have seen these squirrels carrying dead warbler nestlings or just-fledged birds. The squirrels are unlikely to catch birds that fly effectively.

Parents of several species employ extremely elaborate predator-distraction displays. I have seen this display directed at Blue Jays that approached the location of young Black-throated Green Warblers that were just leaving their nest (Morse 1969). Both parents displayed in a way that closely resembled the movements of a nestling about to leave the nest, literally falling from near the top of a spruce tree almost to the ground before breaking their fall by flight. The two parents performed this maneuver approximately two minutes apart, both eliciting an attack from a Blue Jay and leading the jays a considerable vertical distance from the newly fledged young. By the time the jays neared the parent warblers, the warblers had flown into the understory. Later in the day the warbler family was still present at the same site, indicating that they had averted predation. Bent (1953) noted examples of this general display for several species. Individuals at or near ground level are unable to produce this display. Equivalents may include Mayfield's (1960) "rodent run" and other feints—an adult may imitate an injured or a young bird, a female begging for food, or even a female soliciting copulation. Mayfield once observed a male attempt to mount such a female, who repulsed him.

Fledglings. Although the fledgling period does not receive as much attention by students as the nestling period, parents commonly feed their young at least twice as long outside the nest as in it (J. M. N. Smith 1978). Adults sometimes feed their young for a month or more, though progressively less frequently over time. In the Prairie Warbler, young typically remain with their parents until 40–50 days of age (30–40 days after fledging). Nolan determined that they are dependent until an age of 33–35 days. Since the young can barely fly at fledging, the adults still must bring food back to a central place. This restriction diminishes within a few days, as the young follow their parents about, begging. At times, the young follow the parents so aggressively that they compromise the success of their parents' foraging, but they also may learn how to forage themselves. Comparable learning behavior is known in some passerine species (e.g., Krebs et al. 1972; Morse 1978b). Even when the young are separated from their parents, more pecking motions are directed to the substrate as they become older (e.g., Nolan 1978), which may be the major basis for learning how to forage. Kirtland's Warblers can exhibit this behavior the first day out of their nests (Mayfield 1960). Engaging in pecking activities where adults forage probably rapidly improves a fledgling's abilities as a forager—a form of learning enhanced by their parents.

Broods usually split up rapidly after the young begin to fly, often within a few hours (Nolan 1978). Some are cared for by the male and others by the female, a general pattern in warblers (e.g., Mayfield 1960; Pulich 1976). After a few days there is little contact between these two units, although they may remain on or near the territory. Boxall (1983) noted that fledgling American Redstarts that begged from the parent not routinely feeding them were refused food by it. But if the female attempts a second brood, young Prairie Warblers coalesce with the male who takes over feeding them.

Brood division is not uncommon among small passerines (J. M. N. Smith 1978; McLaughlin and Montgomerie 1985), although data are scarce. It appears to be the common pattern among spruce-woods warblers, judging from how many single adults have one or two young. Division of offspring minimizes overlap in feeding areas, distance of carrying food, and predation of an entire brood (Moreno 1984; McLaughlin and Montgomerie 1985). Female Kirtland's Warblers tend to take their young farther from the nest site than do males (Mayfield 1960), a sex-based difference reported in other species.

After a few days the young are often extremely noisy, especially if the parent is nearby. Although such vocalizations might make them vulnerable to predators, they probably are less vulnerable to jays and squirrels than previously. Bird-hunting hawks are their main predators at this stage. Although hawks sometimes kill large numbers of newly fledged passerine birds (e.g., English titmice: Geer 1978; Perrins 1979), this has been uncommon where I have worked. Where present North American hawks also take many prey, including warblers, although even Sharp-shinned Hawks *(Accipiter striatus)*, the smallest of the North American bird specialists, prefer larger birds such as thrushes and sparrows, if available (Meyer, pers. comm., 1986).

The fledglings' loud vocalizing seems paradoxical, but it could simply be a consequence of the adults dispersing with their young to areas of high food density. It would be interesting to compare vocalization patterns in fledglings of the same species at high risk and those at low risk of predation. Young of some small tropical birds, such as manakins, under heavy predatory pressure exhibit far more cryptic behavior (Foster 1974a), but this behavior is associated with slow maturation and extremely low clutch sizes, traits not found among the migratory warblers.

Given the premium placed on the time that adults need to molt before embarking on fall migration, they probably experience pressure to conclude the responsibilities to their young as soon as possible, or, if they are to produce a second brood, to commence it. By contrast, it is to the benefit of the young to retain their parents' services as long and as frequently as possible. The result: parent-young conflict (Trivers 1974), as the young became more incessant and aggressive in their begging as they reach a month in age. Because of their improved flying abilities, they can follow their parents closely, calling loudly much of the time and monopolizing the site where the parent is capturing food. Adults may ignore them or even repulse them aggressively, and the young eventually separate from them (Mayfield 1960; Nolan 1978).

Postfledgling period. Less is known about the period just after the young gain independence than any other time when parulines are on their breeding grounds. Adults are apt to be in heavy molt, replacing various feathers, although not the major flight and tail feathers (Dwight 1900; Bent 1953). After the adult-young bonds have ceased, some adults remain on their territory; others leave. Certain adults return to their territories after the bond with the young has broken in September; some probably leave without returning that fall (Mayfield 1960; Nolan 1978).

Feather replacement requires considerable resources and energy; in the summer these needs are exceeded only during the females' egg-laying period (Payne 1972). Rather than busily hunting for food at all times, molting birds balance their high energy demands in part through low activity (King 1974), which may explain why they are so difficult to observe in this period. Prairie Warblers in molt rarely flew and were even reluctant to move. They remained in perch sites far longer than they did in the breeding season, although they scratched, preened, and stretched more than at other times (Nolan 1978), as did Kirtland's Warblers (Mayfield 1960).

It is also unclear how much most warblers wander locally at this season, although marked shifts in habitat are frequently seen even while adults are feeding fledged young (Morse 1970a). Since some species, such as the Pine Warbler and Kirtland's Warbler, have similar habitat preferences after their young become independent, it is difficult to evaluate and verify the extent of their local wandering at this time.

Departure from the breeding grounds. The amount of time that elapses between separation of parents and young and the beginning of migration differs markedly among species (Bent 1953). For some, the more southerly members of the species are likely to be the earliest fall migrants (Nolan 1978).

Many warblers have already commenced their migration southward—probably in response to selection for favorable migration periods—by the time low temperatures reach their breeding grounds. But some individuals may remain on their territories into September, and thus may be exposed to cold weather then. Investigators seem to assume that warblers are not under strong selection to remain late on their territories. Until we clarify the relation between occupation and reuse of a territory the following year, however, this assumption is unwarranted. Some Kirtland's Warblers remain on their northern Michigan breeding sites into September, when temperatures well below freezing are not unusual; these birds may have climate-induced problems (Mayfield 1960). Their large size should be advantageous in this context.

Conflict during the Breeding Season

It is commonly held that individuals are in conflict via between-male or between-species interactions, but these conflicts can extend to between-sex (mates) and parent-young interactions (see Trivers 1974; Charnov 1982) as well. Warblers provide excellent examples of the latter two kinds of conflicts, which play an important role in some species.

These conflicts peak at certain times. Male-female conflict peaks at the beginning of incubation, when females are confined in space to incubate. Exclusive responsibility for incubation may in its own right result from past male exploitation of females (in contrast, males of many European sylviine warblers incubate: Simms 1985). Regardless of its history, the trait is apparently fixed in paruline warblers. Males can enter into a second mating while the female incubates, a trait that may be widespread among parulines and other songbirds, though it is only recently reported regularly (e.g., Eliason 1986). The cost to first females may be modest: some information suggests that they are less severely affected than second females (e.g., Stewart et al. 1977). If so, pressure on males to remain monogamous may not be especially high. The main drawback of polygamy for the male may result from him spending less time defending the first female's site than he would if monogamous. The decline in effort might allow more disturbance to the female or predation on either her or the young. The male, in turn, might be cuckolded if he switches his attention too soon.

Later in the season, both parents have conflicts with their young. The young attempt to secure as much care as possible, which may not always benefit their parents. If offspring inveigle their parents to divert to them energy or time needed for molting or for additional reproduction, their demands could work against the adults. Since warblers can launch second broods only during a brief period, parents that do and do not complete their critical responsibilities to their first brood by a certain cutoff time should respond to the offspring quite differently: after the cutoff time, the interests and benefits to parent and offspring of some additional care should be much more similar. It would be of interest to determine whether parents are equally solicitous of early and late broods. The parents may be in competition with each other with respect to their contribution to the offspring—and may be in a position to cheat each other. Paruline warblers are particularly interesting in this regard because division of labor after fledging is relatively even. If a second brood is not raised, the parents often separate soon after the young can fly, each with part of the young. In one-brood families, male and female Prairie Warblers take similar numbers of young (Nolan 1978). It is not known which sex feeds the young longer, which provides more food, and which is more successful in rearing young to independence. Presumably after separating, one sex would be unable to manipulate the other.

If parents attempt a second brood, the male typically tends all of the young while the female commences the next nesting (Nolan 1978). Prairie Warbler pairs destined to double-brood initially separate with parts of the first brood in the same way as those not commencing second broods (Nolan, pers. comm., 1987), but comparable data from other species would be of interest. If the males help the females with the second brood, the young of the first brood would not receive care as long or as much (per individual) as would the young of single broods. These birds offer excellent opportunities of evaluating parent-offspring and male-female conflicts. In a similar way, knowing the success of male-maintained total broods, male-maintained partial broods, and female-maintained partial broods is central to understanding the social dynamics of this part of the breeding cycle. Nolan did not find differences of this sort in the few full Prairie Warbler broods he observed; comparative data are again needed.

3 Foraging

The foraging patterns of warblers have been studied more than any other aspect of their biology, with the possible exception of nesting biology at incubation. Studies in the older literature on stomach contents (e.g., McAtee 1912) provide enough general information on food intake to establish many differences between species. Nevertheless, one cannot make precise assertions about how and where resources are gathered from stomach contents alone. Similar foods may be gathered in more than one place, and different foods may be found in the same place. Further, much of this old data on stomach contents is not broken down by season, habitat, or even geographic area. Differential preservation of items in the gut makes it difficult to identify and interpret stomach contents. Soft-bodied food items decompose into unrecognizable form more rapidly than hard-bodied ones—insect larvae become unrecognizable before, say, adult beetles. Therefore, to obtain a balanced picture of resource exploitation for a species, one must supplement stomach-content data with observations of birds in the field.

Still, we have a start. In addition to the early stomach-content studies there are natural historical sketches, such as those of Stanwood (1910) and Mousely (1924, 1926, 1928) in the first third of this century, that provide considerable insight into the foraging patterns of warblers. Kendeigh (1947) performed more systematic studies on the natural history of several warblers. Those and other studies suggested that foraging probably occupies a greater part of warblers' lives than any other single activity, yet I know of no attempt to compile a detailed time budget of a paruline warbler. Foraging contingencies may present warblers with severe challenges, and selection for efficient foraging must at times be quite high. This can occur, for example, when the female spruce-woods warblers incubate their eggs (Morse 1968). During the brief periods that they leave their nests, which total less than 15 percent of the time, these warblers often forage more intensely than any other birds I have ever studied, except the Goldcrest *(Regulus regulus)* (Gibb 1960; Morse 1978b) and its

North American counterpart, the Golden-crowned Kinglet (Morse 1970a), on cold midwinter days. Others have remarked on the high foraging rates of female Black-throated Blue Warblers (Black 1975) and Prairie Warblers (Nolan 1978) during incubation.

It was Robert MacArthur's (1958) study of the foraging patterns of several coexisting, spruce-dwelling *Dendroica* species that stimulated work on the foraging ecology of warblers. MacArthur's warblers all nest in what superficially appears to be homogeneous spruce forest, part of a biome that extends for thousands of kilometers across Canada and the northern United States, dipping southward in the United States along the cool Atlantic Coast and the Appalachian Mountains. Various combinations of these species occupy different parts of this large area. MacArthur's work was pivotal to our understanding of niche partitioning and diversification; that is, testing the proposition that no two species can coexist indefinitely on the same body of resources, if those resources are in limited supply. This theory is called the competitive exclusion principle or Gause's axiom (Gause 1934; Hardin 1960).

MacArthur viewed the spruce-woods warblers—abundant, morphologically as similar as peas in a pod, and nesting in the very same forests—as a key test case for the competitive exclusion principle. Given their morphological similarity, the apparent homogeneity of their habitat, and the large numbers of species (and individuals) within a limited area, they seemed at first glance an exception to the principle, thereby raising strong questions about its universality. By the mid-1950s the principle had been tested only a few times in field situations, perhaps most successfully by David Lack, one of its staunchest advocates, and his students. Lack (1946) found that several coexisting raptors differed markedly in their resource exploitation, which is consistent with, although by no means a test of, this principle. Lack also applied it to a group of English seabirds (Lack 1945), especially cormorants. John Gibb (1954), one of Lack's students, also found support for the competitive exclusion principle in the European titmice (*Parus* spp.), a group of coexisting insectivorous birds sharing many similarities with MacArthur's warbler assemblage. The species studied by Lack and his students exhibited major differences in size, other morphological characters, or habitat usage. That warblers should abide by the same principle was by no means obvious, given their close morphological similarity and abundance in spruce forests. MacArthur's discovery of differences in how warblers exploited a common habitat had a salutary effect on the resource-partitioning concept.

Similar patterns have now been studied so often that we tend to forget the novelty of MacArthur's work when he conducted it. Although Gause, Park (1948), and others had studied competitive exclusion in laboratory microcosms, little effort had been made to pursue the subject with field studies at that time. Many ecologists now assume that similar coexisting bird species will, under close scrutiny, be found to differ in resource-exploitation patterns. Many take this pattern as evidence of competition, past or present, though it remains to be demonstrated for each study that ecological segregation is a consequence of resource limitation.

Substrates and Their Influence on Foraging

On a broad scale one can separate warblers into arboreal, brush, and terrestrial foragers. All of the spruce-forest *Dendroica* (and most other *Dendroica* as well) are arboreal foragers; the yellowthroats *(Geothlypis)* are brush foragers; and Ovenbirds and waterthrushes *(Seiurus)* are ground foragers. Arboreal foragers make up the largest single group—79 percent of the North American species—mainly because the two largest warbler genera are composed of arboreal species (*Dendroica,* with 21 North American species; *Vermivora,* with 9 North American species). Population densities of many of these species are high as well (Robbins et al. 1986). Indeed, most of the highest-density species are arboreal foragers. This result possibly reflects the abundance of that substrate in warbler habitats. Brush foragers make up 13 percent of the total, including a few species, such as the Common Yellowthroat, that attain high densities. Ground foragers are the smallest group, and they make up a smaller part of the ground-foraging guild than do co-occurring taxonomic groups.

Foraging varies along several spatial parameters in the three-dimensional environment occupied by tree dwellers. The extent of an individual's vertical distribution is determined by the structural heterogeneity of a forest as well as its height. Such variation depends on the extent that emergent vegetation, canopy, understory, and herb or ground layers differ in representation and physical structure. In turn, the arthropod fauna harbored by the forest differs concordantly (Morse 1976b). Not only will an insectivore's food supply be affected by the prey's morphological and physiological attributes, but the problems of capturing them will change in response to these attributes. For instance, spruce trees are roughly conical from the lowest point at which needles appear. This shape affects the volume of tree space, and consequently the volume of open space. The ratio of one to the other—a function of tree density, tree height, and the height at which live branches commence—affects how warblers and their insect prey move; in turn, the constraints should affect the birds' morphological and behavioral adaptations.

Coexisting spruce-forest warbler species vary in mean vertical distributions, but their ranges overlap considerably (Figure 3.1). By watching long enough one sees the normally high-ranging Cape May Warbler on the ground, for example, or the low-ranging Yellow-rumped Warbler foraging in the top of a tree.

Arboreal habitats are also diverse in the horizontal dimension, even ones that are extremely homogeneous on a larger scale, such as spruce

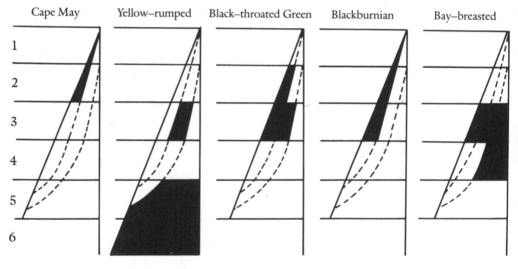

Figure 3.1 Principal feeding positions of spruce-woods warblers in trees, in terms of time spent foraging. Each triangular form is a diagrammatic spruce tree. Zones 1–6 refer to three-meter segments from the top of the tree (Zone 1) to the bottom (Zone 6). Zone 6 is the area below most of the foliage, with few if any limbs. The vertical line at the right side of each diagram is the trunk. The area immediately to its left, until the first set of dashed lines, is the inner part of the tree, which is characterized primarily by large limbs with little foliage. The next area to left, between the two sets of dashed lines, is the middle part of the limbs, characterized by heavy foliage. The area between the outer dashed line and the diagonal line is the outer part of the limbs, including the tips of the vegetation. Sites of the most concentrated activity are shaded until 50% of total foraging activity is mapped. Thus the shaded areas denote where birds are most likely to be foraging. (Modified from MacArthur 1958.)

forests. A bird encounters greater substrate differences when moving about on a horizontal plane (depending on whether it is foraging toward the distal ends of limbs, near the trunk, or even on it) than when merely feeding in the lower or upper part of the crown. Arboreal warblers range from tip foragers, such as the small *Vermivora* or *Parula*, to trunk foragers, the latter, "nuthatch" mode truly mastered only by Black-and-white Warblers. In between lie the majority of species, the foliage gleaners, which concentrate their activity in the midst of the vegetation. Many of the *Dendroica* fit this intermediate pattern (the Magnolia, Cape May, Black-throated Green, and Blackburnian warblers), other congeners concentrate their activities more heavily on the larger, inner parts of the limbs (the Yellow-rumped and Bay-breasted warblers). Although one can order species at points along a continuum of foraging sites, all use most or all of the parts of a tree at least occasionally, similarly to the pattern along the height dimension. The most exclusive of the horizontally partitioned substrates is the trunk. To exploit it efficiently, foragers must work in positions perpendicular (vertical) to their normal horizontal attitude. With the exception of the Black-and-white Warbler, use of the trunk by warblers in the North American temperate zone is confined to pecking from the base of limbs or, less frequently, to clinging to the trunk and pecking, or, at most, weakly crawling about on heavily ridged bark for short distances, as seen in Pine Warblers (Morse 1967c), Worm-eating Warblers *(Helmitheros vermivorus)* (Brewster 1875), or Yellow-throated Warblers in some geographic areas (Bent 1953). In the North American temperate area, other species that exploit trunks freely are largely or exclusively trunk feeders (woodpeckers, nuthatches, creepers). In contrast, Pine Warblers in the Bahamas exploit trunks more adeptly than their North American conspecifics (Emlen 1981; Emlen and DeJong 1981).

How much a species will exploit different horizontal substrates is strongly influenced by the physical characteristics of the trees themselves. The different structure of coniferous and broad-leafed trees (angiosperms) perhaps offers the sharpest contrast. Stiff tips of new spruce growth support 10-gram warblers during late summer, although not earlier in the season. To reach the tips of some large-leaved deciduous trees, however, the bird must perch on the petioles or exploit them in flight. The petioles are unlikely to support even the smallest warblers, although the Parula Warblers and *Vermivora* species, at 6–8 grams, concentrate their activities, usually one species per habitat, on these substrates. The distribution of deciduous leaves also differs considerably from those of conifers: in many deciduous trees that have an umbrella shape most leaves are arranged in a monolayer well off the ground, in contrast to the spruces and other conifers, whose conical shape results in a more even distribution of foliage along the trunk.

The spruce-woods warblers also differ in their directional movements while foraging. This difference can be expressed in the relative frequency of tangential (across branches), vertical (to branches above or below), and radial (outward or inward along branches) movements on the vegetation (MacArthur 1958) (Figure 3.2). My observations suggest that these differences arise from characteristics of the foliage in which a bird forages, rather than independent attributes of the warbler species. Thus, the Cape May Warbler, which forages in treetops, has the largest vertical component to its foraging, and the Yellow-rumped Warbler, which forages widely over trees, has the most even distribution of tangential, vertical, and horizontal movements of any of these species.

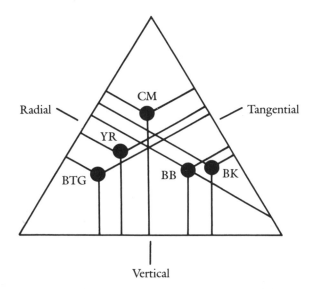

Figure 3.2 Frequencies of tangential, vertical, and horizontal foraging motions. BB = Bay-breasted Warbler; BK = Blackburnian Warbler; BTG = Black-throated Green Warbler; CM = Cape May Warbler; YR = Yellow-rumped Warbler. Lines were drawn from each dot perpendicular to the sides of the triangle. The lengths of these lines are proportional to the total distances that the species moved in the different motions. Thus, about 18% of the Black-throated Green Warbler's motions were radial, 23% vertical, and 59% tangential. (Modified from MacArthur 1958.)

Many warblers are confined to either coniferous or deciduous habitats during their breeding season, although at other times of the year they may have other options. How large a role do structural factors play in habitat use? Some species with breeding ranges concentrated in transitional coniferous-deciduous areas, such as the Magnolia and Canada *(Wilsonia canadensis)* warblers, use both substrates regularly; others differ on a regional basis, even becoming coniferous specialists in one area and deciduous specialists in another. The foraging patterns of the Black-throated Green Warbler differ markedly across its range: in coastal Maine it spends over 90 percent of its time foraging in conifers (Morse 1968); in mixed coniferous-deciduous forests in northern Maine its use of conifers declines to about 60 percent (Morse 1978a, unpublished); and in New Hampshire, where it occupies primarily deciduous forests, it spends only 5–10 percent of its time foraging in conifers (Holmes 1986). In West Virginia, some birds live in pure deciduous growth (Brooks 1947), and the disjunct race *waynei* summers in deciduous and cypress swamps of the southeastern United States (Bent 1953).

In addition to pecking at vegetation under foot as they move along (gleaning), warblers have a wide range of foraging maneuvers. The frequency of different maneuvers also differs greatly among species (Figure 3.3). They include hovering at the tips of vegetation, stretching upward to vegetation overhead, hawking for flying insects ("flycatching"), chasing fallen food downwards, and hanging upside-down from the vegetation. The maneuvers seem related to the characteristics of the habitat that the warblers occupy. Not surprisingly, the Yellow-rumped Warbler (MacArthur 1958; Morse 1968), often concentrated in lower parts of trees, where the habitat is open, is a habitual flycatcher. In contrast, Bay-breasted Warblers, which forage mainly in the dense inner parts of coniferous vegetation (MacArthur 1958; Morse 1978a), seldom indulge in this activity. Yet most species occasionally resort to any one of these techniques.

Search tactics are a third element of the exploitation patterns of arboreal insectivorous birds, along with the differential use of foraging substrates and prey-capture techniques (see Holmes and Recher 1986). Search tactics vary in a continuum from searching for prey over long distances in an open site, as by perchers waiting for large insects in the air column, to searching for prey at close range in dense foliage (Robinson and Holmes 1982). Robinson and Holmes divided search tactics into five categories in their study of arboreal insectivores in a New Hampshire deciduous forest. Different species employed different search tactics (flush-chase, near-surface, open-perch, variable-distance, substrate-restricted), thereby al-

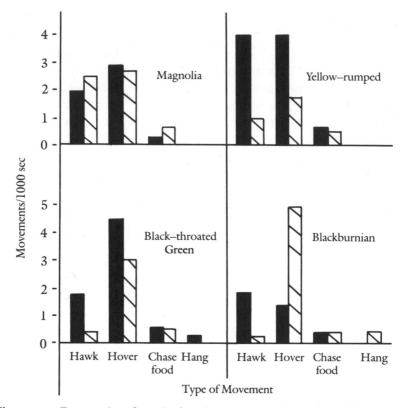

Figure 3.3 Frequencies of certain foraging movements by male (solid bars) and female (striped bars) spruce-woods warblers. (Modified from Morse 1968.)

lowing them a chance at different prey. Species typically excel at one type of foraging tactic and may even be morphologically equipped for maximum efficiency in that tactic, although the same species may sometimes exhibit several tactics. Hence search tactics can constrain the opportunities of a species. In this way, they parallel specialization among warblers to particular foraging substrates and prey-capture techniques.

Search tactics may have phylogenetic constraints. The four warblers studied by Robinson and Holmes (1982)—the Black-throated Blue, Black-throated Green, and Blackburnian warblers and the American Redstart—specialized in only two of the five search tactics (flush-chase, near-surface chase) used by the 11 insectivorous species present. Open-perch search, variable-distance search, and substrate-restricted search tactics were used by such birds as flycatchers, vireos, and chickadees.

Morphological differences (Osterhaus 1962), in part of phylogenetic origin, limit the ability of birds to perform several search tactics efficiently and thereby narrow their repertoires and opportunity for partitioning resources. For this reason they play an important role in determining which species can survive and coexist in a habitat (Robinson and Holmes 1982, 1984). From the tactics of prospective members of an arboreal community, and the physical vegetation structure and distribution of prey, we can predict the composition of the ensuing bird community (Ricklefs and Travis 1980; Miles and Ricklefs 1984). Nevertheless, because morphologically different species sometimes use similar modes of searching (Robinson and Holmes 1982), it is not safe to determine performance from morphology alone, in spite of the blurring of function, and phylogenetic constraints. Search patterns have some generality, since Holmes and Recher (1986) reported a warblerlike pattern in an Australian forest community of unrelated birds.

Generalists and Specialists, Stereotypy and Plasticity

On the basis of the number of foods that they gather, foragers may be divided into generalists and specialists. Generalists exploit a high proportion of the resources present, specialists take only a small proportion of them. The two terms are best treated as end points on a continuum and are most useful when comparing the behavior of two species (or conspecifics) (Morse 1971b, 1980a).

Paruline warblers are, with few exceptions, insectivores during the breeding season, which, on these criteria, qualifies them as resource specialists, although shifts to frugivory and nectarivory by some species at other seasons complicate the matter (Chapters 9 and 10). Further, the high frequency with which shifts have been found at other seasons, once they have been looked for carefully, suggests that these habits are widespread. Even within the breeding season the potential for considerable foraging variation exists, because of the wide range of insect food as well as the variety of sites at which it might be obtained and the maneuvers required to capture it. Although my definitions of generalist and specialist do not cover the latter factors, they are potentially important, for prey available in different places and under different circumstances (flying, hidden, and so on) may qualify as alternative resources.

Although different species typically concentrate on certain substrates, most warblers at least occasionally exploit a wide range of other substrates

(MacArthur 1958; Morse 1968). Some species, however, regularly exploit a wider range of substrates than do others. For instance, the Black-and-white Warbler, a species that might initially be considered highly specialized because it performs much of its foraging by creeping about on vertical surfaces in the manner of a nuthatch, regularly exploits more conventional substrates as well. In addition to the trunk maneuvers it forages frequently on the inner parts of large limbs and on the foliage in typical warbler fashion, sometimes even exploiting the tips of the vegetation (Morse 1970a). As a result, its overall foraging diversity (equality in frequency of use of different kinds of substrate) exceeds that of several other warblers (Figure 3.4). By contrast, small warblers (*Vermivora* and the Northern Parula Warbler) spend a great amount of their foraging time in the tips of vegetation, an area that they exploit much more efficiently than do heavier warblers *(Dendroica)*. Large species often depress the tips of vegetation on which small species forage with ease.

A survey of the warblers as a whole uncovers no strong specialist adaptations approaching those of primarily nectarivorous groups such as hummingbirds. In migratory species this absence could be due to the need to respond to a wide range of conditions over the year, but that hypothesis does not account for the failure of tropical warblers to undergo extreme specialization.

It is common, but mainly untested, wisdom that there is an inverse relation between the number of resources an animal exploits and the efficiency with which it may exploit them (Schoener 1974); that is, the jack-of-all-trades is a master of none (Klopfer and MacArthur 1960, 1961). None of the warbler studies addresses this question directly, although given the diversity and variety of warblers, it should be possible to test this hypothesis. The critical currency with which to gauge efficiency presents a major problem; should we be measuring time or energy? Time is much simpler to measure than energy, but it is so tedious to obtain adequate samples that few time budgets of any passerine birds exist. Time and energy might be convertible, yet it remains necessary to calibrate them, keeping in mind that the relations may not be linear (Wolf et al. 1975). D. S. Wilson (1975) has, for instance, predicted that efficiency of prey capture will not be symmetrical along a gradient such as prey size, but that it will increase rapidly with minimal prey size and decline slowly with increasing prey size. Since spruce-woods warblers, and probably others as well, feed mainly on insects that are considerably larger than the mean

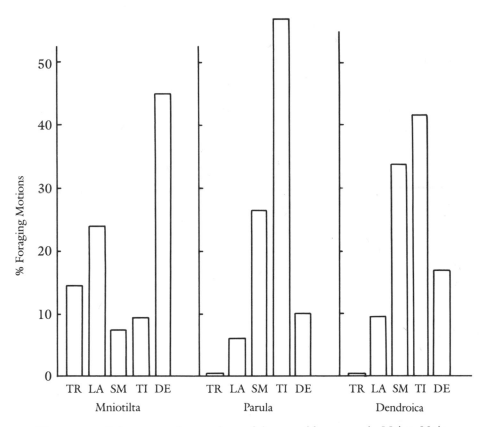

Figure 3.4 Substrate use by members of three warbler genera in Maine. *Mniotilta* = Black-and-white Warbler; *Parula* = Northern Parula Warbler; *Dendroica* = Magnolia, Yellow-rumped, Black-throated Green, and Blackburnian warblers. TR = trunk, LA = large (inner) part of limb, SM = small (outer) part of limb, TI = tip of vegetation, DE = dead limb. (Modified from Morse 1970a.)

insect size in their habitats (3–4 times: Morse 1976b), the shape of this efficiency curve is of considerable importance to them. For instance, in one study the spruce forest contained extremely large numbers of foliage-dwelling insects of 2 millimeters or less (aphids, coccids, etc.), which the warblers seldom if ever took. Instead, they captured caterpillars 15 millimeters or longer in numbers far greater than would be predicted by their abundance in the foliage (Morse 1976b). Recognition of these fundamental relations between abundance and profitability has been incorporated directly into the generation of optimal diet models, which predict the circumstances under which items of different profitability should be exploited or eschewed (Pyke et al. 1977; Pyke 1984). In addition to the possibility that the warblers avoid taking these prey, they may not be detecting them.

Although some foraging patterns may be predictable at specific times, they remain subject to change over longer periods. For this reason it may be important to characterize foraging patterns in terms of their tendency to vary over time, especially since ecological conditions at some time over the lifetime should select for traits, behavioral or morphological. To describe how much birds vary their foraging patterns we use the terms *stereotyped* (birds that exploit resources in the same way regardless of conditions) and *plastic* (birds that exploit resources in different ways as conditions change) (Morse 1971b, 1980a). As with *generalist* and *specialist,* these terms are most useful relative to each other. Although stereotypy and plasticity roughly parallel generalist and specialist, they are not synonymous. Specialists might either be stereotypic or plastic, with the former invariably exploiting only the same narrow range of resources and the latter occasionally substituting others. Generalists may also be stereotyped or plastic, with the former exploiting a wide range of resources with high predictability and the latter using a wide range of resources with little long-term predictability.

Stereotypy has been typically linked to phylogenetic position and to environmental stability, especially the stability that is believed to characterize tropical areas—at least the lowland forests (Klopfer and MacArthur 1960, 1961). The thesis is that phylogenetically old groups possess a low level of neurological plasticity. Given the likely phylogenetic homogeneity of the warblers, it seems improbable that phylogenetic differences among them could be predicted. Yet the proposed relation between environmental stability and stereotypy should be testable.

One would expect to find plasticity well developed in all migratory warblers, since they experience such different conditions during the year. But species vary in their tendencies toward stereotypy during the breeding season, even within a habitat. For instance, among the spruce-woods warblers Black-throated Green Warblers forage much more predictably than do Yellow-rumped and Northern Parula warblers, as can be readily observed when they are isolated on islands with different numbers of other warblers (Morse 1971a) (Figure 3.5). Among these species there is an inverse ranking between dominance (Morse 1976a) and plasticity (Morse

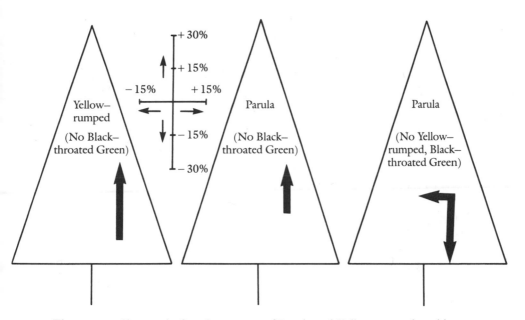

Figure 3.5 Changes in foraging ranges of Parula and Yellow-rumped warblers in different species combinations on small islands along the coast of Maine. The left dendrogram illustrates foraging shifts of Yellow-rumped Warblers when no Black-throated Green Warblers were present (versus when they were present); only Yellow-rumped Warblers and Parula Warblers were present. The center dendrogram illustrates foraging shifts of Parula Warblers when no Black-throated Green Warblers were present; only Yellow-rumped Warblers and Parula Warblers were present. The right dendrogram illustrates foraging shifts of Parula Warblers when neither Yellow-rumped nor Black-throated Green Warblers were present; no other spruce-woods warblers were present. The direction and length of the arrows in the dendrograms indicate the direction and magnitude of the foraging shifts (higher or lower, inward or outward) in the trees. (Modified from Morse 1980b.)

1971a), with the socially dominant Black-throated Green Warbler the least plastic, the socially subordinate Parula Warbler the most plastic, and the Yellow-rumped Warbler intermediate. Social relations produce the same results that one would predict from environmental uncertainty. Indeed, it seems most appropriate to treat the presence or absence of social dominants as an aspect of uncertainty (Morse 1974b). Overall, it would be difficult to separate these three warblers as generalists or specialists, since all three concentrate on certain parts of the foliage: the Parula on branch tips, the others on more proximal parts of limbs. Yet their interspecific interactions provide another potentially vital dimension. Comparable data are needed from several other species, though, before robust comparisons can be made.

Morrison (1981) was impressed with the similarity in foraging patterns of the Orange-crowned (*Vermivora celata*), Wilson's, and MacGillivray's warblers that he studied in Oregon to those of their eastern counterparts (other races of Orange-crowned and Wilson's warblers and the marginally distinct Mourning Warbler). The Oregon populations, probably derived from the eastern taxa during Pleistocene glaciation (Mengel 1964), either confined their activities to habitats that resembled those found in the east, or they foraged in other kinds of areas in a way that resembled eastern foraging patterns, often using only specific parts of such habitats—possibly at considerable expense to their foraging efficiency. Morrison hypothesized that these foraging patterns were in place before these birds' ancestors invaded the west. This explanation would argue for stable behavioral patterns, which exceed those suggested for migratory warbler species.

These species are, interestingly, deciduous foragers. Greenberg (1979, 1984b) found that the Chestnut-sided Warbler, a deciduous-forest species, was much more highly stereotyped in its foraging repertoire than the Bay-breasted Warbler, a coniferous-forest summer resident. He suggested that the marked behavioral differences between these two species could depend on the varying conditions during a year. Chestnut-sided Warblers, which are deciduous specialists in the summer, can expect to encounter broadleaf vegetation throughout the rest of the year, on their wintering grounds as well as their breeding grounds. Consequently, they may not have evolved the level of plasticity seen in species that summer in coniferous forests, such as Bay-breasted Warblers. Such conservatism could also help to explain why within-habitat diversity of western warblers does not match that of eastern species.

Foraging and Optimality Theory

Much of the recent work in foraging has tested optimal foraging theory (reviewed by Pyke 1984), the proposition that animals will forage in a way that maximizes their fitness—in the simplest instance maximizing their resource gain per unit time. Since the relevant measure of fitness is offspring placed in the next breeding generation rather than maximum resource uptake, simple resource maximization is, in a strict sense, not an appropriate measure of optimization. But given the difficulty of obtaining data on reproductive success (direct recruitment measures have not been made in the context of foraging for *any* animal), resource maximization—subject or not to constraints such as predator avoidance, nutrition, and so forth—has generally been used as an estimate of fitness (see evaluation in Morse and Fritz 1987). Many studies of warbler foraging were performed before the spate of research on optimal foraging first appeared in the mid-1970s, and most dealt with questions of niche partitioning. Therefore, they are only secondarily concerned with problems central to optimal foraging theory (see Werner 1977) and have usually provided little direct insight into this burgeoning area of ecology. The important work by Zach and Falls (1976a–c, 1977, 1978, 1979) on Ovenbird foraging is a notable exception.

Optimal diet theory predicts that foragers will choose prey that permit them to increase their rate of intake and eschew prey that will decrease this rate. The prey taken and not taken will be a consequence of the ratio between their energy value and handling time *(E/t)* and will be affected by the size, abundance, distribution, visibility, and problems of processing the prey. Some variables may change as prey are depleted in nonrenewing or slowly renewing systems.

The Ovenbirds usually took large prey in preference to others, as would be predicted. When presented with unfamiliar prey (within a normal size range), the tendency to select large over small prey was especially well marked. With familiar prey, size selection played a lesser role in choice, and profitability, a factor requiring learning, became more important. Ovenbirds responded similarly to simultaneously and sequentially presented prey (Zach and Falls 1978); hence, tests presenting choices available at the same time can be extrapolated to the more frequent field situation of not finding alternatives at the same instant. The level of selectivity decreased with satiation (Figure 3.6). If presented with numerous prey types, the birds typically took more than one, although they shifted to more monotonous diets as experiments drew on. This pattern could result from the birds' need for a mixture of prey to satisfy nutritional requirements, or it could be a consequence of learning prey types and maximizing profitability. Whether insectivorous birds voluntarily mix their diets, or whether the complexity of field situations naturally produces such a mix, is debatable (Tinbergen 1960; Royama 1970). If the former, selecting a mixed diet acts as a constraint that decreases energetic efficiency. Zach and Falls's diet studies suffice to establish that the birds face a complex of factors in even simplified area foraging situations, and that learning and

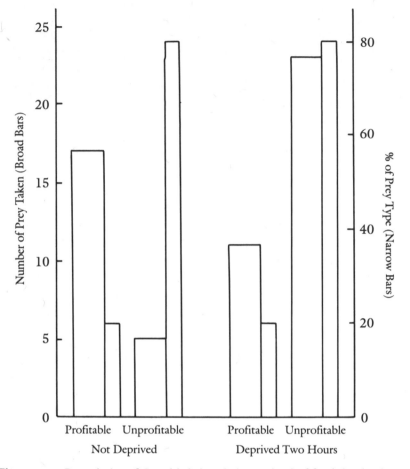

Figure 3.6 Prey choice of Ovenbirds in relation to level of food deprivation. Birds encountered favored, more profitable prey (mealworms, *Tenebrio mollitor*) and less profitable prey (beetles, also *Tenebrio mollitor*) sequentially in a 20%– 80% ratio in experimental arena. (Modified from Zach and Falls 1978.)

memory are often required to ensure that the birds maximize their intake, or even approach that level.

Patch-choice theory similarly predicts that foragers will select hunting sites on the basis of profitability, including them within their foraging repertoire if they permit an increase in resource intake. Ovenbirds also exploited patches of high- and low-density prey in the way predicted by optimal foraging theory, concentrating their activities within areas of high food density (Zach and Falls 1976c). But their fit to predictions was only qualitative because they took prey from low-density sites as well. A recurring tendency to use low-quality sites might result from difficulties of learning the profitability of the areas (Zach and Falls 1976b); it could also be due to some prey in low-quality sites being more profitable than many in the so-called profitable areas (because of high visibility, for example); or it could involve a long-term strategy that permitted the birds to track changing patterns of resource abundance (Smith and Sweatman 1974). Regardless of the reason for this foraging characteristic, it illustrates the difficulty of quantifying profitability if food items are cryptic, as are the Ovenbird's prey. These birds exhibited area-restricted searching (Zach and Falls 1977); that is, they decreased the length of moves between hunting sites and increased the angles of turns made between sites, thereby keeping their activities within high-quality sites and probably enhancing their ability to learn favorable sites. However, they also tended to avoid sites that they had visited recently; this behavior would be advantageous if replenishment time was slow (Zach and Falls 1976c), in this case defined as the period required for other insects to fill the sites recently depleted.

This ground-foraging species resembles arboreal ones, most notably titmice (e.g., Krebs et al. 1983), in that they feed on cryptic or hidden prey. The search for food presents similar perceptual and learning problems to both ground and tree dwellers, and these factors may be responsible for a high proportion of the deviance from the simple models of optimal foraging.

Male-Female Differences

Since male and female warblers are structurally similar, one might not expect to find major differences in foraging patterns between them like those found in strongly sexually dimorphic species, such as the extinct Huia *(Neomorpha acutirostris)* and the Hispaniolan Woodpecker *(Centurus striatus)* (reviewed in Morse 1980b). Yet intersexual differences may be marked. During the breeding season male spruce-woods warblers forage considerably higher in the trees than do their females (Morse 1968), and males alternate foraging with singing in the canopy while females alternate foraging with incubation. Foraging heights of males match their singing heights, as do the foraging and nesting heights of females. It is not known whether these differences remain at other times of the year, but if they are merely due to pursuing an efficient space-restrained foraging strategy (foraging close to display areas by males and near nests by females), one would not predict them at other seasons. A similar distribution has been found in several additional warblers (e.g., Black 1975; Sherry 1979; Franzeb 1983) and vireos (Williamson 1971), suggesting that differences in nonforaging activities between the two sexes will be accompanied by corresponding foraging differences.

Holmes (1986) has explored the hypothesis that sexual differences in foraging result from individuals using the areas closest to the primary sites of other activity (again, display for males and nest maintenance for females). Females of some species foraged higher than males in deciduous forests of New Hampshire. Holmes concluded that in the deciduous forests, with their thick monolayer canopy, the most conspicuous display sites were lower in the trees, beneath the dense vegetation and often lower than the females nested.

Male and female American Redstarts invariably differed in other aspects of their foraging behavior as well (Holmes et al. 1978). Paired individuals that foraged at similar heights differed the most in foraging tactics (relative frequency of gleaning, hawking for insects, etc.), suggesting that each compensated for the other's activity.

Since males in high-density populations seldom if ever feed their females on the nest (Morse 1968), this pattern of spatial separation may be highly efficient for the females. With high intra- and interspecific population densities, a male's time can be completely taken up in territorial defense, leaving him no opportunity to provision the female. Given these circumstances, one might predict that male-female differences in foraging will be less extreme if males contribute to the nesting females' diet than if the females have to fend for themselves. If so, foraging differences could be a consequence of males not feeding females, which in turn results from the probable competition accompanying a densely populated site.

Alternatively, differences in the likelihood of nest predation could account for this difference in male activity about the nest. To choose between these alternative explanations we would need to know whether the males would feed females if these constraints (territorial competition and predation) were relaxed. If they would, a corresponding decline should occur in foraging differences between the sexes. One could test this hypothesis on small islands that have lower levels of potential competitors (intra- and interspecific) and nest predators (Morse 1971a, 1977b) than do larger mainland sites. The sexes tend to forage more similarly on islands than in large mainland populations (Morse 1971a); however, I have no information on males feeding incubating females.

To distinguish the mechanism responsible for differences in foraging patterns, one must conduct observations in situations where only one of the two factors, competition or predation, is low. Studies of low-density Pine Warbler populations in an open, young-growth Loblolly Pine plantation in Maryland, where they were the only species of warbler and one of the few forage-gleaning species, suggest a test. Males and females did not forage very differently from each other, either in height or type of substrate (Morse 1974a). This pattern may have been due to their living in an impoverished and structurally simple habitat with a low volume and diversity of insects (Wahlenberg 1946). In contrast, in a high-density population with abundant food but few similar species present, male and female Yellow Warblers exhibited marked height differences in foraging (Busby and Sealy 1979), like the spruce-woods species.

These observations suggest that a primary nesting-season consideration—display by the male in the presence of many conspecifics—may suffice to engender different foraging patterns in the sexes. Choice of nesting site may fix the female's area of activity as well. Busby and Sealy's results indicate that differences may occur in the absence of significant contact with congeners or similar species, but the presence of other species should enhance this effect. None of these results precludes the possibility of predation playing a direct role.

Foraging Speed

Warblers' foraging speeds, measured in pecks per minute, differ over the breeding season (Figure 2.2). Immediately after returning to spruce-forest breeding areas, both male and female Magnolia, Yellow-rumped, Black-throated Green, and Blackburnian warblers foraged at moderate speeds (Morse 1968). But during incubation and while the nestlings were small, females of each species foraged faster than did their mates. Since females perform all of the incubation, over 85 percent of the daylight hours, and are not fed at the nest by their mates, this difference is not surprising. What is surprising is the extent of their pace, in terms of peck rates and general activity. It is difficult to imagine that these warblers could forage more rapidly, in terms of cost-benefit ratios, without decreasing their efficiency.

As nestlings grow older, the males begin to feed their young regularly and their foraging rates increase correspondingly, while the females' rates decline. The sexual difference in foraging speed disappears, and thereafter male and female speeds remain similar while fledglings are being fed (Figure 2.2). The similarity during fledgling dependency is not unexpected, since males and females wander apart soon after fledging, each with part of the brood. Similar seasonal shifts in foraging speeds have been reported for Black-throated Blue Warblers (Black 1975) and American Redstarts (Sherry 1979); this pattern will probably turn out to be the rule.

Metabolic Demands

Routine metabolic demands can influence foraging routines. We often think of summer temperatures being warm enough for birds to escape the demands of cold weather, but this is not always the case, especially in northern forests. For example, resting oxygen consumption was markedly higher during summer nights for several forest-dwelling passerines, including the Black-throated Blue Warbler (Holmes and Sawyer 1975; Holmes, Black, and Sherry 1979), indicating that the birds increase their metabolic rates at inactive times. Higher metabolic demands raise food demands and might even depress population densities.

The cost and importance of this problem differ according to body size, magnitude of temperature deficit, and size of energy budget. Being small, warblers should be subject to high heat loss per unit mass. The amplitude of the deficit shifts with latitude and altitude as well as with season. Some species are more vulnerable than others. Flycatching birds have high energy budgets because they fly so frequently (Holmes, Black, and Sherry 1979); flight requires a rate of energy consumption roughly 10 times that of the basal metabolic rate (King 1974). In contrast, hopping requires only about half that much.

Bergmann's rule, that size varies inversely with ambient temperature, is usually invoked to account for increases in size with latitude. The numerical dominance of Blackpoll Warblers in montane coniferous zones of northeastern North America may be due to an advantageous area-volume ratio, as may this species' dominance in the northernmost forests across Canada and Alaska. Blackpoll Warblers are considerably larger than most conifer-dwelling species of *Dendroica* at lower altitudes and latitudes

(Sabo 1980). Even so, lipid levels of this species drop during periods of inclement weather subsequent to the birds' arrival on their breeding grounds in June at Churchill, Manitoba (Yarbrough 1970). Of further interest, the other *Dendroica* appearing in these habitats is the Yellow-rumped Warbler (Erskine 1977; Morse 1979), which is larger than the Black-throated Green, Blackburnian, and Magnolia warblers.

Small insectivores routinely operate on tight energy budgets during the breeding season, especially while feeding nestlings (Yarbrough 1971; Holmes 1976). Yarbrough (1971) found that during a year of frequent thunderstorms, breeding birds maintained their metabolisms at such a marginal level that it was often impossible to run standard metabolic tests on them because their condition deteriorated over the holding time prior to measurement. If tested, these birds not infrequently died before morning. Acadian Flycatchers *(Empidonax virescens)* were especially vulnerable, perhaps because of the high cost of a flycatching routine (Holmes, Black, and Sherry 1979), and because of weather-related fluctuations in the availability of flying insects. Habitually flycatching warblers such as American Redstarts should be more vulnerable to these constraints than are other warblers. Although overnight heat losses should be a smaller part of flycatching birds' energy budgets (a consequence of high daytime costs), Yarbrough's results suggest that nighttime heat loss may be more important to them than to birds with more modest energy budgets.

Differences in Island and Mainland Birds

Comparisons of foraging patterns in bird populations on the mainland and adjacent islands or on islands having different species compositions have elucidated the role that other birds, of the same and different species, play in the exploitation of resources. It is usually assumed that foraging patterns differ on small islands because of reduced competitive pressure, or at the very least, because of the decrease in number of potentially competing species. The classic is Selander's (1966) comparison of woodpeckers from the island of Hispaniola, which has only the Hispaniolan Woodpecker, and Texas, which has several woodpecker species. Selander found that the Hispaniolan Woodpecker exploited a wider range of habitats and foraging techniques than did its congener, the Golden-fronted Woodpecker *(Centurus aurifrons)* in Texas. Crowell (1962) and Sheppard et al. (1968) obtained smaller results in comparisons of Gray Catbirds *(Dumetella carolinensis)*, Eastern Bluebirds *(Sialia sialis)*, and Cardinals

(Cardinalis cardinalis) on the faunally impoverished island of Bermuda and the adjacent North American mainland. All of these authors have argued that the island birds responded to a paucity of similar species. This expansion of a niche along one or more dimensions is termed ecological release (Schoener 1967). Unfortunately, mainland and island individuals of all these species are members of distinct, isolated populations, even distinct species (woodpeckers). Thus, although the explanation for the niche differences recorded in these studies is reasonable, the mechanism cannot be directly addressed. Neither can one tell whether niche differences are the consequence of other species' absence, nor whether a species would forage in the same way in the absence of others. If the latter is the case, habitat selection, itself a likely consequence of ecological release in the past, may account for the difference.

One can eliminate the problem of dealing with isolates by comparing individuals separated from a population for a time with members of the same population that are exposed to a large complement of species. Three species of spruce-woods warblers (Black-throated Green, Yellow-rumped, Northern Parula) that I have studied along the Maine coast provide this opportunity (Morse 1971a). These species occupy the mainland spruce forests and nearby islands that are sometimes so small that they support only a pair each of one to three of these species. The islands are close enough to one another and to the mainland that I have observed movements among the sites, yet they are small enough to ensure that only part of the mainland species will occupy them. The pattern of colonization on these islands is highly predictable: if only one species of warbler is present, it is the Northern Parula; if two species are present, they are the Northern Parula and Yellow-rumped; the Black-throated Green Warbler is present only if the other two are present. The Black-throated Green Warbler forages similarly on the islands and in large populations on the adjacent mainland with five warbler species (these three, plus Blackburnian and Magnolia). By contrast, Yellow-rumped and Parula warblers make fuller use of sites exploited by Magnolia and Blackburnian warblers on the mainland. On islands supporting only two species, they also partially fill in the sites that are normally heavily exploited by Black-throated Green Warblers (Figure 3.5) (Morse 1980b). Lastly, if the Northern Parula Warblers are by themselves, their foraging patterns shift even more, and they use sites usually exploited by Yellow-rumped Warblers still more heavily. These shifts closely match the dominance relations among the species (Morse 1971a, 1976b); further, hostile interactions can be seen among these species. Therefore, the evidence for ecological release is much stronger in this comparison than in comparisons between two widely separated populations. If these birds were manipulated by removal or addition of species, they would provide even stronger evidence for ecological release.

In experiments conducted in an observation tent, Emlen and DeJong (1981) demonstrated that neither of two Bahamian populations of Pine Warblers changed their foraging patterns when allowed to forage on the other's normal substrate. Emlen (1981) had previously found that Pine Warblers from Grand Bahama Island foraged quite differently from those on Andros Island. Those on the former island specialized in pine foliage and coexisted with two bark-feeding specialists; those on Andros foraged on both foliage and bark and did not coexist with bark feeders. This experiment demonstrates that the isolated populations have developed distinctly different patterns of foraging, which may fit the same patterns as those seen in ecological release. However, these birds may have lost the tendency to shift their foraging patterns, relying instead on either genetic or cultural inheritance. More studies of this sort are needed to establish the types and extent of differences to be encountered. Although the comparative method and natural experiments may present a persuasive scenario (see Alatalo et al. 1986), they do not provide the controls needed to pursue the argument further. Emlen and DeJong speculated that further studies will uncover other examples of foraging changes like that found in the Bahamian warblers.

The spruce-woods warblers on the islands along the Maine coast serve as an experimental population to test predictions on factors associated with differences in foraging speed. Because insect food supplies on the islands and mainland are similar (Morse 1977b), thereby eliminating food abundance as a variable, differences in species composition and nest predators are the most likely variables to account for differences in foraging speed. Red Squirrels are absent on these islands and Blue Jays visit only sporadically, as opposed to the regular appearance of both species on the mainland. Male and female Black-throated Green and Yellow-rumped warblers on the small islands foraged more slowly during most the season than did their mainland counterparts. This difference in speed between island males and females paralleled the one between mainland males and females (Morse 1981). Given that males make little if any contribution to incubation and care of newborn young, the lower foraging speed of island males probably has more to do with differences in frequency of territorial displays rather than with nest predators. Males on the islands sang much less frequently than did mainland individuals, especially those songs associated with territorial maintenance and encounters between males (Morse 1970b). Island females probably foraged more slowly because they spent less time incubating, as even mainland females almost never made contact with other individuals during their time off the nest. Although no direct evidence was available, island females were observed more frequently than were mainland females (Morse 1981), perhaps because they spent more time off the nests. In that a difference in predation pressure was the most obvious variable, it is tempting to attribute the low foraging speed of females to a change in the time budget occasioned by the absence of nest predators.

The Effect of Vegetation Structure

The studies I have reviewed focused on the effects of breeding phases and interactions with other species, especially congeners, on foraging patterns. But a growing body of literature (e.g., Holmes, Bonney, and Pacala 1979) indicates that foraging repertoires are also strongly affected by the physical composition of the vegetation and type of prey. Studies of species in forests of differing structure further suggest that vegetation may play a role in determining how individual birds exploit an area. For instance, in two forests differing in the presence or absence of an understory, Maurer and Whitmore (1981) found that certain insectivorous birds varied their foraging repertoires markedly, others little if at all. Of the five species they investigated, one of the two warblers, the American Redstart, foraged differently in the two forests, hawking for insects far more frequently in the open understory. The second warbler, the Black-throated Green Warbler, a species that hawks for insects much less frequently, showed moderate changes, intermediate in magnitude between the American Redstart and Red-eyed Vireo *(Vireo olivaceus),* which seldom hawks for insects and which exploited the two forests in a similar way.

Comparing a subalpine coniferous forest and a northern hardwood forest, Sabo and Holmes (1983) found that foliage structure, food availability, and interspecific interactions all played key roles in determining the exploitation patterns of the passerine birds in these two habitats. In general, physical factors appeared to play a greater part in the harsher subalpine habitat; however, that habitat had eight foliage-gleaning warblers in comparison to only four in the northern hardwood forest. Thus, the subalpine warblers countered the overall diversity pattern of birds in the community. This diversity pattern parallels that of the coastal spruce forests where I have worked, which are taller and have a more moderate climate. Sabo and Holmes were at a loss to explain this surfeit of subalpine warblers, but they suggested that it may be due to the rich speciation patterns of the warblers in eastern North America (see Mengel 1964). If so, historical factors are also important in interpreting these relations.

Of species common to both subalpine and northern hardwood habitats, the Black-throated Green Warbler exhibited almost no change in its foraging pattern, which may result from its dominant position in the hierarchy of foliage-gleaning wood warblers; further, Sabo and Holmes (1983) noted that transitional forests appear to be the center of its activity in New Hampshire, such that both types of habitat might normally be included within an individual's foraging nexus. This factor might account for their shifting conditions more in their study than in the study by Maurer and Whitmore (1981), although populations of this species may show clear differences in their foraging repertoires (Morse 1978a). In contrast to the Black-throated Green Warbler, the American Redstart exhibited moderate foraging differences in the two habitats. Sabo and Holmes attributed this shift to the absence of Least Flycatchers *(Empidonax minimus)* in the subalpine forest. Least Flycatchers exhibit many foraging maneuvers in the hardwood forest similar to those of the redstarts and are socially dominant to the latter (Sherry 1979).

The combined results of these studies suggest that a multitude of factors may dictate the foraging repertoires of birds. Thus, simple analyses of single resource-use parameters will not suffice to advance this subject (Sabo and Holmes 1983). This conclusion may come as no surprise, but it does highlight the need to design larger integrated studies that include replicates and experimental manipulations.

Variation over Time

An individual's foraging patterns, as well as its foraging rate, may also differ strikingly over time. Such shifts are obvious in migrating individuals, which drift into a variety of habitats, or in permanent residents that will glean a deciduous forest differently before it has leaves and after they have appeared. Other changes may also occur over time—changes that, though subtle, may be important. Such changes can even occur within a single day. Foraging patterns of Yellow-throated Warblers that have just returned to their breeding grounds on the eastern shore of Maryland

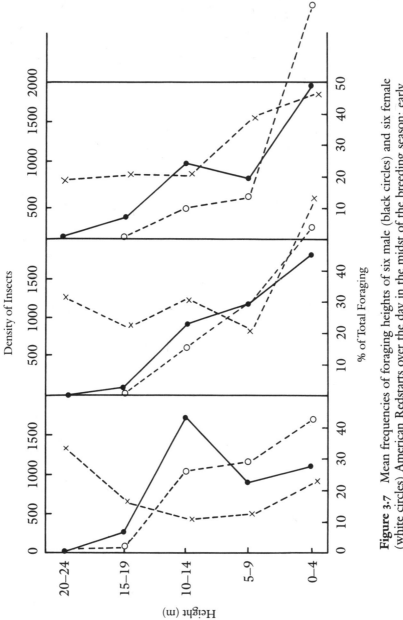

Figure 3.7 Mean frequencies of foraging heights of six male (black circles) and six female (white circles) American Redstarts over the day in the midst of the breeding season: early morning, 06:00–10:00 (*left*); midday, 11:00–15:00 (*center*); and late afternoon, 16:00–20:00 (*right*). The mean numbers of flying insects captured in suction traps per 28,300 cubic meters per hour (crosses) show the varying density of insect food available. (Modified from Holmes et al. 1978.)

differ strikingly with the weather. On cold, early mornings foragers probe their long beaks into the bracts of arthropod-rich Loblolly Pine cones, and later in the day they hawk for insects as the rising sun warms the prey to flight temperature (Ficken et al. 1968).

Holmes et al. (1978) have performed the most detailed study of changes in foraging patterns over a day, on American Redstarts in a New Hampshire deciduous forest (Figure 3.7). Observers simultaneously gathered data on redstarts over an entire day. Certain changes were predictable; for instance, the redstarts, which fill a flycatcherlike niche (Sherry 1979), increased their frequency of hawking for insects as the day warmed up and insects became more active. Further, they foraged progressively lower in the canopy as the numbers of low-flying insects increased. The shifts all appeared to be related to the changing abundance and activity of insects below the top of the forest. Abundance of insects in samples taken by suction traps showed high variance, however; this suggests a patchy distribution, in turn likely to affect the foraging regime elected by the birds. The results of this study must be carefully heeded by students of bird foraging. Simple summations of data from all parts of the day may obscure important variations.

Hutto (1981b) compared the daily foraging patterns of three warblers and a tyrant flycatcher, which differed in their frequency of hawking for insects. Peak foraging of the warblers occurred in the morning, followed by a decrease for the warblers at midday, which Hutto ascribed to the increasing activity of insects (and the difficulty of capturing them). In support of this argument, warbler species with greater overall frequencies of hawking peaked later in the day than those exhibiting little if any such tendency. Thus, Yellow-rumped Warblers showed a foraging peak later than that of the Common Yellowthroat, but even it decreased its rate by early afternoon, while the hawking specialist, the Willow Flycatcher *(Empidonax traillii),* continued to forage at a high rate through midafternoon. As Hutto noted, morning peaks might also be attributed to temporarily elevated energetic demands or to midday thermal stress; but if those were the only factors involved, one would not predict the between-species differences in activity patterns.

The Effect of Inclement Weather

I now believe that periods of inclement weather are one of the most important times for studying activity patterns, though little work has been done at such times. Bad weather can bring on one of the infrequent "crunches" that shape exploitation patterns (Wiens 1977; Dunham 1980). I am as negligent of this oversight in my own fieldwork as anyone else. Although investigations in times of good weather often reveal striking differences among and within species, they present only one aspect, as a study of redstarts (Holmes et al. 1978) revealed for other parameters. Summer storms can drastically affect foraging conditions for the spruce-woods warblers in my coastal study area. Not infrequently, prolonged storms with driving wind and subsequent periods of dense fog and dripping-wet vegetation strike while females are incubating eggs or new-born nestlings. At such times foraging options are severely limited. During heavy rain they, and their mates as well, may become completely inactive. Even when it is not raining, if the foliage is wet, they have to confine their foraging to substrates that they would normally not regular-ly use—usually the inner parts of limbs in areas which contain little if any foliage. Wet foliage may affect foraging even more than light rain itself, judging from bird foraging patterns on wet and dry foliage just after rain has commenced (Morse 1970a). Wet foliage dampens a bird's plumage quickly, thereby destroying its insulatory capability. Wet foliage may even account for variation in abundance of some warbler species, especially the Black-throated Green Warbler in Red and White Spruce forests (Morse 1976b). White Spruce *(Picea glauca)* needles lie at right angles to the branches (Figure 3.8); if wet, they quickly dampen a bird's ventral plum-age. Red Spruce *(P. rubens)* needles lie parallel to the branches and there-fore will not dampen a bird's plumage as rapidly. Birds should thus be able to move about more readily on wet foliage of Red Spruce than of White Spruce. Observations of warblers on both wet and dry foliage are needed from mixed Red and White Spruce forests to test this prediction; my initial data suggest that the warblers do concentrate their foraging on Red Spruce at these times.

Any niche shifts prompted by wet foliage should intensify interactions among the species, since they all increase the proportion of activities un-dertaken within the inner parts of the limbs (Figure 3.9). When vegetation is wet birds perform most of their foraging on bark, a substrate at other times receiving only minimal use by Parula, Magnolia, Black-throated Green, and Blackburnian warblers. In addition to the basic problems of using an apparently suboptimal substrate, arthropods in these sites are likely to be inactive in a storm and therefore less available. These condi-tions should not have as grievous an effect on Yellow-rumped Warblers

Figure 3.8 Red Spruce *(upper left)* and White Spruce *(lower right)* foliage, illustrating differences in the directions of the needles. (Illustrations by Elizabeth Farnsworth.)

since they normally forage heavily on bark substrates; however, their niche may be severely encroached on by the other species at a critical time. The combined abundance of these other species often exceeds five times that of the Yellow-rumped Warbler. Because Yellow-rumped Warblers are socially subordinate to several of the invading species (Morse 1971a, 1976a), such incursions may not be easily resisted and could be partly responsible for this species' low abundance and large territories in the study areas (Morse 1976a, 1977b).

Inclement conditions might even be of central importance in the population dynamics of these species, with nesting success mediated by foraging opportunities providing the vulnerable link. During unusually wet and foggy weather I have observed broods starving in the nest because food is unavailable (Morse 1971a). Unusually wet and foggy weather along the coast of Maine during the summers of 1972 and 1973 correlated closely with the low nesting success of the spruce-woods species in those years (Morse 1976b). Although most anecdotal information suggests that in-

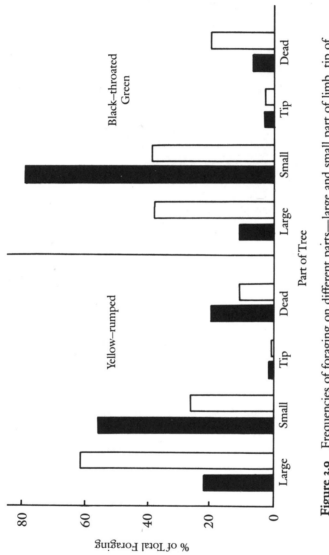

Figure 3.9 Frequencies of foraging on different parts—large and small part of limb, tip of vegetation, or dead limb—of dry (solid bars) and wet (open bars) spruce trees by male warblers. (Modified from Morse 1968 and unpubl.)

clement weather reduces the nesting success of songbirds (e.g., Morse 1980a), little attention has been paid to its effects on foraging.

Wind also modifies birds' foraging positions, probably because of its effect on the thermal environment (Grubb 1975). It lowers foraging effectiveness by presenting the birds with a moving substrate. When winds are high, they may even physically force birds from the preferred substrates on which they are foraging; consequently the birds forage lower in the forests in high winds than at other times. Wind and wet foliage thus force the different sexes and species closer together, thereby increasing competition. During periods of especially high winds the activity declines, leading to less interaction. Such inclement conditions complicate analysis because the birds are less conspicuous when the trees are in motion and the wind is whistling.

Drought might decrease foraging success by its adverse effect on invertebrates that depend on new plant growth for their productivity. In contrast, drought sometimes increases numbers of insects by stressing plants in a way that makes their foliage more favorable than usual to some insects, especially certain defoliating caterpillars (White 1969). Thus, from the viewpoint of paruline welfare, drought may have both negative and positive aspects. Since drought-related insect outbreaks generally occur late in the summer, the warblers' breeding season will have been completed. But high prey abundance may favorably affect newly independent young, since they are then enduring a period of high mortality, resulting from their ineffective foraging (see Lack 1966; Morse 1980a).

Social and Environmental Factors in Niche Partitioning

Studies of paruline foraging habits have demonstrated that the species cohabiting an area may partition resources along one or more of several gradients, and that various factors modify the patterns of partitioning in a particular situation. The potential for partitioning seems to differ strikingly among habitats and situations, however. The very question of multidimensionality in partitioning deserves more attention. Cody (1974) demonstrated a tendency toward reciprocity of foraging-niche dimensions by several passerine species, and Ulfstrand (1976) and Morse (1978b) found similar patterns among European tits and sylviine warblers.

What remains to be seen is how niche relations shift in response to such factors as the social interactions among participants. To what degree are patterns of resource exploitation a consequence of social interactions

among individuals and species and how much do preexisting social relations determine what the exploitation patterns will be? Do intraspecific
interactions play a major role in determining the nature of interspecific
interactions, factors which could influence the niche-partitioning options
available to a species? For instance, does a high species density affect the
nature of interactions among congeners? That is, does the attainment of
high densities generate a syndrome of aggressive behavior, and will aggressive behavior spill over into interspecific interactions? And, if high
densities of a species expand niche dimensions (Svärdson 1949), does a
certain set of individuals from that species use the least typical sites occupied by that species—sites that are usually suboptimal and the last to be
exploited by a species whose population density fluctuates? If so, do individuals that occupy these areas most frequently come in contact with
other species? Are these the individuals least likely to fare well against
other species, perhaps in areas that are optimal for the latter? If so, this
constraint could act as a conservative element in niche partitioning and
interspecific relations.

Factors of this sort are little understood, despite their potential importance in determining the outcome of resource partitioning among species.
Results will be obtained only if students work with individuals of known
history and behavior. The very proposal that social interactions may
strongly affect niche partitioning is a rather new idea, probably because
interspecific interactions and intraspecific interactions have generally fall-
en within the purview of different groups of workers. Interspecific interactions have been studied by ecologists, who generally treat such considerations of niche partitioning as components of a black box, in which
exploitation patterns of species are measured and compared with those of
other species. In contrast, the study of intraspecific interactions has often
fallen within the realm of animal behaviorists, who address these behaviors as ends in their own right.

Although interspecific interactions may be the most important *and*
most frequent ones that an individual will experience, most interactions of
uncommon species may take place with other species, even if they happen
less often than would be predicted from the abundances of species. For
instance, the Blackburnian Warbler, which is two to three times less common than coexisting Black-throated Green Warblers in my study areas
(Morse 1967b, 1976b), has a far different ratio of interspecific to intraspecific interactions than does the Black-throated Green Warbler. Interspe-

cific interactions are quantitatively far more dominant in Blackburnian Warblers' lives on these breeding grounds than in the lives of Black-throated Green Warblers.

What will be the result of these unequal ratios? Will the uncommon species be the least successful ones? If they experience a wide range of other species, the variety in interactions may be so great that their ability to respond adaptively to other species may be weak. Their inability to adapt to interspecific competition could affect their niche-utilization patterns and potentially their distribution.

Random factors can disrupt normal patterns of niche partitioning, and these events are bound to have a tempering effect on exploitation patterns, although most studies have accorded them little attention. If large parts of a habitat suddenly become unusable, particularly when resource demand is high, the fit between the exploitation patterns of the affected populations may be diminished or destroyed. If such an event occurs when young are in the nests, little accommodation may be possible; movement of parents from the area will not save the nestlings. These constraints must play a key role in establishing the ranges of affected species, as well as determining their interactions with coexisting species.

In the coastal spruce forests where I have worked, contingencies such as protracted storms should affect Yellow-rumped Warblers less than the other *Dendroica* species, because of their customary heavy foraging on bark (one of the few substrates on which birds will not quickly wetten their plumage) under normal conditions. The Yellow-rumped Warbler is also the most flexible of the spruce-forest *Dendroica* under normal conditions and the one with the lowest social status. If its flexibility of resource exploitation results from its subordinate position, making the best of a bad situation could serve it well under adverse conditions, giving it an advantage and perhaps balancing its success against that of the others. Measurements of nestling loss by Yellow-rumped Warblers and other spruce-woods warblers during protracted periods of inclement weather would help to evaluate the importance of this environmental variable. If such studies could compare losses in areas of differing storm frequency and severity, we might obtain a better impression of the role of environment in niche partitioning. Environmental factors are normally treated as density-independent phenomena, but if they temper resource-exploitation patterns they can figure into the complex intra- and interspecific dynamics of warbler communities.

4 Habitat Selection and Use

Even a glance at a map of warbler breeding grounds suffices to establish that the birds are not distributed randomly, intraspecifically or interspecifically. This observation tells us little, though, about why a site is occupied by one species and not another. In sorting out possible explanations we must distinguish between habitat *selection,* the choice of a site having certain characteristics in preference to others, and habitat *use,* the act of occupying an area even for reasons other than active choice. Habitat use could be due to lack of a better choice, as happens, for example, when a species is forced by its competitors to remain in a habitat that is less attractive than another (Morse 1980a).

First I will concentrate on bona fide examples of habitat selection. Which factors account for the choices? Attention has been focused on relating bird distribution to vegetational or environmental gradients (reviewed in Morse 1985; Sherry and Holmes 1985) or to structural characteristics of the vegetation, such as foliage height diversity (MacArthur and MacArthur 1961; MacArthur et al. 1962; Cody 1981). Studies of this sort typically measure many variables (for example, characteristics of singing sites or nest sites) and analyze them with multivariate statistics. The resulting correlations are often useful in pointing out future research directions, but they do not establish the underlying cause—though sometimes that distinction is not brought out in research reports (see Morse 1985; Sherry and Holmes 1985).

Sherry and Holmes (1985) have identified several unresolved issues concerning habitat selection, such as the psychological, ecological, and evolutionary mechanisms by which birds select their habitats. Important questions include the following: Why is vegetation structure important, and what environmental factors shape patterns of habitat selection? What is the spatial scale of habitat selection? How do morphology and behavior affect a bird's choice of habitat? Answers to these questions and others like them would greatly advance our understanding of the complex behavior of habitat selection.

Selection According to Features of the Habitat

Structural cues that individuals use in choosing sites must occur predictably within the lifetime of the individual if responses to them are to be genetically encoded; for species such as migratory warblers, this means almost yearly. Thus the response to regularly and irregularly occurring cues should differ.

Structure of the Vegetation

Warblers differ markedly in the types of habitats that they occupy. This is not a trivial observation, for one can precisely place species along gradients of a variety of habitat types. For instance, Cody (1974) found that several western species were spaced along a height gradient that ranged between low scrub (habitat of the Common Yellowthroat) to 4–5 meters or higher (MacGillivray's Warbler), with Yellow Warblers and Wilson's Warblers occupying intermediate positions. Gradients for other habitats, including both deciduous and coniferous forests, have been considered also. Most studies concentrate on the entire passerine bird community of the sites, but since, at least in eastern North America, such a high proportion of the avifauna consists of warblers, the results are of interest. To determine a gradient censuses are taken simultaneously of several habitats. Along one gradient, for example, Magnolia Warblers occupied woodlands that displayed the widest range of disturbance levels—from undisturbed forest to clear-cut areas—in Maine's coniferous habitats (Titterington et al. 1979). But another study by Beals (1960), which incorporated a wide range of mature forest types in Wisconsin, revealed that Magnolia Warblers had little tolerance along this parameter relative to that of several other species. In contrast, Ovenbirds occupied all of the forests in the Wisconsin study, but only two of the Maine sites, the two most mature (that is, undisturbed) areas. Species therefore appear to prosper along gradients of certain types of variables but do not exhibit such flexibility in other gradients: they are generalists along some parameters but specialists along others. Furthermore, the species that occupied most of the sites in both Wisconsin and Maine, the Black-throated Green Warbler, is extremely sensitive to another potential limitation—the minimum size of the area required for a territory. This species needs larger territories than do either isolated Yellow-rumped or Parula warblers, and it thus appears to be excluded from small island forests along coastal Maine (Morse

1977b). It would probably also be absent from small habitat islands of its favored vegetation types. Unfortunately, comparisons are difficult to make in this case because the examples cited involve three different geographic areas. Nevertheless, the studies suggest clear differences among the species, and they also indicate the value of investigations where all variables—size of the area as well as disturbance level—are incorporated in a single study.

Not only the presence or absence but the abundance of birds can be affected by vegetational variables. Light forest fires, for instance, reduced the density of population to a quarter for Blackburnian and Bay-breasted warblers (Apfelbaum and Haney 1981) and a fifth for Prairie Warblers (Moore 1980).

Experimental manipulations of vegetation are rarely done, and those that have been carried out involve few warblers and hence provide limited insight. Kilgore (1971) found that when undergrowth was removed in sequoia groves, one of the three warblers, the Nashville Warbler, an undergrowth species, disappeared, but the two canopy species, the Yellow-rumped and Hermit warblers, were unaffected. Szaro and Balda (1980) also found that removing part of the pine trees in their study area resulted in the disappearance of the Red-faced Warbler.

Another group of workers has used a different methodology, the "habitat-niche" approach, for analyzing determinants of habitat selection. The habitat-niche approach borrows and builds on the ordination methods of plant ecologists and searches for correlations between the vegetational structure of the habitat and the presence of bird species. My review of the subject loosely follows an earlier one (Morse 1985).

Habitat-niche studies measure numerous vegetational variables, which are analyzed by multivariate statistical techniques. For instance, in a seminal study on the vegetational correlates of Arkansas forest birds, James (1971) measured 15 vegetation variables, including percent canopy cover, height of canopy, and number of tree species. The most frequently used methods have been principal component analysis, which reduces a multidimensional array of data to new orthogonal variables ordered in such a way as to explain the greatest possible amount of variance, and discriminant function analysis, which constructs axes from an original data set along which differences among populations are maximized. These methods identify composite continua that represent gradients in vegetation structure.

Using principal component analysis of data gathered from over 400 small (0.04 hectare) plots centered on the singing perches of territorial males of 46 species, including 11 species of warblers, James found that the first principal component, which described a gradient between open country with heavy ground cover and shaded forests with little ground cover, accounted for 65 percent of the variation in vegetation characteristics. Adding the next three components (secondary ordination corresponding to other vegetation features) accounted for 90 percent of the total variance. A plot of the first three components shows that the warblers were randomly scattered among the other species (Figure 4.1). This distribution is consistent with the birds' segregation in space; yet these are correlational results only—they could also be attributed to the direct effect of food, nest sites, or vegetation structure (Collins et al. 1982), among other factors. The conclusion of random spacing derives from the range of habitats censused. If James had continued her analysis to open-country species, the warblers, by virtue of their absence in the latter component, would have been clumped on the plots.

With linear discriminant function analysis, James (1971) further discovered that warblers and other species separated widely on a first axis, which included ranges from xeric to mesic, upland to bottomland, low to high

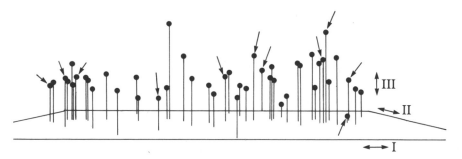

Figure 4.1 Three-dimensional plot of the results of a principal component analysis of habitat variables in birds' occupation of territories. The first component (PC1, represented here by the horizontal axis and labeled I) correlates mainly with increasing number of tree species, percentage of canopy cover, number of small trees, and canopy height; PC2 (the II axis) with increase in medium-sized trees and decreasing shrub density; PC3 (the III axis) with large isolated trees. The three components account for 64.8, 12.5, and 7.7 percent of the variance, respectively, in the features of habitats occupied by the 46 species. The 11 species of warblers are denoted by arrows pointing to data points. (Modified from James 1971.)

biomass, and open country to closed forests. A second axis, ranging from large isolated trees to trees with heavy understory, further separated the warblers. Plotting these two axes against each other, using plots with 3, 10, and 15 vegetational variables, James demonstrated the importance of including several such variables, as revealed by the difference between 3- and 10-variable plots (Figure 4.2). She concluded that different species exhibited clearly differentiated niche parameters in their habitats, and that percent canopy cover, canopy height, and number of tree species per area accounted for the greatest amount of habitat difference in the birds' preferences.

Other multivariate studies have subsequently confirmed and expanded these general results. Anderson and Shugart (1974) and Smith (1977) found that forest-dwelling warblers exploited a narrower range of habitat variables than did permanent-resident ecological equivalents in Tennessee and Arkansas. The warblers nevertheless separated from each other; some species exploited habitat variables at the frequencies predicted by their abundance in the environment, such as the Cerulean *(Dendroica cerulea)* and Hooded warblers, and others exhibited strong habitat selectivity, for example the Pine Warblers, Yellow-breasted Chats, and Ovenbirds. Most species with similar ecological requirements were separated by habitat; where such separation did not occur, vertical separation occurred within the habitat (e.g., Cerulean Warblers are canopy feeders and Hooded Warblers are undergrowth specialists).

Most of these studies have, unfortunately, differed in details of methodology, and direct comparisons are not possible. But Noon et al. (1980) have compared habitat-use patterns of birds in eight study areas in the northeastern and north-central United States. Their results, using discriminant analysis, revealed much consistency of habitat use over a broad geographic range: only two of 25 valid comparisons showed habitat shifts. Both were warblers: Black-and-white Warblers in the southern part of the region selected forests with more tall trees and a more open understory than did those in the north, and Ovenbirds in the east were more confined to deciduous forests than were those in the western part of the study range. But the availability of several habitats varied over the geographic range of the study, and it appeared that both of these shifts were responses to differences in habitat availability. Several other warblers—Cape May, Yellow-rumped, Blackburnian, Kentucky, and Canada warblers—were consistently associated with certain habitats, even when their availability differed strikingly.

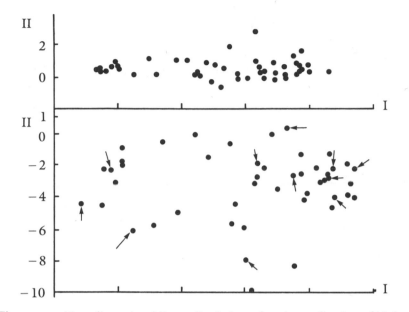

Figure 4.2 Two-dimensional linear discriminant function ordination of bird habitats. The horizontal axis (Axis I) of both plots expresses a continuum from xeric to mesic conditions, from upland to bottomland, from low to high biomass, and from open country to forest associations. The vertical axis (Axis II) of both plots describes a continuum from areas having large isolated trees with relatively open understory to areas with biomass concentrated in the lower strata (dense shrubs, small trees, etc.). The upper plot incorporates 3 vegetational variables: percentage canopy cover, canopy height, and number of tree species. The lower plot incorporates 10 variables, including the three noted above plus canopy height × number of trees 3–9 inches in diameter at breast height (DBH), canopy height × number of trees larger than 9 inches DBH, shrub stems per 0.02 acre, canopy height × number of shrubs, number of trees 3–6 inches DBH, percentage ground cover, and number of trees 3–9 inches DBH squared. A 15-variable ordination (not pictured here) achieved additional separation, but it did not account for as much additional variance as did the 10-variable ordination from the 3-variable one. The 11 species of warblers are denoted by arrows pointing to data points. (Modified from James 1971.)

Whitmore (1975) conducted studies in Utah that were designed to match James's (1971) Arkansas study. Unfortunately, the two studies had only two warblers in common, the Common Yellowthroat and Yellow-breasted Chat, but in discriminant function analyses these species occupied similar positions in the two studies, indicating that the birds make similar vegetational choices over wide geographic areas. If these studies had all used a standardized procedure, more such comparisons could be made, and they would be much more valuable in unraveling the variables of habitat selection than the piecemeal results presented.

In an effort to counter this problem, Collins (1983) compared the habitat-niche of Black-throated Green Warblers in five geographic areas, from Maine to Minnesota, using similar data-gathering techniques at each locality. In contrast to the results obtained from the James (1971) and Whitmore (1975) studies for the brush-frequenting Common Yellow-throat and Yellow-breasted Chat, Collins found that the geographically separated Black-throated Green Warbler populations differed markedly in their selection, a conclusion that I reached independently from different data (Morse 1976b). Although this species is highly stereotyped, at least at my study site (Morse 1968), Collins's result emphasizes that it is not safe to characterize an entire species' repertoire on the behavior of a local population.

In studies on the habitat variables of spruce-woods warblers occupying small islands along the coast of Maine, Hendricks (1981) made the interesting discovery that several of the outlier points in her analysis involved either males that did not attract females or pairs that were not successful in rearing young. Her result stresses the point that the birds' habitat choices are not random, and that habitat selection is subject to immediate and major fitness payoffs. These results do not, though, reveal the basis for the favorable or unfavorable nature of these sites.

Although practitioners of multivariable techniques have argued the virtue of this methodology and the resulting approach to habitat selection (e.g., James 1971; Collins et al. 1982), the methods are subject to problems unless used with care. In spite of the attempt at objectivity, the choice and number of variables can affect the results, as James demonstrated (Figure 4.2). Sensitivity analyses on data sets would be useful in assessing the number of dimensions needed and the importance of the choices made, yet only James has presented work in this context. Her result, as revealed in the differences between 3- and 10-dimensional analyses, points out the value of such analyses.

The technique of obtaining samples from a focal point, typically a male singing perch or a nest site, subjects the method to difficulties, especially if a bird's choice of habitat depends on more than one critical factor or if that factor is not the one chosen for the sampling. If a species' territory is heterogeneous, the problem can be serious, as James (1971) warns. Nevertheless, analyses in the literature have used different criteria for establishing the sampling site within the territory. Collins (1981) found that data on singing perches and nest sites of open-country species often differed, establishing that these two should not be mixed. Comparing, rather than confounding, such data might help to unravel the critical factors associated with habitat selection; to the best of my knowledge, however, this technique has not been exploited.

Although multivariable techniques are useful for discovering correlates between bird presence and vegetation characteristics, the investigator is left to interpret the resulting axes; that is, to determine how the original habitat variables contribute to the new axes. Some studies have been successful in separating species with single habitat factors (e.g., James 1971; Whitmore 1977), others were unable to do so (e.g., Smith 1977). Smith suggested that species respond to more than one habitat factor collectively; alternatively, James and Whitmore may have classified vegetation factors in a way different from Smith.

A major focus on the habitat-niche approach is to explain species-specific habitat affinities without resorting to competitive displacement (Collins et al. 1982). Although a healthy alternative to inferring *only* a competitive basis for observed distributions, it is important to emphasize that, by itself, the habitat-niche technique cannot preclude the possibility that competition is responsible for some, or even all, of the species-specific patterns observed (Sherry and Holmes 1988). The methodology involves correlations only, and although sometimes the results are treated as if the variables caused the outcome, such conclusions are unwarranted in the absence of other information. The answer can be eventually resolved only by experiments. Statements such as that of Collins et al. (1982), "We think that the distribution of warblers along habitat gradients is the result of individualistic responses by the species to characteristics of vegetation structure," are no more admissible as causal explanations of community structure than the assumptions of competition-mediated differences espoused by MacArthur and others. Indeed, the habitat-niche approach has been a useful method for learning about bird-habitat relations, but its greatest virtue may lie in pointing the way to experimental tests of habitat selection in the field.

The habitat-niche approach in some ways reverts to earlier efforts of Robert MacArthur and his colleagues (MacArthur and MacArthur 1961; MacArthur et al. 1962) to characterize bird habitats on the basis of the vertical distribution of vegetation (foliage-height diversity). They designed this methodology to predict bird species diversity and eventually the presence of individual species in habitats on the basis of easily measured variables. They initially calculated the proportion of vegetation at ground (0–0.6 meter), understory (0.6–4.6 meters), and canopy (4.6 meters to top) levels. The demarcation line between understory and canopy was later raised to 7.6 meters. The investigators enjoyed reasonably good fits to predictions based on their studies of eastern deciduous and coniferous forests, including many habitats with a dominance of warblers. Triangular, three-dimensional plots of vegetation in territories occupied by the species in question can be superimposed on plots of the three vegetational components based on sampling within a habitat. The degree of overlap should predict the presence and abundance of the species at another site. Later efforts, such as applying the method to tropical forests (MacArthur et al. 1966), were less successful. Perhaps the method's initial success was due to the characteristics of the eastern forests in which the researchers first worked and developed the technique. Reference to the results of multivariate analyses, especially the study of James (1971), who compared the patterns obtained by using different numbers of variables in her analyses, suggests that MacArthur and his colleagues were working with so few variables that they had a high error factor, especially when species with special requirements (for example, those seeking special nesting sites such as hollow trees) were included. MacArthur's effort was laudable, however, in being designed to search for general patterns and testable predictions. The habitat-niche approach, by contrast, has often been confined to the description of bird species within a range of vegetational variables so broad that elements of the vegetation potentially responsible for differences in habitat selection have often been obscure.

Special Features of the Vegetation

Sometimes circumstantial evidence strongly suggests that warblers respond directly to habitat cues. Blackburnian Warblers are common as breeding individuals in tall spruce forests along the coast of Maine, but they were invariably absent on small islands lacking tall spruces, a pattern that held consistent over a period of nine years (Morse 1977b). The only one of ten such islands that Blackburnian Warblers regularly explored (but not to remain and breed) was the sole one containing several tall spruces. Yet the volume of tall vegetation on this island did not match that of territories on the adjacent mainland and large islands. The frequency of inspections of this one island suggests that these birds nevertheless respond positively to tall trees (Morse 1967b, 1971a, 1977b). Further, Black-burnian Warblers were present in all the large forests except a young, even-aged stand with no emergent trees (Morse 1976b).

Kendeigh (1941) noted that Yellow Warblers do not nest around prairie marshes unless the vicinity contains a tall exposed site, such as a dead tree. This prominence is used as a singing perch, and apparently it is a necessary feature of a territory. Although its function and importance was not tested by Kendeigh, one could readily do so, simply and elegantly, merely by erecting artificial promontories at a range of heights. The height required may not be great. In Maine Yellow Warblers frequently occupy marshes with Speckled Alders *(Alnus rugosa)* no more than 3 meters in height and no other emergent vegetation (Morse 1966b). Of course, the prairie and eastern birds may have different requirements.

Yellow Warblers also occupy riparian vegetation, apparently because of the extensive undercover, in addition to the tall vegetation. Ficken and Ficken (1966) found that 8 of 10 male Yellow Warblers obtained a mate when they established territories in sites with a dense understory, but not one of 5 males obtained a mate when they established territories in sites differing only in the lack of an understory. Their study has the interesting aspect of suggesting that the females exhibit stronger habitat preferences than do males. Males that established territories in sites without understory might have been unable to find territories with an understory, but the Fickens' paper omits that point.

Food Supply

Some species appear to establish sites as a consequence of the food supply present. This condition is perhaps most evident with the Spruce Budworm specialists: the Bay-breasted, Cape May, and Tennessee warblers. These species often do not even nest in large geographic areas for long periods of time unless a Spruce Budworm outbreak occurs, and then they leave when the outbreak has run its course (Brewster 1938). The abundance of these species often becomes extremely high during major outbreaks (Kendeigh 1947; MacArthur 1958; Morse 1978a). Ironically, numbers of these warblers may have stabilized over the past 45 years because of budworm control programs in northeastern North America. These programs may have kept budworm numbers at a low to medium epidemic level over long periods of time (Blais 1973), when otherwise they would have run their normal course within a few years, killing trees en masse and making formerly wooded areas unable to support another outbreak for 35 years or more (Royama 1984).

Although the relation between these three warblers and budworms is rather well known, a stronger test of the cues associated with settling involves the establishment of Bay-breasted Warblers at an outbreak of Forest Tent Caterpillars *(Malacosoma disstria)* on deciduous vegetation, primarily aspens, in southern Manitoba—well south of their normal breeding range (Sealy 1979). This choice of habitat suggests that the warblers respond directly to their caterpillar prey rather than to the vegetation. The deciduous habitat is unusual for this species, which as a breeder is perhaps as strongly associated with conifers as any other warbler species; however, Greenberg (1979, 1984b) has remarked upon the high level of foraging plasticity of Bay-breasted Warblers at other seasons.

Social Stimulation

The presence of other individuals might in its own right serve as a measure of site quality. For an individual lacking detailed information about a habitat, this cue might be a good one, although previous arrivals must have made the decision to settle the site in the first place. The establishment of tradition (philopatry) would account for the retention of such patterns once they are established. Little attention has been paid to this problem in warblers as compared with nesting seabirds. Considerable attention has been accorded the Kirtland's Warbler, however, simply because, as an endangered species, it has been surveyed and color-marked exhaustively (Ryel 1979). These birds confine their breeding-season activities to Jack Pines *(Pinus banksiana)* 2–5 meters in height and with extensive branching at ground level. Within a local homogeneous pine stand, their breeding territories are clumped. Social factors may influence their pattern of occurrence within these stands.

Several other warbler examples exist. Sealy (1979) noted that the Bay-breasted Warblers attracted to Forest Tent Caterpillars in southern Manitoba nested closer to each other—a small colony in the midst of a much larger outbreak area—than predicted by chance. In a third example, Meanley (1971) reported that Swainson's Warblers often occupy only contiguous territories in the extensive southern canebrakes, frequently in the midst of much larger unoccupied expanses of similar habitat. Lastly, Blue-winged *(Vermivora pinus)* and Golden-winged *(V. chrysoptera)* warblers not infrequently assume colony-type distributions in the midst of much larger areas of similar habitat (e.g., Ficken and Ficken 1968d; Gill 1980). This pattern may turn out to occur regularly in colonizing situations.

One can characterize most of these examples in relation to food or a related factor such as vegetation—a good indicator of present or future food abundance. The Kirtland's Warbler may be an exception. Few examples seem unequivocally associated with nonfood variables, though the two hole-nesting warblers, the Prothonotary and Lucy's, may be limited by nest holes (Bent 1953). But the Prothonotary Warbler could have additional criteria for choosing sites because they usually inhabit low-lying swamp forests. They should be amenable to nest-box experiments that simultaneously explore the importance of associated habitat factors. Comparisons of their habitat selection with that of conventional nesters can elucidate the effect of a limiting factor such as holes on the "choosiness" of a species for other variables.

Case Studies of Habitat Selection

Black-throated Green Warblers in different habitats. Black-throated Green Warblers are more abundant in Red Spruce forests than in White Spruce forests along the Maine Coast. Further, birds in mixed areas concentrate their activities in the Red Spruce trees (Morse 1976b), where they can forage more easily. Red Spruce needles lie flat and parallel to the branch; White Spruce needles lie approximately at right angles to a bird hopping over them (Figure 3.8). Consequently, the birds must hop higher, and expend more energy, when foraging in White Spruce than in Red Spruce. Wet White Spruce foliage is especially unfavorable (Chapter 3). The biomass of insects (standing crop) in the two species of trees does not differ much during the breeding season, including the most critical periods of the breeding cycle—the incubation and early nesting phases (Figure 2.2). The only significant difference in insect biomass occurs at the beginning of the breeding season, when territories are being claimed: White Spruces then have more insects than Red Spruces. If territories were determined on the basis of insect prey available when they are chosen, Black-throated Green Warblers should be more abundant in White Spruces than in Red Spruces, the opposite of the pattern observed. This analysis assumes that standing crop is an appropriate estimate of energy flow in both habitats, which seems a legitimate hypothesis given that birds did not drive the standing crop to low levels and that most insect and bird species were the same in both habitats.

Shifts in density that accompany general population declines also favor the argument that Black-throated Green Warblers prefer Red Spruce foliage. Their numbers were extremely low in 1974–75, following two poor breeding seasons (Morse 1976b) and high mortality in spring migration (Finch 1975). Population densities nevertheless remained stable in two Red Spruce study plots but declined markedly in White Spruce plots, both in large forests and on small islands (Morse 1976b, 1977b). That this shift was associated with habitat choice, rather than differential survival of the population, is strongly suggested by the behavior of members in a White Spruce study area. This population initially settled at nearly normal density in 1974, but its numbers then declined to the low level that characterized all Black-throated Green Warbler subpopulations exploiting White Spruce that year. Most likely these individuals shifted to Red Spruce forests, although I have no data that explicitly verify this suggestion. The pattern matches other studies in which unsaturated preferred habitats are occupied by conspecifics that immigrate from neighboring suboptimal habitats (Brown 1969a; Fretwell and Lucas 1970; Krebs 1971).

Not all Black-throated Green Warblers are confined to spruce forests in the breeding season, however. A well-differentiated geographic race *(waynei)* breeds from the Dismal Swamp of southeastern Virginia and northeastern North Carolina south to eastern South Carolina in an exclusively deciduous area (including cypresses). Moreover, less than 150 kilometers from my study area in coastal Maine, Black-throated Green Warblers inhabit mixed coniferous-deciduous forests. In fact, both in northern Maine (Morse 1978a) and in the mountains of New Hampshire (Morse 1979) this species occupies mixed coniferous-deciduous areas in highest densities, even when pure coniferous forest (spruce-fir) grows nearby.

The pure coniferous habitats in northern Maine and the mountains were numerically dominated by *Dendroicas* not currently represented in the coastal spruce fauna—the Bay-breasted Warbler in northern Maine (Morse 1978a) and the Blackpoll Warbler in the mountains of New Hampshire (Morse 1979; Sabo 1980). Though neither species overlaps Black-throated Green Warblers in foraging sites and methods, their densities are high, both in actual numbers and percentage of the entire avifauna. Such high densities might curtail the Black-throated Green Warbler's abundance despite the modest foraging overlap between them and the Blackpoll or Bay-breasted warblers; this is reinforced by Kendeigh's (1947) and Morris et al.'s (1958) observations that some *Dendroica* species (including the Black-throated Green Warbler) declined during outbreaks of the

Spruce Budworm, a phenomenon accompanied by a striking increase in Bay-breasted Warblers. Bay-breasted Warblers were present on Hog Island in the 1930s (Cadbury and Cruickshank 1937–1958), but it is not known whether they exerted such an effect then.

In Maine, immediately inland from the study area, Black-throated Green Warblers inhabit sites of white pine and hemlock where spruces are sparse, but they do not breed in areas that are not conifer-dominated. This pattern seems qualitatively different from the one in northern Maine and at moderate elevations in New Hampshire; if so, it would be of interest to explore the choices of the transitional populations settling between these geographic sites.

The discrimination patterns of Black-throated Green Warblers are fine-tuned, given their differential response to Red and White spruce; further, they change over a small geographic area, judging from their performances in coastal and northern Maine and New Hampshire. But it is uncertain what cues they use, though some potential cues do not play a major role. For example, cues such as food abundance do not appear to be of primary importance, if we judge from their responses to Red and White spruce forests upon return. And if one can extrapolate their position in the social hierarchy from my study areas, one can conclude that food abundance did not play a role in their settlement patterns in other areas.

Pine Warblers in pine and mixed habitats. The Pine Warbler, a species characteristic of pine forests throughout its range, presents one of the clearest responses to a habitat type. In Maryland there was a close relation between the proportion of pine vegetation and the density of nesting Pine Warblers: in plantations of Loblolly Pines, densities exceeded those of adjacent areas containing few Loblolly Pines in the midst of deciduous trees. The warblers treated mixed areas as if they were dilute pine forests, almost invariably moving from one pine tree to the next and ignoring the deciduous trees between them (Morse 1974a). These results suggest that Pine Warblers respond directly to the pine foliage.

Sherry and Holmes's (1958) studies in a mixed deciduous-coniferous forest in New Hampshire indicate that the Blackburnian Warbler, a species dwelling in coniferous habitat over its breeding range, concentrates its activities in patches of hemlock and Red Spruce, exhibiting a pattern that resembles that of the Maryland Pine Warblers. Habitat specialists such as these routinely seek out favorite sites within heterogeneous habitats.

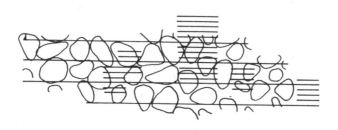

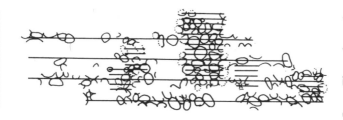

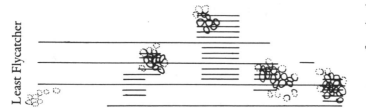

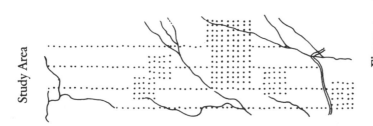

Figure 4.3 Dispersion patterns of territorial Least Flycatchers, American Redstarts, and Black-throated Blue Warblers in the Hubbard Brook Experimental Forest, New Hampshire, June 1981, plus map of the study area. The four vertical transect lines and the more extensive grids were marked at 50-meter intervals. In the map of the study area double lines are roads and single lines are streams. In the species maps, curved solid lines are territory boundaries, and dotted lines approximate territory boundaries that were not precisely mapped. (Modified from Sherry and Holmes 1985.)

Swainson's Warblers in lowland swamp and mountain habitats. The Swainson's Warbler, which frequents low, dense deciduous cover and often forages on the ground, was traditionally known from the extensive wooded swamplands of the southeastern United States. It is therefore surprising to find it as a breeding bird in the mountains of West Virginia (Brooks and Legg 1942; Meanley 1971). Although these habitats initially appear quite different, the canebrakes of the southern swamp forests and the rhododendron-laurel thickets on the mountainsides in West Virginia bear a rough similarity, especially in their foliage-diversity patterns (MacArthur and MacArthur 1961; MacArthur et al. 1962). Studies to date do not reveal whether Swainson's Warblers use both habitats in similar ways (Meanley 1971), but the birds' behavior, with most of their activity near to or on the ground in dense growth, is consistent with their similar exploitation patterns in both places. In neither habitat do they frequent the densest undercover; instead, they exploit areas where small openings appear among the vegetation (Meanley 1966).

Use According to Opportunity

Even if individuals cannot occupy preferred sites, the pattern of compromise in habitat use may not be random. This aspect of habitat selection has been little examined, but it may be appropriate to think of alternatives as second or third choices, and so on, rather than the absence of selection, unless there is evidence to the contrary. The responses of individuals with these restraints should provide useful insights into their choice of sites in the absence of competitors, as well as into the mechanisms of habitat selection.

Dispersion Patterns

As I noted earlier, a few species (Kirtland's, Bay-breasted, and Swainson's warblers) exhibit social stimulation; they settle in clumped territories in the midst of large expanses of unoccupied but favorable breeding habitat. In other words, they assume a colonial distribution in homogeneous habitats. Each species attains only small population sizes in the regions, and available habitat is unlikely to be limiting.

To the best of my knowledge, however, only Sherry and Holmes (1985) have systematically evaluated the spatial distribution of an entire guild, a group of co-occurring species sharing a common life-style within a community. The members of this guild—five warblers, a flycatcher, and a vireo—occupy old second-growth northern hardwood forest in the Hubbard Brook Experimental Forest in New Hampshire. Intraspecific spacing differs greatly from species to species (Figure 4.3). Only the Least Flycatcher, which has by far the smallest territories and is aggressive in its interactions with some warblers (Sherry 1979; Sherry and Holmes 1988), exhibits a clumped distribution. Although clumping could be simply due to the size and heterogeneity of the sites sampled, Sherry and Holmes's study encompassed a large enough area to permit testing at several scales. The flycatcher has mean territory sizes of 0.18 hectare with little variance, and clumps strongly at 1- and 4-hectare levels. The level of clumping is much lower at the 16-hectare level, reflecting the scale of the clumping; that is, groups of birds tend to be evenly distributed in space and have a clear colonial patterning, but this may be on a smaller scale than it is for the Kirtland's and Bay-breasted warblers.

By contrast, American Redstarts at Hubbard Brook are randomly distributed, even though they are abundant and highly aggressive in their territorial interactions. Sherry and Holmes conclude that the birds distribute themselves in this way to avoid parts of the habitat unlike their own, and that they spread out in the rest of the space in ways typical of territorial species. The first factor leads to clumped distribution, the second to even distribution. In anything other than a homogeneous habitat, one might expect such a distributional pattern. Black-throated Green and Blackburnian warblers, being at much lower densities than redstarts, probably have a similar pattern. Black-throated Blue Warblers defend large territories and are evenly distributed over virtually the entire study area. Ovenbirds and Red-eyed Vireos also have evenly distributed territories, only partly filling the study area, and their territories are smaller than those of the Black-throated Blue Warblers.

In subsequent studies Sherry and Holmes (1988) removed Least Flycatchers from certain areas and in a subsequent year also observed a natural decline of Least Flycatcher numbers (Figure 4.4). In the absence of this species, adult-plumaged redstarts spread quickly into the clumped sites previously occupied by the flycatchers, establishing that these sites were highly acceptable habitats. In turn, second-year redstarts, which sometimes coexisted with the Least Flycatchers, declined in these areas, presumably in response to the socially dominant older redstarts. Least Flycatchers recolonized in subsequent years, and numbers of redstarts correspondingly decreased in these sites. Sherry and Holmes thereby dem-

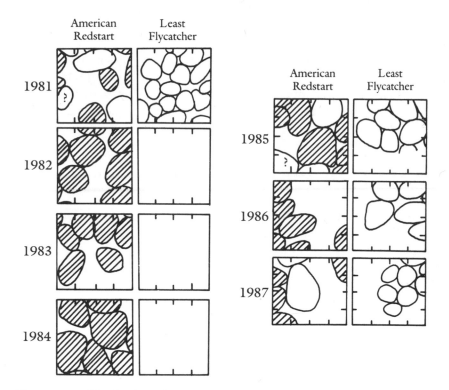

Figure 4.4 Distribution of American Redstarts and Least Flycatchers on a 4-hectare plot before removal of Least Flycatchers (1981), after removal but before return of Least Flycatchers (1982–1984), and after return of Least Flycatchers (1985–1987). Cross-hatched redstart territories denote adult-plumaged males (after second year), unhatched redstart territories denote immature-plumaged males (second year), ? denotes birds for which age information not recorded. Data for Least Flycatchers not broken down by age. (Modified from Sherry and Holmes 1988.)

onstrated that part of the basis for the "random" distribution of redstarts at Hubbard Brook resulted from the normal occupation of sites by the aggressive and socially dominant Least Flycatchers.

Thus, even within a single community, species can have different patterns of intraspecific dispersion. Distributions are probably due to intra- and interspecific factors, as well as to structural ones. The frequency with which interspecific interactions affect scaling patterns needs to be investigated more broadly. It is not enough to assume that territoriality is strictly an intraspecific phenomenon if resource exploitation patterns overlap.

Effects of Intraspecific Density

Population density can affect the types of habitats that individuals use. Birds may avoid crowded sites or be forced into sites not of their choosing. Svärdson (1949) noted that high population densities often broadened the niches of species, but only when a certain density is reached relative to the characteristics of the habitat does expansion occur. Favored habitats are exploited exclusively, or nearly so, until crowding and its consequences (such as intraspecific fighting) reach a point where it becomes more profitable to occupy an otherwise inferior habitat (Fretwell and Lucas 1970).

Individuals obtaining favored habitat sites may differ from others in several ways. First, they may hold these sites simply because they were the first ones there. Others may prevail because of superior size, age, or previous experience. These traits may be related, since old individuals usually arrive on the breeding grounds before young ones.

Habitats exploited by second-year and older male redstarts differ markedly (Ficken and Ficken 1967). In deciduous-coniferous forests of the northeastern United States, second-year males often established territories in coniferous habitats, older males in deciduous habitats. I have reported a similar pattern (Morse 1973), and Howe (1974) found old and young males occupying different deciduous habitats in Michigan. Second-year males also used areas frequented by Least Flycatchers more often than did older males (Sherry and Holmes 1988), perhaps because adults seldom occupied them. No one has investigated more southerly populations to determine whether analogous partitioning happens in primarily deciduous formations.

The difference between second-year and older males is unlikely to be a consequence of age-related changes in habitat choice. Since old redstarts usually dominate young ones in interactions, second-year birds were more numerous on small, largely spruce-clad islands than on larger islands with lush deciduous growth (Morse 1973, 1977b).

Missing thus far is a study of known individuals, including second-year males, to establish whether they use the same sites, or habitats, that they used the preceding season. Such a study would provide insight into how site occupation is affected by return date, age, plumage, site tenacity, and a tendency to return to the birthplace (philopatry). Redstarts should be studied in woodlands that contain adjacent adult and second-year birds nesting in deciduous and coniferous habitats. Experimental removals of adults and second-year birds, nesting in deciduous and coniferous habitats, would determine the effect of the two classes on habitat-use patterns of others. Thus far one cannot eliminate the possibility that natal habitat plays a role in choices of second-year birds, and it is worth exploring this. Early learning figures into habitat selection of other warbler species (Greenberg 1983, 1984a), but recovering enough locally reared passerine fledglings to test the idea would be a difficult task.

Age-related differences in habitat use may be widespread among warblers; however, the lack of conspicuous plumage differences in most species complicates investigations of individuals that are not banded. Not surprisingly, among warblers such studies are confined to redstarts, although second-year birds of most warbler species can be distinguished in the hand (Dwight 1900; Bent 1953). Age-related differences in plumage patterns are associated with hormonal titers in other passerine birds that have age-related plumage differences, and these differences are also linked to competition for suitable nesting space—several blackbirds of the Icteridae family provide examples (Orians 1980). It thus seems appropriate to see if the redstart's prime habitat is at a premium more often than other warblers' habitats are.

Intraspecific abundance also affects the distribution of Black-throated Green Warblers in adjacent Red and White spruce forests in Maine. These individuals appear to operate in accordance with an ideal free distribution (Fretwell and Lucas 1970); that is, when the best habitat is already crowded, it is most profitable to settle in a secondary one; if numbers decline in the primary habitat, however, invasion there becomes profitable.

The Role of Other Species in Habitat Distribution

Much circumstantial evidence suggests that congeners and other species regularly affect the niche partitioning of co-occurring species (Chapter 3). Less evidence supports the role of exclusion in the segregation of warbler species between habitats, although Cody (1985) noted several examples from other bird groups that may be explained in this way. Two examples with which I am familiar could be accounted for by exclusion.

In the northeastern part of its range, the Northern Parula Warbler frequently is found in conifers, along the coast of Maine most often in spruces. In these spruce-dominated areas, Northern Parula Warblers usually confine themselves to the edges of the forests, and in my study areas (Morse 1977b) they mainly occupy the edge of the shore. In 1965, 30 of 34 pairs breeding on Hog Island, a large, spruce-clad, near-shore island, nested along the shore. Two others nested along the edge of large blowdowns created by hurricanes in 1954, and two nested on the edges of large natural openings (created by extensive ledges) in the middle of the island. These small warblers, which forage mostly on the tips of vegetation, regularly encounter another similarly foraging common breeder in the area, the Golden-crowned Kinglet, a species that is not usually found on the edges of the forests. The kinglet frequently initiates and prevails in aggressive encounters, even though it is slightly outweighed by the warbler.

Northern Parula Warblers also interact with *Dendroica* warblers in these spruce forests (Morse 1971a, 1976a), but these interactions appear to generate niche shifts rather than habitat exclusion. Northern Parula Warblers occupy more of the small offshore islands in the vicinity than do any of the *Dendroica* species (Morse 1971a, 1977b), and they shift their foraging patterns in the absence of other species, most notably the Yellow-rumped and Black-throated Green Warblers. Only twice in 70 island-years did Golden-crowned Kinglets nest in the small island spruce forests, which range in size from 0.16 to 1.65 hectares.

Further south (Morse 1967a), Northern Parula Warblers course through mature deciduous forests more widely, rather than being edge specialists. There, they are not in the company of Golden-crowned Kinglets, nor of other species with similar size and foraging patterns. They are perhaps the only tip-gleaning warblers in the vicinity; *Vermivora*, the other common tip gleaners, do not frequent those forests (*Vermivora* are represented in the southern part of the Northern Parula Warbler's range only by the extremely rare, possibly extinct, Bachman's Warbler).

A second example of exclusion as the cause of habitat segregation is the Yellow Warbler. A riparian and swamp-edge species over much of its range, the Yellow Warbler often occupies suburban habitats. In none of these situations does it use coniferous forests, yet it regularly occupies these areas on islands off the Maine coast (Hebard 1961; Morse 1973, 1977b). One assumes its presence there to be associated with the absence of congeneric spruce-woods warblers. It also occupies more mature habitats, including deciduous growth, on these islands than on the mainland. It may be using the sites in the absence of Chestnut-sided Warblers and American Redstarts, two species that occupy successional continua from swamps to mature deciduous forests on the mainland. Chestnut-sided Warblers almost never occur on these islands, but the redstarts also occupy coniferous forests in this region and colonize small islands with spruce forests. Either Yellow Warblers or redstarts may be present on the small islands, and the species may even differ from year to year. On slightly larger islands they may coexist (Morse 1973), and there they have the same pattern of segregation as on the mainland.

The tropical Yellow Warblers (Golden and Mangrove warblers) have a similar pattern in some places. In continental regions they confine themselves to coastal fringing vegetation, usually mangroves, but on many smaller islands, especially the Lesser Antilles, they occupy forests over much or all of the landmass, a pattern correlated with the small number of other passerine insectivores on these islands (Morse 1966b).

Predators

I have no clear evidence that warblers avoid habitats because of the presence of predators. Where accipiters have been studied in close apposition to parulines, there has been little evidence that these hawks affect the distribution of resident passerines. Mueller et al. (1981) and Meyer (pers. comm., 1986) report attacks by accipiters on warblers, but the raptors prefer larger prey than warblers. In fact, Mueller et al. noted Yellow-rumped Warblers and American Redstarts within 20 meters of a Sharp-shinned Hawk nest in Ontario.

Hildén (1965) has suggested that the mere presence of predators could dissuade individuals from using an area, thus altering the distribution of individuals. Yet, little attention has been paid to this effect of predation, and Hildén cited only one doubtful example of a small passerine, a sylviine warbler. In that instance, Amann (1949) reported that Wood Warblers *(Phylloscopus sibilatrix)* avoided areas with high densities of the field mouse *(Apodemus flavicollis)*, a species which robs the nests of small birds, but he presented no convincing quantitative data that eliminated alternative explanations. Tits in Wytham Wood, England, settle in the vicinity of European Sparrowhawks' *(Accipiter nisus)* nests, even though this predator may take as many as one-fourth of the tits yearly (Geer 1978; Perrins 1979). Examples of seabirds avoiding nesting on islands inhabited by humans, and their pet dogs and cats, are more widespread, although here again the actual mechanism seems to have received little attention.

Balancing a Multitude of Variables

Habitat selection is not a simple response to a common set of physical variables by different species; instead, a variety of causes come into play. To categorize warblers' responses in many communities, one must understand both their biology and their social interactions. These biological factors vary with the severity of the climatic factors of the habitat (Sabo and Holmes 1983). Nevertheless, the physical environment, typically the prevailing vegetation, is the governing variable to which all species respond. Whether birds respond directly to this factor or to a dependent variable, such as potential competitors or the food supply, differs among species. Evidence for response to vegetational structure, to other species, or to food supply exists in abundance (e.g., Morse 1978a; Sabo and Holmes 1983), but most investigators have considered only selected aspects of the general problem, thereby precluding a systematic assessment of how variables differ in importance. Sabo (1980) argued that inflexible species depend on vegetation in making their choices, but they may become inflexible in the first place simply as a consequence of favorable dominance relations with co-occurring species or by unusual adaptations, thereby ensuring themselves of a predictable environment. Even birds of the socially dominant species are sensitive to the density of their own species (Morse 1976a).

Not only is there variation in sites exploited locally, but patterns of habitat selection sometimes change over distance, as seen in the Black-throated Green Warbler's habitats. Unfortunately, the basis for this difference is not entirely clear, for these birds may occupy their favored sites in some areas but not in others. The Black-throated Green Warbler is an interesting subject for evaluating regional patterns, however, since sites chosen in one locality may receive little if any use in others. For instance, although boreal spruce forests are the center of their abundance along the coast of Maine, Black-throated Green Warblers are uncommon in lowland spruce-fir forests in northern Maine (Morse 1978a) and montane coniferous forests in New Hampshire (Morse 1979). Rather, they mainly occupy mixed coniferous-deciduous forests and even range into forests with few conifers (e.g., Holmes et al. 1986).

The geographic differences in the Black-throated Green Warbler's habitat use run counter to the results for several other warbler species arrived at by Noon et al. (1980). This suggests that most species of the eastern North American forest are conservative in habitat use over wide ranges;

several species exploited similar habitats in different regions, even if those habitats were uncommon. Regional differences in habitat use appeared to be a response to the absence of habitats typically used. Habitat availability thus appears to be a major influence in determining presence and abundance of certain species.

Collins (1983) noted that the analysis by Noon and colleagues used habitat measurements gathered at a community level rather than at specific activity sites of the individual birds, which means that the characteristics of the area only were evaluated; so it may not accurately reflect the selective processes of the individuals. Yet in lieu of information to the contrary one would not expect the two to differ. The tendency for most species to exhibit as many similarities as they do over substantial geographic areas suggests that either they belong to guilds of quite similar pattern from one region to the next, or that interactions among species do not strongly affect their habitat-choice patterns. Since many species are flexible on a within-habitat (between-niche) level (Chapter 3), that flexibility may have the most effect on interspecific accommodation and on maintaining their ecological discreteness. One would not predict this if interspecific territoriality occurred or if species were segregated by community, as the European sylviine warblers often are (Simms 1985).

It is of interest that Collins's concerns arose from the example of Black-throated Green Warblers, which had the lowest level of within-habitat flexibility of any of the species that I studied in depth. It might be enlightening to determine whether within- and between-habitat flexibility are negatively correlated in these species. The Black-throated Green Warbler is, curiously, one that Sabo would not have predicted to change its selection of habitats.

Experiments that permit rigorous quantitative assessment of these variables (Sherry and Holmes 1985; Morse 1985) must be designed if we are to sort out the variables affecting habitat selection and use. Despite the difficulties associated with such testing, certain—perhaps most—environments provide an opportunity for at least partial tests with relative ease. Landres and MacMahon (1983) suggested several simple manipulations that could be carried out with minimal difficulty or environmental damage. Further, possibilities for collaborative work with researchers studying modern forestry techniques have hardly been explored (Morse 1985). Opportunities to test habitat selection theory are already available, and conceptual advances will profit greatly from taking those opportunities.

5 Territorial Behavior and Other Aspects of Resource Exploitation

Interactions among warblers most often occur in the context of acquiring or retaining a resource—a certain space, a mate, food, or a nesting or roosting site. The ability to occupy distinctive habitats may be due to how successful an individual is in repelling conspecifics and other ecologically similar species. The outcome of contests of this type can determine how the individual exists (whether dominant or subordinate, for example) in a territory, which, following Noble (1939), I define as any defended area. Defense can take many forms, from physically aggressive displays to an actual confrontation. Such interactions are generally concentrated at specific times, especially when territories are being established. Consequently at other times territories may seem merely to be exclusive areas, although challenge will readily bring on defense.

The Advantages of Holding a Territory

Warblers studied during the breeding season exhibit territorial behavior. In that most of their singing appears to be associated with territorial defense or maintenance, we can infer that territoriality occupies much of the time and energy of males during the breeding season. Warblers defend Type-A territories (e.g., Mayr 1935; Hinde 1956), that is, all-inclusive territories. Thus, they usually are evenly spaced within favored habitats. All warblers I know with large breeding territories thus obtain all or most of their food on these sites.

Although occasional short-term temporal overlap doubtlessly occurs between adjacent territories (Morse 1976b; Ford 1983), the frequency of intrusion is low relative to the amount of time that the owners spend on their own territory. Warblers illustrate all-inclusive and exclusive territoriality as well as any species living under comparable conditions. Some intrusion may take place simply because of dense cover—a bird may not immediately see all interlopers, but even so most territory holders are remarkably adept at detecting intruders. On several occasions an observer's first notice of intruding males comes from an attack by the resident on the intruder (Morse 1966b). And even if the ability to intrude brings into question the inviolability of territorial boundaries, the intruder's usual behavior leaves no doubt about its status. An encroaching bird will be silent and take flight immediately upon discovery, even though it may be the rightful owner of a territory along the very boundaries of the territory it has invaded.

The functions (or at least benefits) attributed to territoriality are many, and they differ so much from one situation to another that no simple theory of function has emerged (see Hinde 1956); indeed, such efforts have generally been unproductive (Brown 1969). We do know, however, that during the breeding season males (and probably females, too: e.g., Nolan 1978) invariably defend their territories if possible.

Brown's concept of a territory as an area that is economically defendable seems to differ little from Noble's definition of a territory as "any defended area," but Brown's emphasis on economics enables one to predict when territories will be held. This is a major conceptual advance. The area in question must have resources rich enough to make it worth the effort; for example, more calories must be gained by maintaining a territory than by not doing so, or defending a territory must enhance the probability that the defender will have access to a minimal level of resources. Situations that favor this type of space use occur when resources are rather rich but are only loosely clumped. If they were strongly clumped, so much energy might be expended in defending an area that even if defense were physically possible, it would lower the defender's chances of survival—or at least its chances of reproduction. By contrast, poor areas will not sufficiently reward the effort put into defending them.

This concept of the economics of defendability seems consistent with the situations in which most warblers nest. They are insectivorous during the breeding season, and sources of superabundant food either do not exist or do not last long enough to ensure a sustained resource throughout the nesting period. Spruce Budworm outbreaks are an exception (Kendeigh 1947; MacArthur 1958; Morse 1978a); yet these are not always exploited heavily by all the species that encounter them (Morris et al. 1958; Morse 1978a).

I have implied that the food supply assumes a primary role in territoriality, a concept Brown advances. Food availability is frequently of major importance among breeding warblers. The high foraging rate of female warblers during incubation (Morse 1968) indicates that food is abundant enough for the birds to acquire adequate resources in a limited time. The problem may turn out to be more complex, however. The poorer the resources, the longer a parent will have to forage, and the greater the period that its nest will be vulnerable to climatic vicissitudes and predation. In studies of gulls, Hunt (1972) showed how, during times of scarcity, parents might find enough food for their young, but they required so much time that although the young did not starve, the loss of unattended nestlings to predation was high.

An alternative hypothesis for explaining why territoriality is beneficial regards predation as a direct factor: if nests are spread out in territories, search by would-be predators will be unprofitable. Horn (1968) proposed predator avoidance to explain why Brewer's Blackbirds *(Euphagus cyanocephalus)* spaced their nests evenly within a limited area of terrestrial habitat rather than clumping them, as do several closely related species that nest in inaccessible marshes. No direct evidence for this type of pattern has been reported in warblers.

An explanation often advanced for all-inclusive territories is that territoriality minimizes the interference of other individuals to the point that successful reproduction becomes possible. By keeping intrusions of other males low, territorial males should minimize the probability of cuckoldry. Given the period of vulnerability and the length of defense over the season, the latter explanation inadequately accounts for territorial behavior, even by males, but it could be a contributing factor.

Territory Size

The size of a site a warbler will defend differs according to the species in question, the site's quality, and the number of conspecifics and ecologically similar species present. Spruce-forest warblers have interesting differences that illustrate these patterns well.

Use of territory space changes during the breeding season, although specific information on this pattern is lacking in most species. Territories are often largest at the beginning of the breeding season, but later in the season individuals may not regularly exploit parts of territories they set up earlier. Declines in aggression and in utilized areas (*sensu* Odum and Kuenzler 1955) often denote a decline in territory size (e.g., Lack 1968), but these data cannot in themselves be taken as unequivocal evidence for a decrease in maximum territory size. For example, spruce-woods warblers, which decrease their use of territory over the season, are still intensely territorial when offspring begin to leave the nests; this is apparent when young wander to the edge of, or even out of, the territories (Morse 1976a). Nolan (1978) found that male Prairie Warblers used their entire territory throughout the breeding season, but those with the largest territories confined most of their activity to only part of their territory.

The size of a territory held by warblers varies from habitat to habitat and is related to how individuals use sites, how many resources are available, or both. Territory size of Black-throated Green Warblers varies between Red and White spruce forests; territories are half-again as large in White as in Red spruce forests, probably because of the greater difficulty of foraging on White Spruce foliage (Morse 1976a).

Variations in territory size can be related to the parts of the habitat that are used. Within dense populations we find close correlations between foliage volume and territory size of Black-throated Green and Blackburnian warblers. The Blackburnian Warbler concentrates its activities in treetops, where there is less foliage volume per unit area than in the areas frequented by Black-throated Green Warblers. Blackburnian Warblers concentrate their foraging in areas with only 40 percent as much foliage as Black-throated Green Warbler sites (Morse 1967b), and their density also averages about 40 percent that of the Black-throated Green Warbler in places having many Black-throated Green Warblers. This relation between territory size and foliage density occurs in a broad context of situations (MacArthur and MacArthur 1961). For example, Blackburnian Warblers never nested on small spruce-clad islands under 1 hectare that contained only a few high-elevation spruce trees. Nevertheless, this species inspected the one island in my study that had tall trees more than it inspected 10 other islands combined (Morse 1977b). Black-throated Green Warblers, by contrast, occupied several islands, but none with less than 0.35 hectare of forest. These forests were tall enough to incorporate the range of heights that Black-throated Green Warblers frequented in large mainland forests (Morse 1967b). Thus, if we wish to know whether a species will find a site satisfactory, we need to know the usage pattern of the species in question.

Territory sizes of deciduous or coniferous specialists may differ primarily as a function of which vegetation is commoner in the areas occupied. The coniferous-favoring Blackburnian Warblers sometimes occupy principally deciduous habitat, but they hardly use the deciduous vegetation and traverse especially large areas in search of conifers (Holmes 1986). Odum and Kuenzler (1955) referred to such areas as maximum territories, and the areas actually used within them were termed utilized territories. Maximum territories of Pine Warblers occupying mixed forests can also be several times that of their utilized territories. Such birds occupy forests where the proportion of conifers falls as low as 10 percent (Morse 1974a).

In other circumstances, special structural features of the environment may play an important role in territory size. Nolan (1978) compared foliage height with territory size of Prairie Warblers. The mating success of birds on territories of varying sizes suggested that the largeness of a site was not a reflection of poor quality. Display and surveillance sites are necessary elements of territories, and their absence can make the sole difference in whether the territory is satisfactory. If satisfactory display and surveillance sites are limiting, more tall trees should increase the number of territories and possibly decrease territory size as well. Kendeigh (1941) suggested that lack of display and surveillance sites limited the number of territories held by Yellow Warblers around the edges of prairie marshes.

Abundance of food alone can affect territory size, but such relations are hard to document because food availability is seldom measured. Even within a species, the relation between food supply and territory size may differ. For instance, both Stenger (1958) and Zach and Falls (1975) reported an inverse correlation between food supply and territory size of Ovenbirds, but Zach and Falls (1979) could find no such relation in an area determined to be less favorable for this species. Neither could Smith and Shugart (1987) find it among Ovenbirds in another geographic area, instead reporting a significant correlation between territory size and structural features of the environment. Yet one can easily infer a relationship between Spruce Budworm outbreaks and the appearance of Bay-breasted Warblers as breeders (Kendeigh 1947; Crawford et al. 1983), whose densities can reach 5 pairs per hectare (Erskine 1980). This is a higher density than that of any other warbler species except for the Tennessee Warbler, another budworm specialist on the breeding grounds (Erskine 1984).

Numbers of conspecifics can affect territory size, which fits Huxley's (1934) elastic disk hypothesis. This hypothesis proposes that as numbers of birds increase, the size of territories decreases to an irreducible minimum. Population density and territory size of Black-throated Green Warblers usually follow this pattern closely. In 1974, after two unsuccessful breeding years, the population density of Black-throated Green Warblers was extremely low. In the White Spruce habitats density declined to little more than half that recorded either before or in subsequent years. More individuals settled in the White Spruce forest initially, but left before nesting. For 10 to 14 days the remaining Black-throated Green Warblers retained territory sizes comparable to those observed during other years, but then they spread out; territories reached sizes twice that of other years, but size remained consistent with predictions from the elastic disk hypothesis. Population and territory sizes remained stable in Red Spruce forests, however. Other White Spruce forests nearby had the same decline as the intensively studied site. The number of Black-throated Green Warblers on small islands dropped sharply, too (Morse 1977b). All these results suggest local movements into favored sites. Nolan (1978) found that Prairie Warbler territories bounded by territories of several conspecifics were smaller than those bounded by few or no territories, but the strongest relation was between the proportion of the boundary bordering other birds' territories and the territory size.

Since so many species of congeneric warblers coexist in the spruce forests, it is important to determine whether they affect one another's territory size. In a competitive context, the birds should compensate by increasing their territory size, which in turn decreases the density of conspecifics. If, however, the species do not interact much, territory size should not be affected.

Territories of Northern Parula and Yellow-rumped Warblers nesting in spruce forests along the coast of Maine differ greatly in size (Figure 5.1). In diverse assemblages of warbler species, Yellow-rumped Warblers have the largest territories of any warblers and are widely dispersed through spruce forests at a low density. Yet, on smaller islands, where many of their customary associates (especially Black-throated Green Warblers) are absent, they occupy areas as much as 2.5 times smaller than their territories on the adjacent mainland. Only Northern Parula Warblers (the smaller of the two species) inhabit the smallest island areas. Not surprisingly, being smaller than the Yellow-rumped Warbler (and at least as plastic in the areas they will exploit), they occupy sites even smaller than the smallest inhabitated by Yellow-rumped Warblers (Morse 1977b). Both species are socially subordinate to the Black-throated Green Warbler (Morse 1976b) and change their foraging pattern markedly in its presence (Morse 1971a). In comparison, the Black-throated Green Warbler changes neither its foraging pattern in the presence of other species nor its territory size. Black-throated Green Warblers never nested on islands with spruce forests smaller than their customary mainland territories (Morse 1977b). Though

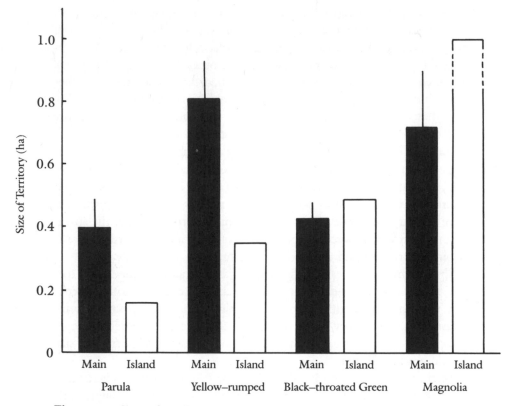

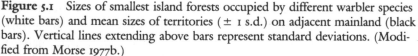

Figure 5.1 Sizes of smallest island forests occupied by different warbler species (white bars) and mean sizes of territories (± 1 s.d.) on adjacent mainland (black bars). Vertical lines extending above bars represent standard deviations. (Modified from Morse 1977b.)

factors such as foliage structure and numbers of conspecifics affect territory size in Black-throated Green Warblers, the presence of other warblers, at least Yellow-rumped and Northern Parula warblers, does not.

Magnolia Warblers resemble Black-throated Green Warblers in not occupying island habitats smaller than their normal territory size on the mainland (Morse 1977b), which may account for them occupying only the largest of the small spruce-clad islands studied (Figure 5.1). That they nested only one year in nine on this island is less in line with these relations, but most Magnolia Warblers that I observed on small islands (4 of 5) were found on the largest one. The Magnolia Warbler attains much higher population densities in some parts of its geographical range. In fact, it even reaches much higher densities in mainland areas of disturbed coniferous-deciduous forests adjacent to the spruce forests. In young Red Spruce forests in the Appalachian Mountains of West Virginia, it attains densities several times that of the coastal coniferous areas in Maine (Hall 1984).

Dominance Relations between the Sexes

Males of most passerine species are dominant to females, with certain notable exceptions (Morse 1980a; Smith 1980). Since males usually set up territories before females arrive on the breeding grounds, this pattern could be a consequence of territorial males being dominant in sites by the time females first arrive. Female Prairie Warblers and American Redstarts have this relationship; the response of a territorial male on first encountering a female on his territory is to attack (Ficken 1963; Nolan 1978).

In interactions between male and female spruce-forest warblers over the course of a summer, males usually dominated females (Morse 1976a). I based this conclusion on a small number of observations, however, perhaps a reflection of the low frequency of interactions between males and females other than mates, with whom aggressive encounters may be infrequent. If males and females partition the habitat during the breeding season, as these spruce-forest species do, they automatically ensure that females encounter females more often than males other than their mates, and vice versa.

Smith (1980) has proposed, though, that this pattern of male dominance is reversed in many monogamous birds, at least during the early parts of the mating period. She attributes the failure to report this pattern to the lack of attention to early periods of the pairing sequence. Nolan (1978) has reported that female Prairie Warblers rapidly become more dominating soon after pair formation, to the point that they routinely turn on aggressive males that attack them and force the males to retreat. Nolan asserts that females never completely dominate males, though several females attacked by their males during nest building retaliated by chasing the males. Smith's hypothesis is based on a few observations of reversal and scant information supporting other dominance patterns; thus it requires extensive testing. If it is valid, it could elucidate the resource-exploitation patterns and social relations between a pair during the rest of the breeding season. Female dominance could serve as the basis for intersexual partitioning, although no data support this.

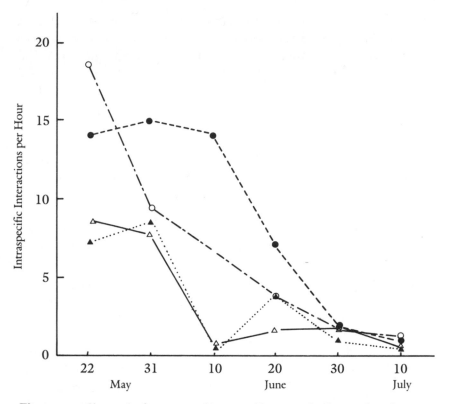

Figure 5.2 Change in frequency of intraspecific aggressive interactions between warblers over the breeding season. White circles, Black-throated Green Warblers; black circles, Yellow-rumped Warblers; white triangles, Blackburnian Warblers; black triangles, Magnolia Warblers. (Modified from Morse 1976a.)

Territorial Defense over the Breeding Season

Overt territorial interactions are most frequent during the early part of the breeding season and then decline precipitously during the incubation and nestling stages (Figure 5.2). This change in frequency is probably a consequence of territorial establishment, after which vocal or visual display usually suffices to proclaim ownership of a territory. Decline in frequency results either from the lowered importance of overt defense or from the substitution of auditory display for visual or physical displays. Ovenbirds, for example, do not respond as much to the songs of adjacent territory holders during the middle of the breeding season as they do to unfamiliar individuals—even birds a few territories away (Weeden and Falls 1959).

This may be explained by the song-recognition hypothesis. The investment in territory-related singing nevertheless may be great in dense populations, judging from the difference in time spent singing songs by Black-throated Green Warblers in large populations and those isolated on small islands (Morse 1970b). However, the frequency of singing in the large populations also declines over time (Morse 1970b).

Territorial defense does not end with decrease in song by the spruce-forest warblers, as can be seen after juveniles begin to move about. If young birds wander off their parents' territory, the owners on which they trespass may attack them. Prolonged encounters take place, with the territorial owner attacking either the young or their parents. In contrast to spruce-woods species, Mayfield (1960) reported that male Kirtland's Warblers with young sometimes encroach on territories without challenge, and Nolan (1978) noted that trespassing male Prairie Warblers with young were occasionally tolerated by the territorial male.

Use of the Territory at the End of the Season

As feeding the young becomes more demanding, warblers seem more inclined to wander off the territory. This tendency is probably enhanced by the random movements of the young (Mayfield 1960), even in the face of attacks from adjacent territory holders. In that many birds are fledging young at nearly the same time during a good breeding year (Morse 1976b), however, the effectiveness of territoriality probably declines at this time. The tendency to maintain territories late in the summer differs among species. Black-and-white Warblers aggressively attacked wandering birds, both conspecifics and other warblers, after this behavior had disappeared among other warblers (Morse 1970a).

Family groups of spruce-forest species often move from their nesting territories to areas of temporary food abundance, such as White Birch (*Betula papyrifera*) forests adjacent to spruce forests when the birches support infestations of lepidopteran larvae. Nolan found that about half of his male Prairie Warblers (31 of 63) remained consistently on their territories until migration and that only 13 percent (7 of 54) of the females remained. The males' behavior could be a remnant of territorial behavior, if the food supply were adequate for them and their young. It would be useful to know whether males that remained or left were feeding young and, if they were, whether the numbers being fed differed among those that remained and those that left. Other variables, including number of young per parent and feeding and predatory regime on and off the territory, might influence the results. Nolan (1978) tried unsuccessfully to relate the tendency to remain on territory or to leave with reproductive success, age, or characteristics of territories.

Juvenile American Redstarts remained on territory until migration (Ficken 1962). Comparison of philopatry between individuals whose young remained on territory until migration and those that did not remain would help researchers evaluate the importance of time spent on the territory in selecting their first breeding territory.

Although territoriality may completely break down when the young can fly well, and parents of many species often leave the vicinity of their nesting territories, Nolan (1978) reported a return to the territory by some departed parents in September, shortly before migration began (14 more of the 63, or 22 percent, of the males, 3 more of 54, or 6 percent, of the females). The purpose of return is not clear, but it may be associated with a mild increase in territorial behavior and song (Nolan 1978), accompanied by gonadal recrudescence, as has been recorded for several passerine species (Andrew 1969). It might be linked to reclaiming the territory. Whether the returning bird refamiliarizes itself with features of the territory, thereby facilitating recognition, or whether it reestablishes territorial relations with neighboring territory holders, one can only guess. Since two-thirds of Nolan's Prairie Warblers either remained on their sites throughout the summer or returned, late-season occupation is not a casual behavior. Nolan did not report whether late-season occupants were more likely to reclaim territories the next year than those not known to return at the end of the season.

Use of the Territory in Successive Years

Being migratory, individuals must vacate their territories at the end of the season and reclaim them anew at the beginning of the next season. After establishing themselves on a territory, warblers (especially males) have a strong tendency to return there the next season. Two-thirds of returning male Kirtland's Warblers (Mayfield 1960) and nearly three-fourths of returning male Prairie Warblers (Nolan 1978) used the same territory in successive years. Including individuals that disappeared, Nolan found that just under half of the males reoccupied their territory of the preceding year. Male Prairie Warblers' invariable success in reclaiming territories occupied the previous year may in part result from a greater willingness, compared with the usurper, to escalate their aggressive behavior (Maynard Smith 1982). The returning individual has the benefit of "knowing" that the site is acceptable. If the returnee reared young on the site, it might perceive the territory as being even more valuable. In several bird species, successful breeders are more likely to return to their former site in the succeeding year than individuals that survived but failed to rear young (Morse 1980a), although Nolan found no difference of this sort in the Prairie Warbler. An estimated 65 percent of the territorial male Prairie Warblers survived yearly, as calculated by the number of males returning to their previous nesting colony. Survival of young birds between fledging and return was about 39 percent, a much lower figure but one that is high in comparison with other young passerines (Lack 1954).

In contrast to the precise philopatry of established territory holders, there is no evidence of young returning to their exact home sites. A few first-year birds do return to the vicinity of their birthplace (Mayfield 1960; Nolan 1978) and establish territories. Even accounting for high first-year mortality, their return rate is much lower than that of older birds. Other bird species follow this pattern too. Nolan estimated that 11 percent of the young returned to the area where he worked (although none returned to their natal territory), which suggests that even if the young do not home in with the bull's-eye precision of their parents, they may reappear in a small area.

Most new sites are colonized by young birds. These are probably the birds that form new Kirtland's Warbler populations in Jack Pines that have become large enough for nesting. In contrast, older birds sometimes remain in sites that have become unsatisfactory. Mayfield (1960) and Walkinshaw (1983) reported territorial Kirtland's Warblers in old colonies of overaged Jack Pine forests. Numbers of birds at the sites declined precipitously, sometimes to the point where only a single or few unmated males remained.

Female warblers are generally less apt to return to a site than males are, again in common with other species of birds (Morse 1980a). Nevertheless, Mayfield (1960) and Nolan (1978) found that females regularly return to their previous year's breeding site, sometimes remating with their mate of the preceding year. Nolan found that 38 percent of returning females used their previous year's territory; when individuals not again found were included in the count, 10 percent of the females nested on their territory of the previous year. This difference in philopatry may be due to lower return rates, but it also is affected by whether the female returns soon enough to establish herself on last year's territory. Late-arriving females not infrequently settle near their last year's site, however. Although both males and females appear to prefer their previous site, they do not select last year's mate in preference to another; rematings occurred no more frequently than would be predicted by chance in adjacent territories, a pattern observed in other groups of birds (Morse 1980a; Morse and Kress 1984). Thus, apparent mate fidelity may be no more than the consequence of site fidelity.

Territories claimed in successive years are not invariably of the same size or shape, as each site is affected independently by such factors as population density, food supply, and whether adjacent territory holders return. Nevertheless, year-to-year overlap of territorial boundaries, even in locations not determined by the physical configurations of the site, such as a forest-grassland edge, showed a high degree of continuity in Prairie Warblers; Nolan (1978) reported that in about 90 percent of the reoccupations, year-to-year overlap exceeded 80 percent.

Interspecific Territoriality

Given the complex interspecific relations among some warbler species, one might wonder if warblers in dense multispecific aggregations exhibited interspecific territoriality; that is, do they defend exclusive territories against other species? What are the functions and advantages of interspecific territoriality? Is it primarily adaptive, directed between species that resemble each other and hence compete for resources and space (Orians and Willson 1964; Morse 1980a), or is it the result of mistaken identification (Murray 1971)? Regardless of its basis, participants in interspecific territoriality are similar enough that they are likely competitors for common resources, and the niche partitioning and niche shifts of these birds indicate that it may be an adaptive behavior.

I know of no good examples of interspecific territoriality by paruline warblers, even in the diversity of spruce-woods warbler assemblages. Most interspecific territoriality in other groups occurs when the habitat is low in height, as with marshes, or when similar species (sibling species) have recently come in contact. One might thus predict that Blue-winged and Golden-winged warblers, closely related and frequently hybridizing species that probably only recently have come into regular contact with each other, would exhibit interspecific territoriality. When they do interact, however, these birds do not form interspecific territories (Ficken and

Ficken 1968d; Gill 1980). The pattern of interaction among parulines is thus diametrically opposed to the widespread incidence of interspecific territoriality among their Old World counterparts, the sylviine warblers (Cody and Walter 1976; Cody 1978).

Hierarchy and Territoriality

Coexisting wood warblers, such as spruce-forest species, exhibit well-defined dominance patterns that are unlikely to be completely unidirectional, even in a given area (Morse 1976a). Reversals could be a function of the part of the territories where birds encounter each other, or they could be influenced by the birds' reproductive status. Reversals in dominance relations can be seen in interactions between the small Northern Parula Warbler and the larger *Dendroica* species. In a few instances Northern Parula Warblers were observed to initiate and successfully attack a *Dendroica* warbler, and each of these events occurred close to its nest (Morse 1971a).

Sometimes clear interspecific dominance does not occur. Yellow Warblers and American Redstarts exploit different habitats, with Yellow Warblers most frequently occupying younger or wetter sites than redstarts. But sometimes they are in close contact with each other, as on islands along the coast of Maine covered with mixed spruce–mountain maple growth. There redstarts generally inhabit more conifers than do Yellow Warblers, although on small islands the two species seem randomly distributed, probably a consequence of which species colonized first (Morse 1973). The number of hostile encounters initiated against birds of the other species is similar (Morse 1973). In adjacent mainland areas the two species are differentiated along a successional gradient—for example, on the edge of old beaver ponds in the northeastern United States—but the Chestnut-sided Warbler often inhabits sites between them. The relations between Chestnut-sided Warblers and the other two species have not been adequately investigated, but it would be profitable to do so in light of the tendency of Chestnut-sided and Yellow warblers to countersing when holding adjacent territories (Morse 1966b).

Territoriality and Population Size

Territoriality and aggressive behavior might limit populations, or at least the size of the breeding population. Brown (1969b) has pointed out, however, that unless one finds reproductively competent nonbreeding birds of both sexes at a breeding site, one cannot assume that territoriality limits population. Few experiments have tested this hypothesis, and they have yielded mixed results: some indicating an effect, others not (reviewed by Morse 1980a). Yet, massive removals by Stewart and Aldrich (1951) and by Hensley and Cope (1951) of several species of spruce-woods warblers during an outbreak of the Spruce Budworm in northern Maine provide striking evidence for the presence of nonbreeding birds. These workers removed a few hundred warblers from a 16-hectare area—more birds than those initially present (1.7 to 3.4 times the original number of territorial birds of the five commonest species were removed). Although they did not know the origin of the incoming birds, or their previous breeding status, or whether they would have bred under these circumstances, the magnitude of the influx implies that many individuals were previously floaters prevented from breeding because satisfactory habitats were saturated by other birds. Both males and females entered, although by far more males were collected, possibly because they are more conspicuous and hence more often collected.

Field observers frequently note wandering individuals, which routinely have been assumed to be nonbreeding males. Studies with marked individuals, however, have demonstrated that some are territorial. Most frequently the wanderers are males (Nolan 1978), which is a consequence partly of their being more conspicuous and partly of their not being bound to the nest by responsibilities. On the small coastal islands where I worked, it is easy to identify foreign birds; new singers are conspicuous, and often the change in behavior of the resident birds signals their presence (Morse 1970b). Most of these visitors were also males. If this apparent male bias is real and not just the consequence of males being more conspicuous, it raises the question of whether the size of the breeding pool is controlled by territoriality or by the number of the less common females. To complicate the issue, a necessary resource for males is a territory, which, if satisfactory, will allow the male to obtain a female, and which will not be shared. But females may enter into a polygynous relationship if there are no female-free territories, so what would otherwise be a female surplus can be accommodated. These unions could be the females' best option, but since second broods are less successful than broods of the first females, the fitness of these "surplus" females is intermediate to those of females that have either a monogamous mate or none at all. In light of Brown's concern, the birds are mated, and thus will minimize the frequency of territoriality limiting population size.

Interactions between Summer Residents
and Permanent Residents

Relations between the populations of resident and migratory species have received far too little attention, in spite of their implications for the structure and diversity of both and for the total diversity as well. Parulines that have taken up life-styles other than that of the numerically dominant leaf gleaners seem especially likely to make contact with residents of other species. Trunk and large-limb foragers, represented by the Black-and-white Warbler, are an obvious case in point. Their foraging habits bring them into sites exploited by nuthatches, creepers, and woodpeckers. They interact with these birds regularly, and in my study areas Black-and-white Warblers had many fights with Red-breasted Nuthatches *(Sitta canadensis)*, the most frequently encountered member of the bark-foraging guild (Morse 1970a). Although Black-and-white Warblers also exploit foliage and interact with gleaners, relations with bark foragers are probably more important to them than they are to other warblers.

Interactions between migrants and residents play an important role in the lives of both groups in Old World gleaners (Ulfstrand 1976, 1977). Ulfstrand was struck by the similarities in resource exploitation of twig- and foliage-gleaning species: the resident Coal Tit *(Parus ater)* and Goldcrest and the abundant migrant sylviine, the Willow Warbler *(Pylloscopus trochilus)*. Both permanent residents used parts of their foraging areas much less often in summer than in the winter, the Coal Tit by narrowing its niche and the Goldcrest by shifting its exploitation to other parts of the habitat. Additionally, many of these birds were migratory themselves; their densities were much lower in the summer than in winter, in contrast to other titmouse species that did not overlap heavily with summer residents. Ulfstrand proposed that the Willow Warbler excludes the Coal Tit and Goldcrest. Specific nesting or feeding requirements of the latter two species could also explain the shift, which in turn might facilitate the invasion of the Willow Warblers.

A comparable situation may exist in North America, at least in northern coniferous areas where spruce-forest warblers comprise up to 70 percent of the breeding bird fauna and residents less than 20 percent (Morse 1977b). In my study areas along the Maine coast the majority of the resident population is made up of two species whose foraging patterns do not substantially overlap those of *Dendroica* warblers: the Red-breasted Nuthatch, which forages extensively on large limbs and trunks as well as foliage, and the Golden-crowned Kinglet, a branch-tip specialist. One of the other commonest species in the coniferous forests during the winter, the Black-capped Chickadee *(Parus atricapillus),* deserts the areas during the summer for sites in adjacent deciduous or mixed coniferous-deciduous forests, perhaps in response to its preference for dead birch stubs, in which it excavates its nests. By contrast, during the winter it extensively uses spruce foliage (Morse 1970a) exploited by numerous *Dendroica* warblers of several species during the summer. These feeding habits are consistent with the permanent partitioning between migrants and residents, similar to the one reported by Ulfstrand. The main interaction among migrants and residents in this habitat is between the Golden-crowned Kinglet and the Parula Warbler, the smallest warbler species in the area (Chapter 4).

Where large forested areas have recently been reduced to remnant, islandlike habitats by deforestation, permanent residents, including exotics that frequent edge-type habitats (e.g., Starlings and House Sparrows), have invaded; and many migrant species, prominent among them certain warblers, have declined precipitously in recent years (Whitcomb et al. 1977, 1981). Whitcomb and his colleagues argued that these opportunities are opened up for residents because migrants are unable to recruit to these small areas. This, too, indirectly implies that under stable conditions migratory species can exclude permanent residents from niches. The islandlike areas also become more edgelike, which must increase the opportunity of permanent residents to invade. Lewke (1982) found that most permanent residents in a floodplain habitat in Washington were edge species, and the forest branch and foliage foragers were usually summer residents or winter residents that replaced each other. In these deciduous habitats, adaptations for winter foraging on leafless trees (largely by kinglets and chickadees) are not the same as those appropriate for summer leaf gleaners (the Yellow Warbler and Red-eyed Vireo). Thus, in Lewke's study the three groups of species exploited distinct habitat characteristics in a manner similar to the complementarity of resident and migrant foraging types in spring (Herrera 1978). This suggests that interactions between residents and migrants, if they occur at all, should be confined to certain foraging repertoires. For instance, aerial prey capture is the province of migrants, but foliage gleaning of both. In general, resident species have a broader and more variable repertoire, and migrants fit in where residents have not exploited resources. The migrant group's diverse repertoire as a whole happens more through changes in species composition than in repertoires of individuals or single species. Lewke's and Herrera's studies indicate that residents monopolize constant resources and migrants capitalize on temporary ones. Nevertheless, temporary summer food supplies constitute the major resources at these sites over the period of a year, and given the migrants' success with them it is by no means appropriate to consider them inferior to the permanent residents.

The dominance of spruce-forest warblers in their habitat could be due to yearly fluctuation of food supply in the habitat, which will support few birds during the wintertime but considerable numbers during the summer. An uneven food supply selects either for migratory species or species able to hibernate on site, an option not open to these birds. Resident-migrant interactions should change as one moves southward into regions with larger percentages of permanent residents, sites at which higher densities of birds can survive a winter. If the environment supported more birds during the winter, migratory species would presumably be less suc-

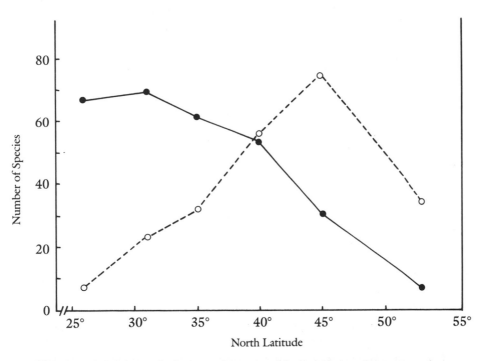

Figure 5.3 Numbers of winter or permanent (black circles) and summer resident (white circles) passerine bird species in a latitudinal gradient along the eastern coast of North America. (Data from AOU 1983.)

cessful. A latitudinal shift of this ratio along the eastern seaboard of North America is consistent with this argument (Figure 5.3).

Interactions among Summer Residents

The parulines impinge on other groups of migratory species as well as on permanent residents of the breeding grounds. The vireos, sluggish twig and small-branch foragers, are often encountered by warblers. Vireos are larger than most paruline warblers, however, and the slow, deliberate type of foraging generally associated with large size seems to be related to what kind of prey is sought. The largest foliage-gleaning warblers, the Bay-breasted and Kirtland's warblers, are also sluggish and deliberate in their foraging and they, as well as the vireos, take different prey from that taken by smaller, more active species (Robinson and Holmes 1982, 1984). Slow movement may permit a predator to focus on cryptic prey, which are also larger but would not be spotted by more rapidly moving birds. Vireos are usually less abundant than warblers, and many habitats in the east support one species of vireo, or two at most, per habitat (Williamson 1971; Rice 1978), although the Red-eyed Vireo attains high densities in mature deciduous habitats.

Several species of warblers are ground dwellers, especially the Ovenbird and waterthrushes, and to a lesser extent the Swainson's and Worm-eating warblers. Thrushes are common in this guild; though larger, they exploit the same substrate. Sparrows feed heavily on the ground, although they usually take a wider variety of foods than warblers do, including some that warblers would not use.

Another adaptive zone entered by warblers is flycatching, or hawking insects in the air column. This life-style is dominated by tyrannids, even in the high latitudes; this subgroup is the only set of tyrannids that has successfully penetrated well into the north. Even so, American Redstarts, though foliage gleaners too, spend much of their effort flycatching, and several other parulines, including Yellow-rumped Warblers, regularly flycatch. The redstart's life-style places it in competition with Least Flycatchers, and probably other *Empidonax* species elsewhere. Sherry (1979) and Sherry and Holmes (1988) found aggression between the two species in New Hampshire, probably because they exploit northern hardwood forests in a similar way. Their relationship is governed, at least partially, by the Least Flycatcher's dominance and aggression directed at the redstart and the redstart's ability to exploit parts of the habitat not used by flycatchers. Although the redstart lives in other forests where the fly-catcher is not present (Morse 1973), members of the genus *Empidonax* occupy habitats throughout North America, so investigators should be on alert for other interactions such as these.

Flexibility in Spatial Patterns

Their reliance on insects as a food supply is a major influence in setting warblers' social and spatial patterns. Their resources are usually only mod-erately patchy in favored areas, though they select territories with superior resources. The direct responses of many species to certain physical struc-tures in the environment probably reflects a resource potential in the environment that is constant enough to be interpreted by the birds as a stable set of cues, thus programming the birds to a predictable territorial pattern. Although resources may be superabundant during a caterpillar outbreak, the retention of aggressive behaviors at such times suggests that these behaviors are basic adaptations to an environment not undergoing such an outbreak, and that the outbreak conditions are not important enough to suppress them. From the viewpoint of food acquisition, "inap-propriate" aggressive behavior comes into play when it makes the least difference—when resources are not limiting and food-gathering con-straints are minimal. Alternatively, at these times other functions of terri-toriality, such as spacing to minimize nest predation, may assume an im-portant role.

At an interspecific level warblers illustrate well how space- and aggression-related factors determine guild relations and membership. Normal patterns of territory size, dominance relations, and flexibility in space or resource use indicate that space and resource use are greatly influenced by interactions with other birds and by habitat constraints. These conclusions contrast with those based on habitat selection, which reveals a high consistency in performance for most (though not all) spe-cies from one geographic area to another. Perhaps exploitation patterns take place at a within-habitat level and then spread rapidly.

Flexibility is consistent with the rarity of interspecific territoriality in this group. But one might view the spatial patterning (niche partitioning) of spruce-forest warblers as a mosaic in which habitats are partitioned in compartments, each occupied or dominated by one species to the partial but not total exclusion of others. Quantitative differences among species in exploiting compartments seem consistent with territorial precepts but

also resemble home ranges. Thus, if two species have a dominance relationship and their niche preferences overlap, the dominant species can
exclude subordinate species simply through exploitation. The result will
quantitatively resemble a low-grade interspecific territory, which, however, is an attractive option because costs of defense are low and the potential for overlap is substantial. Ecological release exhibited by subordinate
species in the absence of the dominant species speaks to the effect of the
dominant's presence and the behavioral nature of the shift. Whether the
effect is due to the dominant's presence or whether exploitation competition plays a role is not clear.

The advantage of the spatial relationship among the spruce-woods warblers lies in its low defense costs. Most interspecifically territorial songbirds in North America occupy habitats whose vertical component is inadequate for vertical position—grasslands or marshes, for example:
(Orians and Willson 1964)—or exploit resources in the air column itself,
as when *Empidonax* flycatchers feed on flying insects. Niche partitioning
among parulines in eastern North America differs from that of their Old
World ecological counterparts, the sylviine warblers, in western Europe
(see Chapter 15). Members of sylviine genera usually are less diverse within
habitats, in spite of their rich regional diversity, and there is substantial
evidence for interspecific territoriality among congeners where they meet.

6 Predators and Parasites

Temperate-zone bird populations are often considered in terms of the availability of food and the subsequent competition for it. Hypotheses about predators or parasites have attracted little attention, but this aspect of bird biology deserves serious attention. Not only might it play an independent role in determining the abundance and composition of the species present in a population, but its effects may be interwoven with those of food and competitors.

In this chapter I focus on two types of predation: predation of adults and fledged young, and predation of nest contents. Even though nest predators may not endanger adults, if they consistently prevent successful nesting they will diminish the would-be parents' future fitness to zero, which results in genetic death as surely as if the parents themselves were killed. Thus, responses to the two types of predators may differ, but the eventual consequences of failure are the same. Avian brood parasites are another threat to nesting birds (Lack 1968; Payne 1977), as are invertebrate parasites (Clay et al. 1985).

Predation of Adults

We lack adequate information about the predation of adult warblers during the breeding season. Nolan (1978) reported only two acts of predation of adult Prairie Warblers, both about the nest; and I have never observed predation of adults during my 15 years of fieldwork with breeding warblers. Yet it would be a mistake to assume that it is unimportant. Most of my observations were in areas devoid of the usual avian predators of small birds in North America: Sharp-shinned and Cooper's (*Accipiter cooperii*), hawks. These hawks take numerous small birds, and warblers probably form a large part of their diet because of their abundance. Mueller et al. (1981) in Ontario and Meyer (pers. comm., 1986) in New Brunswick noted that although warblers were frequent prey of nesting Sharp-shinned Hawks, the hawks appeared to prefer larger birds such as thrushes. During my fieldwork on spruce-forest warblers, from the early sixties to the mid-seventies, hawks were very low in number, probably as a result of heavy use of persistent chlorinated hydrocarbon pesticides (Moore 1967; Ratcliffe 1970). Only in 1986 did Sharp-shinned Hawks again nest in the vicinity of my Maine study areas.

From studies of the similar European Sparrowhawk in Wytham Wood, England, where the species has recently returned, we know that the species takes 20–25 percent of the numerically dominant group of small breeding birds, the titmice, during the breeding season (Geer 1978; Perrins 1979). Although the titmouse populations remained stable after the hawks' return, the number of foreign-born breeders increased, which implies that the local population was maintained through immigration. High predation rates such as this must also affect predator-avoidance behavior, and in turn may dictate what type of foraging is acceptable. One would assume that the effects of Sharp-shinned Hawks would be similar for paruline warblers, were these hawks a common nesting species.

Nest Predators

It is commonly believed that many eggs or young in nests are lost to predators (Lack 1954) when observers lead predators to them—with their odor trails, for instance. But only a few tests of this hypothesis have been done. Willis (1973) found no difference in predation of Bicolored Antibird *(Gymnopithys bicolor)* nests that were regularly visited or of nests surveyed from a distance: both instances revealed many predators. If Willis's results are representative, nest predation may have a large effect on nesting success. Yet, little is known about nest predation of warblers or other species in relation to such variables as population density and species composition.

Studies of spruce-woods warblers and other passerine birds in mainland or large island habitats and on small islands provide insight to the effect of predators on population density. As the size of the island decreases, the population density increases (Morse 1977b). The isolation and even the smallness of the islands may account for an absence of Red Squirrels and a scarcity of Blue Jays, the most common nest predators on the mainland. Therefore, the high density of small-island birds may be due to a general absence of nest predation.

Nest predators in other areas include a wide variety of birds, especially other corvids; small mammals, including various carnivores; and snakes. The relative importance of these predators probably varies with the geographical region and habitat. For example, snakes are believed to be more prevalent nest predators in tropical areas than in temperate ones (Skutch 1976), and obligate egg-eating snakes inhabit only tropical areas (Gans 1974). Nolan (1978) concluded that snakes, along with Eastern Chipmunks *(Tamias striatus)* and Blue Jays, were significant nest predators of Prairie Warblers. Mayfield (1960) attributed nest predation of Kirtland's Warblers mainly to birds, especially Blue Jays and crows, although he also listed several mammalian predators. The part of the habitat, the nest location, and the nest characteristics may also enter the picture. Ground nesters will be exposed to small predators that are unable to climb, for example, and Prothonotary Warbler nests, because they are placed in holes, will not be vulnerable to certain types of predators, such as jays and crows.

Martin (1988a) has recently addressed the question of nest predators as a factor affecting the number and distribution of species in a habitat. Numbers of species in high-elevation Arizona forests correlated with the density of foraging substrates and nesting substrates, but correlations with nesting substrates were stronger than correlations with foraging ones. Further, ground and understory nesters, groups of species especially vulnerable to nest predators, showed the strongest correlations between substrate density and nesting density, further suggesting the possible importance of nest predation in breeding distribution. Martin (1988b) then determined experimentally that a predator's intensity of search rises with increased occupation of potential nest sites and that predator pressure is significantly higher when nests are distributed within a single type of nest site than if they are distributed among more than one type of site. Thus, equal numbers of MacGillivray's Warbler and Black-headed Grosbeak *(Pheucticus melanocephalus)* nests should generate less hunting pressure than twice the nests of either species alone. These predator pressures should select for territorial behavior among their prey species, especially intraspecific territoriality, in a way that also resembles resource-based territoriality. Accordingly, interspecific interactions should be stronger among species with similar nesting habits than among those with dissimilar nesting-site preferences. Although Martin did not conclusively eliminate the alternative of foraging substrates as an explanation for his results, his study emphasized the importance of addressing the role of nest predation as an alternative hypothesis for community structure.

Brood Parasites

The Brown-headed Cowbird is the only passerine brood parasite in the breeding range of most temperate-zone warbler species. It lays eggs in nests of various passerines, but its effect varies with the host species and is probably undergoing major change at this very time. Brown-headed Cowbirds were originally inhabitants of the Great Plains of central North America and were associated with the giant bison herds of that region. Their range, now encompassing most of North America (AOU 1983), expanded when Europeans arrived with their livestock and removed the forests (Friedmann 1929; Mayfield 1965). Only recently has it overlapped with many host species, especially those of the eastern forests, the western montane region, and the West Coast. For instance, Mayfield estimated that cowbirds parasitized about 50 species before European colonization, but now they may parasitize 200 species. Newly exposed species usually have no defenses against cowbird parasitism, and if the hosts are attractive the cowbirds could drive them to extinction. The Kirtland's Warbler, initially a rare and localized species, could soon become extinct if there were no cowbird control (Walkinshaw 1983). Even so, its numbers dropped almost threefold over 10 years of heavy parasitism (Chapter 11).

This same pattern of host decrease is known in other instances where brood parasites invade new areas (May and Robinson 1985). Rothstein (1975a,b) presented the problem of would-be hosts accepting or rejecting cowbird eggs; although newly exposed species or populations are likely to suffer severe depredations, many long-established associates do not reject these eggs, either (Rothstein 1975a). The hardest-hit species or populations have presumably become extinct. Some species that do not routinely reject cowbirds eggs are, nevertheless, poor hosts (May and Robinson 1985). Probably all accepting warblers are desirable because of their food and rearing habits, and size, but species with unusual diets—such as American Goldfinches *(Carduelis tristis)*, which feed their young a diet of regurgitated seeds (Rothstein 1975b)—may be unacceptable hosts.

Rothstein (1975a) placed artificial cowbird eggs in the nests of many North American birds. He concluded that few species remain in a dimorphic condition (both accepters and rejecters in a population) for long, which he attributed to selective pressures accompanying different responses once they are expressed. However, May and Robinson (1985) suggested that signs of dimorphism do exist and are maintained by frequency-dependent selection.

The cowbird has extended its range into the boreal spruce forests, where it parasitizes even such species as the Black-throated Green Warbler, though not often. Forest-interior species are protected because of the cowbird's dependence on open-country feeding areas (Brittingham and Temple 1983; Rothstein et al. 1984).

Young cowbirds emit begging notes superficially similar to fledgling calls of several warbler species. Although it is unlikely that their call is characteristic of warblers only, a survey of other potentially parasitized species would be of interest to determine whether the call notes of young cowbirds have converged in the direction of common host young. If there has been no selective premium on specific distinctness in the begging response of passerines, as Hamilton and Orians (1965) suggested, however, cowbird begging calls may function most effectively in the absence of any such directional selection, past or present. Woodward (1983) described cowbird begging calls as characteristic of passerine begging calls, although louder (not surprising because of their usually superior size) and more persistent than the calls of any of their hosts. Their calls may thus function as supranormal stimuli.

Many host species recognize full-grown cowbirds, and both the adults and large, begging young are apt to be attacked or scolded by passerine birds when in the vicinity of their hosts' nests (Ficken 1961; Folkers and Lowther 1985). Robertson and Norman (1976, 1977) have demonstrated experimentally that a wide range of passerine birds, including Yellow Warblers and Yellowthroats (but not Yellow-rumped Warblers) respond more strongly to cowbird models than to other species. These results and other observations run counter to those of Hofslund (1957) on Common Yellowthroats, as well as Hann's (1937) on Ovenbirds, who recorded no such hostile behavior. In some instances Robertson and Norman's accepting species responded more strongly than rejecters, though their result was not statistically significant. Were this pattern to prevail, it would suggest alternative means of reacting to the cowbirds, also a mechanism of defense that might counter parasitism in the absence of rejecting behavior.

The cowbird–Yellow Warbler relationship has been well publicized because a warbler may bury cowbird eggs laid in its nest, sometimes along with its own eggs (Bent 1953). Bent reported several tenement nests, a consequence of cowbirds' repeatedly laying in the warbler nests. The record is a six-"story" nest (Berger 1955)! In a study using artificial cowbird eggs, however, Rothstein (1975a,b) concluded that the Yellow Warbler and other warbler species were accepters, finding no evidence that it

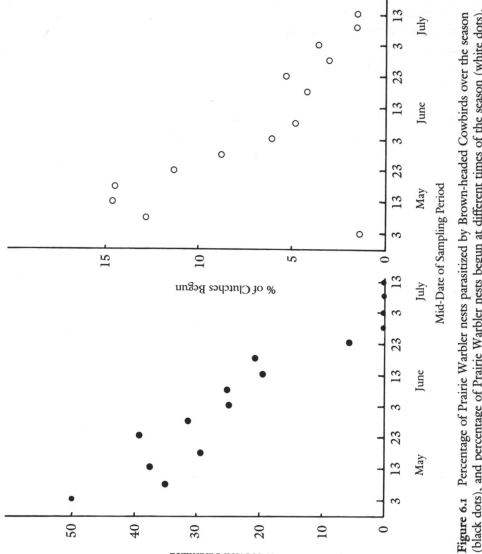

Figure 6.1 Percentage of Prairie Warbler nests parasitized by Brown-headed Cowbirds over the season (black dots), and percentage of Prairie Warbler nests begun at different times of the season (white dots). (Modified from Nolan 1978; used by permission of the American Ornithologists' Union.)

would selectively bury cowbird eggs without burying its own. If a cowbird laid before a nest was completed, the warbler might cover the cowbird eggs as a foreign object. This behavior could protract nest construction so that construction and egg-laying would occur simultaneously, and thus cowbird eggs laid with warbler eggs might be buried. Rothstein concluded that neither Yellow Warblers nor any warblers he studied were adapted to respond to eggs deposited in their nests.

In contrast, Clark and Robinson (1981) suggested that Yellow Warblers possess a finely tuned antiparasite strategy that rejects cowbird eggs by egg burial or by nest desertion. Cowbird eggs are probably too large for Yellow Warblers to eject, a problem for other warblers as well. Alternatively, puncturing the egg would foul the nest, so Yellow Warblers bury cowbird eggs if they have not laid more than one of their own eggs; if they have laid as many as three eggs before the nest is parasitized, they are more likely to accept the cowbird eggs. The success of nests in which warblers buried eggs was as high as that of unparasitized nests, and reuse of the site did not increase vulnerability to nest parasitism; however, the success of parasitized nests was lower. Clark and Robertson pointed out several advantages to warblers of accepting cowbird eggs laid after at least three of their own. The accepters' energy investment is greater by that time, and, further, the delay of laying another clutch is longer. Also, if they have started incubating, their own eggs are older than the cowbird's eggs. Cowbird eggs hatch in 1–2 fewer days than Yellow Warbler eggs, but if the warbler gets a head start so that its eggs hatch as soon as the cowbird's, its nesting success is higher than if the cowbird hatches before the warblers. Desertion is an alternative response to parasitism early in the season; it may be a more drastic response because additional time is required to build another nest, as well as to lay eggs. Nests might also be deserted if after parasitism they became unstable because they were built higher.

Other warblers cover cowbird eggs much less frequently than do Yellow Warblers. Nolan (1978) reported only three such eggs in his extensive studies on Prairie Warblers, and Mayfield (1960) noted that Kirtland's Warblers occasionally embed cowbird eggs when they are laid before the nest has been completed.

Other strategies might be preferable under different circumstances. Deserting the site would be a more advantageous strategy if a single cowbird frequently laid more than one egg in a single nest. This makes the nest a natural target, and thereby greatly reduces the opportunity of the host to bring off young. Cowbirds probably lay a limited number of eggs, and the length of their reproductive season is correspondingly curtailed (see Bent 1958). Thus, host species might succeed in waiting out the parasite by renesting. Indeed, incidence of parasitism on the Prairie Warbler declined markedly over time (Figure 6.1).

The host's tendency to rebuild or abandon a nest should, accordingly, shift over the season, as Clark and Robertson (1981) found for the Yellow Warbler. Success of renesting birds should be lower than that of birds rearing young on their first attempt, for several reasons. The season has drawn on, so opportunities for young may not be as good. Even if fledging success were comparable, survival after fledging might be lower, simply because periods of peak food availability do not coincide as well with the demands of interrupted parents as with those of parents of an undisturbed brood. In response, later broods are smaller. Clark and Robertson (1979) pointed out that broods that are out of phase with conspecifics and other species about them—Red-winged Blackbirds *(Agelaius phoeniceus)* in their study—may suffer more predation than other nests. Lastly, late-fledging broods may have less success simply because they will be younger at migration. Unfortunately, data on the survival of young are scant (May and Robinson 1985); nothing seems to be known about the success of young from pairs that reared their first broods and those from pairs that renested because of parasitism early in the season. We do know that cowbirds lower the fledgling production of Prairie Warblers by 13 percent, the decrease about equally divided between nests failing to produce any young and nests producing fewer young (Nolan 1978).

Ectoparasites, Endoparasites, Disease, and Commensals

Warblers appear to carry a normal share of parasites and disease-producing organisms, yet the literature on the subject is incomplete and scattered. Bent (1953) listed ectoparasites taken from several species of warblers. Anyone who has banded birds is familiar with some of them, especially louse flies (Hippoboscidae), the peculiarly flattened insects that hide in birds' feathers and feed on their blood. But a much longer list could be made, and some could play a larger role in the welfare of passerine birds than do hippoboscids. It includes chiggers, ticks, and mites, as well as representatives of several insect groups: chewing lice (Mallophaga), swallow bugs (Cimicidae), and blowflies (Calliphoridae). Many species of these parasite groups attack a wide range of host bird species (Boyd 1951), but specificity is much higher in such groups as the Mallophaga (Foster 1969). Some species concentrate on nestlings, and their life cycle may be centered on nest sites. They constitute the greatest problem for hole nesters and other species that use sites more than one year. Warblers and other open-nesting passerines seldom if ever reuse a nest in the next year, or even for a subsequent brood (Bent 1953). The numerous nest parasites often increase mortality of the young, or at least decrease their growth rate, thus potentially prolonging their nesting period and depreciating their condition at fledging, thereby adversely affecting their survival (e.g., Nolan 1955).

Nolan (1978) recorded parasites that his Prairie Warblers came in contact with, and it was clear that these organisms cannot be ignored. Larval blowflies (*Protocalliphora* sp.) caused the greatest mortality in these nests. Yet mites and chiggers occupied all 23 Prairie Warbler nests that Nolan analyzed, and one or more of these species was responsible for blood or scabs on the tarsi of the nestlings. Ticks also were found on the warblers, and along with the mites accounted for loss of blood, serum, and feathers, at an unrecorded cost. These acarines attack both adults and nestlings, but nestlings carry many more, probably because the young are not mobile and cannot groom themselves (although the parents occasionally remove ectoparasites from them), and because the nest environment is a good microhabitat for growth and reproduction of parasites.

Species that reuse the same sites probably suffer ectoparasitism more often than most warblers do. One wonders if the only paruline hole nesters, the Prothonotary and Lucy's warblers, experience greater parasitism than do other paruline species since they use cavity nest sites that have been occupied previously. This added burden may be due to nest sites being limited, or to nest sites offering offsetting advantages that compensate for the many ectoparasites encountered. For instance, Eastern Phoebes *(Sayornis phoebe),* which nest under bridges or in buildings, use the same sites year after year and are often infested with blood-sucking mites. Klaas (1975) reported infestations of Northern Fowl Mites *(Ornithonyssus sylviarum)* that resulted in up to 8 percent nestling losses over 4 years in Kansas and that weakened others to the point of premature fledging and hence high mortality. Nolan recorded numerous *Ornithonyssus* in a few nests and suggested that it is the most injurious mite to Prairie Warblers, although he did not document deaths. Cliff Swallows (*Hirundo pyrrho-nota*) at large colonies suffer debilitating juvenile mortality from swallow bugs that infest established colonies; their presence may prompt periodic moves to new sites (Brown and Brown 1986).

A wide variety of endoparasites, roundworms, tapeworms, flukes, and the like, are reported for passerine birds (e.g., Davis et al. 1971; Rausch 1983), but the reports are generally only descriptive. The data are adequate to indicate, though, that small passerine birds have their own representative endoparasites, as do other groups of vertebrates, and we can only assume that their effects parallel those of more heavily studied host groups. Nolan reported finding nematodes in Prairie Warblers killed at television towers, but has no more to say about them.

Birds harbor many "traditional" diseases from protozoan, bacterial, and viral infections. Sometimes Avian Pox (*Poxvirus avium*) lesions are obvious, especially on the feet. Serious lesions lead to the loss of digits and easily debilitate birds to the point where they starve or are easy prey. Nevertheless, Nolan relates an instance of a female Prairie Warbler who, over a few years, had lost her left hallux and all but the proximal elements of the toes, yet the bird nested and helped in feeding her young until a predator ate them. Less than 1 percent of Prairie Warblers examined in Nolan (both breeding birds and television-tower kills) had pox lesions. Karstad (1971) also noted pox foot lesions on *Dendroica* and other warblers.

Malarial diseases are widespread in birds and involve at least three genera of protozoan parasites transmitted by dipteran groups that attack birds, including mosquitoes, black flies, and hippoboscid flies. The effect on wild bird populations is not clear, although fatal cases have been observed (Herman 1955) and high levels have been found in game birds and in island populations of passerines (Warner 1968; Riper et al. 1986). Coccidiosis (protozoan), cholera (bacterial), aspergillosis (fungal), psittacosis, and ornithosis (viral) are all important diseases of domesticated birds that are known from wild populations of galliforms and other species. But the ranges and frequencies of these types of diseases (to passerines, or even to parulines), and their significance, are unknown. Equine encephalitis (both Western and St. Louis) virus apparently does not measurably affect the birds carrying it (Herman 1955).

The ecological literature is deficient about the impact of parasites and diseases on birds, at least on nongame species—a deficiency extending to the role of disease and parasites in population processes (Anderson and May 1979; May and Anderson 1979) that is only beginning to be redressed (Dobson and Hudson 1986). It was not always so; in old ecology textbooks, such as the classic *Principles of Animal Ecology* by Allee et al. (1949), one finds more attention given to disease and parasites than is seen at present, except for recent efforts by Anderson, May, and their colleagues. More attention to these topics may also be found in the literature from wildlife biology, which includes journals devoted to the problem, such as the *Journal of Wildlife Disease*.

The one clear sense we have from a survey of ecological literature is that little is known about the effects of parasites and disease on the dynamics of warblers, other passerine birds, or many nongame animals. This does not imply that their effects are negligible, and the topic warrants attention in field population dynamics. Nolan's efforts are a welcome start, though they are confined to ectoparasites. His approach differs from the earlier one of Mayfield (1960), who dismissed the potential ectoparasites as "seldom seen on Kirtland's Warblers," an attitude common to most workers.

Although nests present warblers and most other birds with their most concentrated exposure to parasites, many arthropods in these nests are not parasitic on the birds; they are probably best considered commensals. The nest's construction provides food—from the nesting materials, unremoved feces and other debris, and dead baby birds—and protection. Other species may parasitize or prey on the parasites of the birds. It is thus appropriate to think of nests as small, temporary communities. A majority of the species extracted from the 23 just-abandoned Prairie Warbler nests that Nolan (1955, 1959) analyzed with Berlese funnels (a drying treatment that forces arthropods and other organisms from the nests) were not parasites of the birds, though parasites were well represented by parasitic mites and calliphorid flies. Other residents included spiders, snails, collembola, lacewings, bark lice, thrips, leafhoppers, aphids, predaceous bugs, beetles, moth larvae, several other dipteran families, ants, and wasps. Some of these species, such as aphids and leafhoppers, probably were only in transit across the nest at the time it was collected. Others, such as the moth larvae, may have been feeding on the dead plant material. Some, such as the collembola and dermestid beetles, probably were scavenging on dead animal material. Pirate bugs (Anthocoridae) may have been feeding on other arthropods, and scelionid wasps may have been parasitizing insect or spider eggs in the nests.

Predation, Parasitism, and Population Dynamics

The literature on the effect of predation or parasitism (including disease) on warbler populations is virtually nonexistent, as is true for other songbirds as well. In fact, the attention accorded the subject in terrestrial communities as a whole is deficient. This state of affairs is probably related to the concentration in research on competition-related measures. These may indeed turn out to be the dominant factors regulating populations; but the inattention to alternatives does the field a disservice. A conclusion that other factors determine population levels does not necessarily mean that predation or parasitism is inconsequential. If prey species must take action to avoid predators, the resultant behavioral patterns may in their own right affect the population dynamics. For example, predator-avoidance behavior can affect how and where prospective prey safely exploit food resources, which could have a damping effect on these populations. This factor could even dictate the outcome of competition between prey species: if one species escapes from predators more economically than others do, or under a wider range of circumstances, it might enjoy a competitive edge. Although I cannot provide a paruline example, the phenomenon may not be uncommon. Kangaroo Rats *(Dipodomys merriami)* and Pocket Mice *(Perognathus penicilatus)*, seed-eating rodents of the deserts of southwestern North America, may coexist because the Kangaroo Rats' highly developed cursorial and auditory capabilities allow them to exploit areas farther from cover than can the less cursorial Pocket Mice, which would be vulnerable to hawks and owls at comparable distances from cover. In contrast, Pocket Mice require fewer seeds per individual and may be more efficient in gathering seeds under cover than are the Kangaroo Rats (see Rosenzweig and Winakur 1969; Rosenzweig 1973). In this instance, predator-avoidance behavior may facilitate coexistence.

Even if predation often receives inadequate attention, its role in basic population dynamics is generally recognized as a structuring force in some communities, and development of theory has therefore proceeded apace. The situation for parasitism and disease is very different, with rare exception (e.g., Anderson and May 1979). Yet the literature from wildlife management and wildlife disease journals illustrates that these factors are likely to be of major importance. There is good reason to suggest that they operate, at least in part, in a density-dependent way; some diseases or parasites depend on high population densities to maintain themselves.

Not only that, but parasites may lower population densities when food is scarce and diminish the vitality of the competing individuals, partly by diverting for their own use a substantial part of the resources that their hosts have gathered. Different levels of parasitism or different sensitivity to parasites could also affect competitive outcomes (e.g., Park 1948). An adequate evaluation of these possibilities would require analysis of parasite loads at a community level. Predictions are by no means clear at the outset, since some parasites attack a wide range of hosts (effectively all of the warblers, or songbirds, for instance), and others may be nearly host-specific.

At a local level, parasite loads can determine nesting strategies of warblers and other birds. More specifically, the ability of many nest parasites to inhabit continually used sites could select for birds not limited by type of nesting sites. Established nests, likely to be occupied yearly, would be convenient places for overwintering parasites. Under these conditions parasites may have an important role in the population dynamics of their hosts. One would expect the two hole-nesting warblers to harbor a larger parasite load, especially as nestlings, than their open-nesting relatives. Since hole nesters have larger clutch sizes than their closest open-nesting relatives—a pattern that holds for the Prothonotary Warbler but apparently not for the Lucy's Warbler—it would be of interest to determine if nestling losses to parasites cancel out this initial advantage.

7 Display

Warblers are highly vocal and often brightly colored birds. They may sing more than one song—a repeated, recognizable pattern of notes—and these songs are sung in different contexts, making them of interest to students of animal communication (Kroodsma 1983). Other vocalizations come into play in aggressive, sexual, and defensive circumstances. Songs of warblers are largely or totally confined to the males. With their distinctive coloration and conspicuous flash patterns, male warblers engage in frequent visual displays also. Visual displays might assume special importance in a taxon with many coexisting, closely related species. The combination of auditory and visual displays allows for great possible diversity in signaling. I will examine that diversity and how it provides isolating mechanisms and efficient communication in various habitats. I also relate the number of warblers to the wide array of songs and coloration.

Song Types

Warblers' song repertoires, ranging from one to many songs, vary widely among species. Certain species sing only two songs, which differ in context and may vary from individual to individual; still others have more than two songs, which are sung in two or more contexts (Lemon et al. 1987).

A clear example of a two-song bird is the male Black-throated Green Warbler, whose songs can be referred to as Unaccented Ending and Accented Ending Songs, following Ficken and Ficken (1962a) (Figure 7.1). These songs are named simply because of their distinctive last notes. On their breeding grounds, male Black-throated Green Warblers typically sing Accented Ending Songs when near females, low in vegetation, or foraging (Morse 1967b, 1970b). They seldom sing them when stationary on exposed perches. Lein (1972, 1978) proposed that birds sing Accented Ending Songs in the absence of stimuli to sing Unaccented Ending Songs—usually sung when advertising a territory or close to other males.

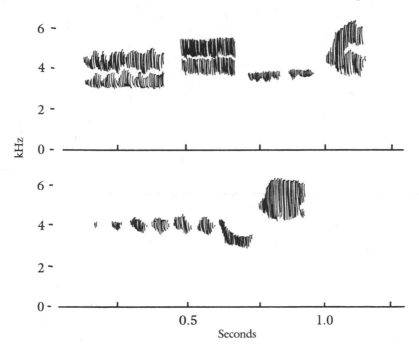

Figure 7.1 Sonograms of Black-throated Green Warbler songs: *above,* Unaccented Ending Song; *below,* Accented Ending Song.

Most Unaccented Ending Songs are sung while stationary on exposed perches, such as the top of a tree or the end of a limb high in a tree. The contexts of the Black-throated Green Warbler's two songs are so constant that an observer who hears a bird sing can predict its activities.

The total amount of time spent singing peaks following the period of most frequent overt fighting (Morse 1976a). Before males have solidified the boundaries of their territories, they may enlarge or switch territories at the expense of individuals that do not sing regularly. Göransson et al. (1974) and Krebs (1977b) found that songs minimize trespassing in other songbirds. Over most of the season Unaccented Ending Songs predominate among Black-throated Green Warblers, but shortly after they return from migration Accented Ending Songs predominate (Figure 7.2). Birds establish pair bonds when Accented Ending Songs are frequent; males and females are often in close contact with each other at such times. Accented Ending Songs may thus play a role in pair maintenance.

The singing patterns of isolated Black-throated Green Warblers and those in large populations are strikingly different. Although Accented

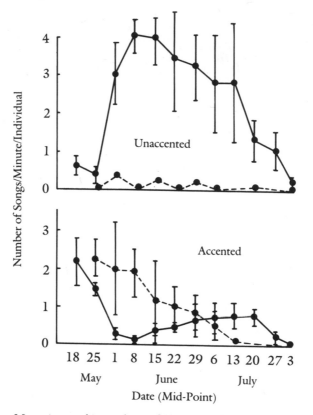

Figure 7.2 Mean (± 1 s.d.) numbers of Unaccented Ending Songs (*above*) and Accented Ending Songs (*below*) sung by male Black-throated Green Warblers on large mainland habitat (solid lines) and small islands (dashed lines). (Modified from Morse 1970b; used by permission from *Nature*. Copyright © 1970 Macmillan Magazines Ltd.)

Ending Songs were sung frequently by both mainland and island birds at the beginning of the season, inhabitants of small islands lacking more than one conspecific male sang Unaccented Ending Songs only irregularly. Further, most of these songs were heard when birds countersang against other conspecifics on a nearby island or the adjacent mainland (inevitably singing on the side of the island nearest to the other singer) or when they responded to a wandering conspecific male that had temporarily landed on the island.

Birds sang much more frequently on the mainland than on the islands, but their rates declined in both locations as the season progressed. In large populations songs became more frequent in the early part of the season, reaching a peak after males and females were together. By comparison, isolated males sang the most at the beginning of the season, a consequence of seldom singing Unaccented Ending Songs. Accented Ending Songs declined quickly in mainland populations after pairing, but slowly and evenly in small-island populations. Accented Ending Songs rose late in the season in large populations as Unaccented Ending Songs declined.

Other spruce-woods warblers vary from this scheme. Blackburnian Warblers also have two song patterns in similar contexts, but I cannot predict them as accurately as those of the Black-throated Green Warbler. In a similar sense, the Northern Parula Warbler sings one type of song in contexts similar to the Accented Ending Songs discussed above, and a second song type analogous to the Unaccented Ending Songs, yet the latter are more variable than are the Unaccented Ending Songs of Black-throated Green Warblers. Yellow-rumped Warblers usually sing songs on one pitch and either ascending or descending songs. The one-pitch song occurs in contexts similar to those in which an Unaccented Ending Song is sung by other species, and the ascending or descending songs in contexts similar to those in which the others would sing Accented Ending Songs. These songs, especially the one-pitch song, may vary in trill speed among individuals. Magnolia Warblers' songs are superficially divisible into three major types, but I have not determined how many types a single bird will sing.

The Yellow Warbler, Chestnut-sided Warbler, and American Redstart provide an interesting variation on this basic theme of song types. Each species has one or more species-distinct song pattern and numerous other songs, some resembling those of the other two species. Species-distinct songs are given in contexts similar to those of the Accented Ending Song in Black-throated Green Warblers, and at least some of the nondistinct songs in contexts similar to Unaccented Ending Songs (Morse 1966b; Kroodsma 1981). Lemon et al. (1985, 1987) referred to these as repeat and serial modes. Although these species differ in breeding habitats, there can be considerable interaction among them.

Nondistinct songs might convey interspecific communication (Morse 1966b). Some songs are similar enough to the human ear, at least, that they may be difficult to distinguish (Kroodsma et al. 1983). But Lein (1978) felt they are easily distinguished, and Kroodsma (pers. comm., 1988) noted that Unaccented Ending Songs of Yellow and Chestnut-sided warblers are tonally distinct. If the birds did experience the difficulty that some humans do, such vocalizations might serve the singers well, for a single song pattern might act as a territorial proclamation against conspecific and interspecific competitors. The songs associated with agonistic interactions sound similar to the human ear; those associated with courtship and other interactions with females sound dissimilar, hence minimizing the likelihood of hybridization. Kroodsma (1981) noted that Unaccented Ending Songs of Blue-winged and Chestnut-sided warblers vary according to geographic region and have distinct regional dialects, whereas Accented Ending Songs are much more similar in different areas.

Conspecific tropical Yellow Warblers ("Mangrove" and "Golden" warblers) may sing primarily or entirely Unaccented Ending Songs (Ficken and Ficken 1965; Morse 1966b), although Nedra Klein (pers. comm., 1988) has found that members of at least two populations occasionally give what I suspect to be Accented Ending Songs. Tropical Yellow Warblers usually are confined to mangrove swamps or small islands, both of which have depauperate breeding faunas and few if any closely related species. These birds might thus seldom encounter a situation where they were isolated from conspecifics, or where they were confronted with similar species. It would be interesting to know what songs tropical Yellow Warblers usually give to females.

The first songs in the morning and the last songs in the evening should be predictable since contacts (and therefore external stimuli) among individuals should be minimal, if only because it is difficult to engage rivals or mates in the semi-darkness. First and last songs of the day should be Unaccented Ending Songs because the birds are scattered but not in close contact with females. At those times females are on their nests. Lein's hypothesis would predict Accented Ending Songs in the absence of external stimuli, although for unstated reasons he included low light levels among strong external stimuli.

For Black-throated Green Warblers, Unaccented Ending Songs are almost always sung at dawn and dusk—a pattern that holds throughout the season, even when Accented Ending Songs predominate in large populations during the day (Morse 1989). Unaccented Ending Songs also predominate in Blackburnian and Northern Parula warblers at this time, suggesting similarities in function. Lein also reported Unaccented Ending Songs at dawn and dusk, both in Black-throated Green Warblers and Chestnut-sided Warblers. Kroodsma et al. (pers. comm., 1988) reported this pattern for Chestnut-sided Warblers, and Lemon et al. (1987) found the same relation in redstarts. Thus, the condition appears to be widespread. Song types cannot be adequately explained according to whether Unaccented Ending Songs are sung under conditions of strong external stimuli.

Lastly, these species sometimes sing on their wintering grounds and while migrating. I have heard Black-throated Green Warblers singing in migration and on their wintering grounds in elfin oak forests in the high mountains of Costa Rica. Migrants in Maryland usually sing Unaccented Ending Songs, the exceptions being late migrants. By the time they have reached Rhode Island, 550 kilometers to the northeast, their predominant song is the Accented Ending Song. Here there is no difference between the dates of individuals singing Accented and Unaccented Ending Songs, but these dates average 4 days later than the Maryland birds and occur when the earliest individuals have already reached their breeding grounds. Birds on their breeding grounds sing mostly Accented Ending Songs.

Birds in the high Cordillera de Talamanca of Costa Rica in early April almost always sang Accented Ending Songs, although the songs often were incomplete. This pattern also appears late in the summer, at the end of the breeding season. The roles of song on the wintering grounds and in migration have not been carefully studied and are not known. E. S. Morton (pers. comm., 1987) suggested that singing on the wintering grounds may be confined to the period immediately preceding migration.

Many observations of Black-throated Green Warbler song patterns are consistent with Lein's hypothesis that Accented Ending Songs are given in the absence of strong external stimuli to sing Unaccented Ending Songs. But it does not explain why birds sing on migration and on their wintering grounds. Two explanations might be proposed. The first, that they are defending a territory, does not fit what happens on the breeding grounds. Although reporting some intraspecific aggressive behavior of this species in Costa Rica, Tramer and Kemp (1980) found scant evidence of territoriality. Accented Ending Song is the wrong song for this explanation anyway, unless one follows Lein's argument to its extreme and hypothesizes that they should be sung when territorial motivation is weak. It seems unlikely that Black-throated Green Warblers set up territories on migration, although Rappole and Warner (1976) have shown that Northern Waterthrushes do so temporarily. If they did set up territories on migration, the song type again seems to be inappropriate in one of the two sites (Rhode Island).

These may be merely testing periods, in which the bird practices its songs for the upcoming breeding season; alternatively, they may use it to repel nearby individuals (Morton 1986). The "testing" hypothesis would predict that the Accented Ending Song should be sung since it will be most important at the beginning of the breeding season. Out-of-season songs may eventually provide useful insights into the function of Accented and Unaccented Ending Songs.

The picture of song patterns of warblers that I have described thus far is oversimplified. In addition to the well-developed Accented and Unaccented Ending Song types given by many species, birds sometimes sing muted or incomplete songs at territorial boundaries after prolonged chases, or near the female or nest (Ficken and Ficken 1962a). Either type may be sung; redstarts usually gave muted Accented Ending Songs, but Chestnut-sided and Kirtland's (Mayfield 1960) warblers gave Accented and Unaccented Ending Songs that were muted. Muted songs of both Yellow and Black-throated Green warblers usually were Unaccented Ending Songs when the birds were engaged in behavior associated with conspecific males, Accented Ending Songs when associated with females. It is generally agreed that close contact may result in low-volume songs or even silence in both male-male and male-female encounters, a situation where visual signals may play the major role. Lein described similar results and listed a separate category, jumbled song, for the Chestnut-sided Warbler, which was virtually confined to territorial encounters. Though distinct, jumbled songs might be combined with Unaccented Ending Songs or interspersed with incomplete and muted Accented Ending Songs and call notes. The jumbled songs were also muted in relation to full-volume songs of other types. Black-throated Green Warblers sing similar songs, which nevertheless are most like Unaccented Ending Songs and seem best included with that song.

Regardless of motivation, the Accented Ending Song is associated with pair formation. Once that act is completed, Accented Ending Songs rapidly decline. Lein observed a threefold decrease in Accented Ending Songs associated with pairing, but attributed it to territory formation, which requires support. Nolan (1978) found that male Prairie Warblers who lost their mates reverted to the early-season song type (Accented Ending Song), presumably in an attempt to remate; he further observed that unmated Prairie Warblers fighting at territorial boundaries in the presence of unattached females sang Accented Ending Songs in each of 10 instances observed. It is difficult to ascribe either of Nolan's reports to the Lein hypothesis, but they do support an argument of separate functions. Kroodsma et al. (pers. comm., 1988) have experimentally removed female

Chestnut-sided Warblers and obtained a similar result, further supporting an intersexual role for Accented Ending Songs.

McNally and Lemon (1985) found that male redstarts responded similarly to playbacks of Accented and Unaccented Ending Songs early and late in the season. But they played the songs from the middle of the redstart territories, which, as they note, may not have been appropriate.

The song repertoire is small in Black-throated Green Warblers, with only one Accented Ending Song and one Unaccented Ending Song, but it is considerably larger than this in some species. Lein reported two distinct Accented Ending Songs and two Unaccented Ending Songs within a population of Chestnut-sided Warblers, each used by most of the birds and alleged to differ in motivational context. At the very least, it seems appropriate to assign Chestnut-sided Warbler songs to two main classes (Payne et al. 1984). Bankwitz and Thompson (1979) recorded 4 distinct Accented Ending Songs and 60 Unaccented Ending Songs of Yellow Warblers in two areas of southern Michigan, although they did not report on the repertoire sizes of individuals. Lemon and colleagues (1985) found that individual American Redstarts in New Brunswick sang, on average, 4.4 songs, one Accented Ending Song (which they called the repeat mode) and 3 or 4 Unaccented Ending Songs (the serial mode). Local redstart populations had considerably more songs than did any individual, however. In spite of the high stereotypy of Black-throated Green Warblers, we would profit from a detailed sonographic analysis of their songs.

Individual repertoires convey information about the singer's identity and help to distinguish among neighbors, with which singers presumably have established territorial boundaries, and strangers, which may attempt to set up territories at their expense. In several species of warblers, territory holders respond more strongly to unfamiliar songs than to those of adjacent territory holders (e.g., Weeden and Falls 1959).

Several explanations could account for the among-species variation in numbers of Unaccented Ending Songs. Krebs (1977a) has suggested that a large repertoire conveys the impression that more individuals are present, thereby dissuading other individuals from attempting to settle, but this seems an unlikely explanation of repertoires of the redstart's size and variability in size (Lemon et al. 1987). Repertoires of many songs may convey an advantage through sexual selection, if this trait correlates with fitness. The latter explanation also finds little support in redstarts, because birds with mature plumage do not have larger repertoires than do first-year birds (Lemon et al. 1985).

Why does the Black-throated Green Warbler have only a single Unaccented Ending Song? If Unaccented Ending Songs help to maintain territories, members of dense populations should be under pressure to develop repertoires that present critical information efficiently to surrounding individuals. Black-throated Green Warblers may accomplish this act differently in that their Unaccented Ending Songs vary among individuals in such a way that one can distinguish between them on the basis of sonograms, and often with the human ear. It seems probable that the birds can do so, also. To determine whether levels of discrimination differ, one might compare responses to playbacks of songs by adjacent and nonadjacent Black-throated Green Warblers with responses to redstart songs. The two species may have two methods of communicating individuality (within-song vs. between-song variability).

Evolution of Repertoires

The song repertoires of the wood warblers raise questions about how they evolved. The frequency with which parts of repertoires are sung is highly dependent on context. Male Black-throated Green Warblers completely isolated from other males seldom or never sing Unaccented Ending Songs and Accented Ending Songs are scarce among tropical Yellow Warblers in contact with many conspecific males, but no congeners. Whether tropical Yellow Warblers sing more than one song type, there may be a correlation between alternate song types and the diversity of the community, although direct evidence for this relation is generally lacking (Kroodsma et al. 1987). By contrast, this pattern says nothing about why this type of repertoire has been adopted, rather than, say, the type that one finds in certain wrens or thrushes, which can have more than a hundred vocalizations (Verner 1976). The latter kind of repertoire has not been reported in warblers, although the Yellow-breasted Chat might have such a pattern.

Stein (1962) has studied the song patterns of the *Dendroica virens* complex (the Black-throated Green Warbler and closely related species) and offered an instructive introduction to the question of the evolution of song. Although the Black-throated Green Warbler occupies a species-rich community of parulines, other species (Townsend's, Hermit, Black-throated Gray, and Golden-cheeked warblers) share their habitats with many fewer warbler species. Stein reported that the songs of the three western species (Townsend's, Hermit, and Black-throated Gray) were more variable than the songs of the Black-throated Green and Golden-

cheeked warblers. He interpreted the difference between the western species and the highly stereotyped Black-throated Green as a consequence of
the differences in warbler species diversity—that species discrimination
may not be as critical in depauperate western communities as in eastern
populations. This does not account for the high stereotypy of the Golden-
cheeked Warblers, since this species, which breeds only in the oak-juniper
forests of the Edwards Plateau in Texas, also occupies a low-diversity
habitat. Because Stein recorded only a few songs from these species, his
sample may not adequately categorize their song relations.

The Ovenbird, in contrast, has a single song type (Lein 1981) but rich
variability within members of a population, thus providing the basis for
individual recognition. Ovenbirds respond differently to familiar and unfamiliar songs of conspecifics (Weeden and Falls 1959), and they also have
a flight song, much less frequently sung than the basic song; its function is
unclear (Lein 1981). Several warbler species sing flight songs, but with rare
exception species with two regular song types do not sing flight songs.
No *Dendroica* are known to give them (Ficken and Ficken 1962a).

The low geographic variation of Accented Ending Songs suggests that
they are primarily sung by individuals under intersexual pressure, and the
regional variation of Unaccented Ending Songs, justifying the term "dia-
lect," suggests that they are primarily sung by individuals under intrasexual pressure (Kroodsma 1981). Consistent with Kroodsma's argument of
inter- and intrasexual pressure is Ficken and Ficken's (1965) report that
Chestnut-sided Warblers and American Redstarts singing only Unaccented Ending Songs were less successful in securing mates than were birds
with a full repertoire, although birds with just the one song were not
totally unsuccessful. More systematic measures of success by birds with
different repertoire sizes would help in testing this question.

A cautionary note is in order. Playback studies have measured male
responses to song, even if the questions of interest in the studies involved
female response. Although male response is the most appropriate measure
for assessing the role of song in establishing and maintaining control of
territories and the mates within them, the female's response is the most
appropriate measure for determining the role of song in mate choice.
Even though female discrimination of song is central to species discrimination and pair formation, it has been assumed that male responses to
songs in analogous contexts will be an accurate substitute. The focus on
male response has probably arisen because of the lack of a simple measure
of female response comparable to the countersinging or aggressive re-

sponse of males. Further, it is reasonable that male and female responses are comparable, for it may be important for males to respond to the same songs sung by individuals that could compete for a territory or mate.

Yet, female and male responses to a song might in fact differ. Females would stand to lose a great deal more from a response to a song that resulted in a heterospecific mating than would a male that attacked a heterospecific male. Criteria for response to intraspecific songs should also differ; females should respond to differences in the merits of various territories and males, but males should be under pressure to respond to other males similarly: they all are threats. Thus, females should be more discriminating in their response to song than males. This conclusion may not invalidate the results of playback studies, but it suggests caution.

King and West (1977) have devised a test of female response by injecting estradiol into females and measuring solicitation displays of females in response to male song. Although not a convenient measure in the field, this assay has promise in the laboratory. Using this method, Searcy et al. (1981) demonstrated that female Swamp Sparrows *(Melospiza georgiana)* were sensitive to differences in both syllable type and temporal patterns of songs; in contrast, males were sensitive only to differences in syllable type. These results are consistent with the hypothesis of enhanced discrimination in female response to song.

Song and Environment

Ficken and Ficken (1962a) correlated acoustical frequency of warblers' songs with the birds' usual foraging positions in the forest; high-pitched songs are most often given in treetops, low-pitched songs in lower, dense vegetation. Two high-foraging species averaged 8,100 hertz; seven medium-foraging species, 6,207 hertz; and seven low-foraging species, 3,136 hertz. Lemon et al. (1981) obtained similar results from a group of 19 warblers nesting in New Brunswick. The Fickens attributed this relationship to the characteristics of sound waves: high-frequency sounds attenuate more quickly in dense vegetation than do low-frequency ones (see Morton 1975). Thus, high-dwelling species, such as the Blackburnian Warbler, have high-pitched songs, while species that often sing from lower in the forest, such as the Yellow-rumped Warbler, have lower-pitched songs. Birds singing in treetops, at least of spirelike conifers, encounter fewer obstructions between them and their receivers than do those in dense underbrush. The correlation is of particular interest in that many species' songs are sung from prominent treetop singing perches, rather than the height where most foraging is done.

Warblers living in vegetation much denser than the understory of a mature spruce forest fit this pattern. *Oporornis* warblers (the Connecticut, Mourning, MacGillivray's, and Kentucky warblers) live in dense vegetation and often sing in it. Their songs are low-pitched, loud, and ringing. One of the Magnolia Warbler's songs shares this trait with *Oporornis* warblers more closely than do any of the other spruce-woods warblers, perhaps because most of its populations live in dense second-growth forests. The Magnolia Warbler and *Oporornis* songs seem to the human ear to have a strong ventriloquistic quality. It is unclear whether these songs affect their intended receivers in this way, but songs, particularly ones at low frequencies with repeated elements, are subject to distortion by reverberation when sung in forest vegetation. Although Richards and Wiley (1980) discussed these aspects of vocal transmission in terms of "honest" communication (i.e., in communicating location accurately), vocalizations might be used in a manipulative way—to make the singer appear to be where it is not (see Krebs and Dawkins 1984). In that way it would be possible to minimize movement in patrolling a territory, although intruders might learn such a ruse if it was frequently not backed up by action. If this hypothesis is valid, Magnolia Warblers singing their ventriloquistic song should move less while singing this song than when singing others, but if they used this song in territorial situations (in the context in which the Unaccented Ending Song is sung), one would predict low mobility anyway.

Morton (1970) and Chappuis (1971) analyzed the sound-propagation properties of several habitats in relation to the inhabitants' vocalizations and found a connection between the transmission properties of different environments and the physical properties of the vocalizations. Subsequently, Marten and Marler (1977) analyzed sound propagation in deciduous and coniferous forests of the northeastern United States and found patterns in the temperate-zone forests similar to those in tropical forests (Figure 7.3). Attenuation of vocalizations is more severe at high frequencies than at low frequencies, except that low-frequency vocalizations (below 2 kilohertz) close to the ground were also excessively attenuated. Height of vocalization is an important variable near ground level; all frequencies are attenuated more between the ground and one meter than

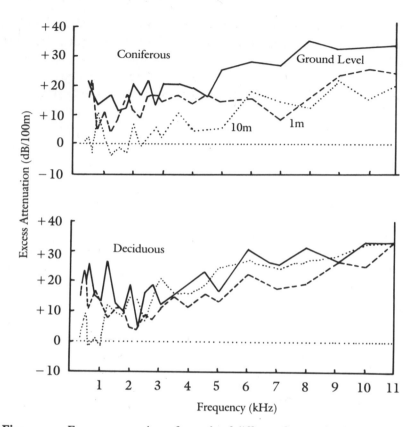

Figure 7.3 Excess attenuation of sounds of different frequencies in coniferous forest *(above)* and deciduous forest with leaves *(below)*. Measurements were made at ground level (solid line) and at heights of 1 meter (dashed line) and 10 meters (dotted line). (Modified from Marten and Marler 1977.)

at greater heights. Attenuation is also likely to be high in the midst of the canopy because of the many obstructions. Not surprisingly, the louder the vocalization, the farther it carries.

Birds can maximize sound propagation by singing loud songs of low frequency at certain heights. Why, then, are not all songs sung in this way? Several explanations might apply. Songs of different frequency should maximize the distinctness of a species, which is an advantage in a diverse community; the benefits of specific distinctness could even select for singing at frequencies that do not maximize transmission distance (see Wiley and Richards 1978). This explanation could account for the wide range in frequency in vocalizations of spruce-woods warblers. Further, since problems of attenuation are low in the open, the higher the singing bird is in the spruce forest, the less severe the problems of attenuation. Because the problem does not pertain merely to the singer's location but to the entire distance from it to its intended target—if the intended target, typically a conspecific, is also located high in the forest—selection against high frequency should be weakest.

Birds singing in forests should, in the absence of other complications, favor low-frequency songs. If birds of high-density species communicate over distances that are on average shorter than those birds of low-density species communicate over, they might use high-frequency vocalizations more easily than their low-density counterparts (see Lemon et al. 1981). On this basis, though, one would predict that Black-throated Green Warblers, the most abundant spruce-woods species and the one with the smallest territory sizes, would sing higher-frequency vocalizations than Blackburnian Warblers, but the reverse is actually the case. Physical limitations of the vocal apparatus might constrain pitch. Since members of a single genus such as *Dendroica* sing over a wide range of frequencies (Brand 1938), however, this factor is unlikely to play a role.

The distance a song will travel should differ in coniferous and deciduous forests because of variable growth patterns. Many deciduous forests have open areas below the dense canopy in which physical obstructions are sparse in comparison to the canopy. The deciduous canopy itself is dense and the resultant possibilities for attenuation high; this means that, all things being equal, songs should not be sung there. This pattern contrasts with the spruce forests, which have more vegetation low to the ground and less open space below the canopy than deciduous forests have. Conversely, the "canopy" level of the spruce forests is more open than the canopy in a deciduous forest, as a result of the conical shape of spruce trees.

Patterns of sound attenuation nicely explain the foraging and singing patterns of two warblers. The Ovenbird, a ground forager, seldom sings from the ground (Lein 1981); it has a loud, low-frequency song, which it typically sings after flying to a singing perch more than a meter above the ground (Lein 1981). This behavior does not completely solve the problem of communicating efficiently to individuals on the ground, but if most of the distance between the singing perch and the receiver exceeds one meter in height, the difficulty should be minimized. Male Black-throated Blue Warblers usually sing at a lower height than females forage (Holmes

1986). Being deciduous-forest birds, they can take advantage of low foliage density at song perches. The link between foraging and singing sites of male warblers may be associated with the use of favorite sites for communication.

Attenuation is by no means the only complication associated with sound transmission. At least two other factors, amplitude fluctuation and reverberation, determine whether a song is recognizable. They modify or degrade vocalizations, robbing them of their species-specificity. Birds can minimize degradation by selecting times and locations for singing that diminish effects, by producing vocalizations with little similarity to environmental acoustical perturbations, and by introducing redundancy into vocalizations. Inability to exploit these possibilities can circumscribe the singers' opportunities.

Amplitude fluctuations—that is, fluctuations in loudness—arise mainly from wind-generated turbulence. When vocalization has attenuated nearly to the level of background sound, amplitude fluctuations may even exceed the natural modulation. Reverberating sound waves, primarily generated from scattering surfaces such as leaves, interfere with rapid amplitude modulation and repetitive frequency modulation, such as trills and buzzes (Richards and Wiley 1980). The high likelihood of reverberations may explain why few forest birds give such vocalizations. Reverberations also decrease the bird's ability to localize sound, a problem that becomes acute above 8 kilohertz. Omnidirectional signals reverberate much more than do unidirectional ones; this could affect vocalization transmission (see Witkin 1977).

Intermediate-range frequencies are most suitable for long-range communication, as they minimize amplitude fluctuation, reverberation, and attenuation. Most songbird species use the 2–8 kilohertz range. Some warblers are among the few species that produce energy above that level, for example, the Blackpoll Warbler. High-frequency vocalizations are favored in short-range communication, and many call notes fall into this range; vocalizations used in longer-range transmission would be disadvantageous because of the danger of predation or other factors.

The acoustical environment is, then, a complex, multidimensional variable as vocal animals experience it. Most critical variables remain obscure and unintegrated, although they figure into communication and social interactions. Vocalizations thus could be under selective pressure to respond to environmental factors, and it is of interest to speculate on how acoustical constraints affect social systems and population biology.

Morton (1982) put forth the "ranging hypothesis" to explain the acoustical constraints of the sound environment and songbirds' adaptations and responses to them. The ranging hypothesis proposes that animals assess the location of singers by the quality of their song relative to a learned template of it. Comparison of the two indicates how much the song degrades as it passes through the environment, which can provide clues to the distance of signals in the habitat. Signals could be varied by other means, perhaps through amplitude control, thereby obscuring the normal message. Richards (1981) demonstrated that in song playbacks Carolina Wrens respond to degradation rather than to amplitude. Gish and Morton (1981) have also shown that the songs Carolina Wrens sing in their own territories do not undergo as much degradation as do songs of birds chosen randomly from other territories. This implies that they match their songs to their habitats. Shy and Morton (1986) demonstrated that these birds could therefore discriminate between familiar songs given on and off their territory—which is important to territorial defense. It might behoove an intruding male to sing in a way that would mask its intrusion; simultaneously, it would be advantageous to the territory holder to decode messages from the others, including deceptive ones, to gauge threats to its own territory.

The problem of selective degradation is not unique to Carolina Wrens. It is conspicuous to the human ear in Black-throated Green Warblers, though it remains to be seen whether it operates effectively at levels relevant to territorial maintenance and defense. Warblers provide an alternative system for study, however, by virtue of their wide within-habitat diversity and differences in their use of auditory and visual cues. The appropriateness of the ranging hypothesis to warbler songs is also by no means clear, in that the hypothesis is based on the ability of a receiver to compare a song with a learned template. Although song learning is not widely established in parulines (but see Kroodsma et al. 1983), since species with one or two songs will share the same songs, parulines may indeed be able to compare a song with such a template. Further, the Kentucky Warbler, a species with a single song type, modifies that song when countersinging against playbacks—and probably in natural situations as well—to match songs of contenders in a way analogous to song matching, as studied by Morton and colleagues in Carolina Wrens. Kentucky Warblers raised and lowered their frequency ranges and altered low and high frequencies of songs in response to song playbacks (Morton and Young 1986).

Competition for Singing Time

Singing birds encounter other problems besides the physical environment. It may be difficult to use particular ranges of wavelengths if they already are heavily loaded (the "noisy channel"). Ovenbirds minimize acoustical overlap with adjacent territory holders by singing in the intervals between their neighbors' songs. This phenomenon, frequently referred to as countersinging, is generally taken to be a response to other individuals, potential competitors. Nevertheless, spacing of these songs is more precise than would be predicted on the basis of this consideration alone. In fact, given the length of an Ovenbird song, one could even commence a countersong before the initial singer had finished its song, if response were the only consideration. Yet Ficken et al. (1985) demonstrated experimentally that far more songs were given in the first tenth of the intersong interval than predicted by chance, and there was almost no overlap. Wasserman (1977) and Schroeder and Wiley (1983) found that White-throated Sparrows (*Zonotrichia albicollis*) and Tufted Titmice (*Parus bicolor*) have similar singing patterns, probably a common characteristic of songbirds.

Channel sharing should become more problematic as density increases. Members of dense populations in homogeneous habitats could experience sound environments far more complex than the pairwise interactions considered by Ficken and colleagues. In maximum territorial packing, with hexagonal territories (Grant 1968), six individuals will be equally nearest neighbors. Although packing seldom achieves this level of symmetry, in dense, homogeneous forests as many as four equidistant nearest neighbors would not be unusual (e.g., Morse 1976a). At their highest densities, Black-throated Green Warblers sing at close range to each other, and given their resulting small territories, the potential for acoustical interference would be substantial.

Other studies have explored similar interspecific problems. Popp et al. (1985) found that four species of passerines, including Ovenbirds, tended to avoid overlapping another's songs, a pattern earlier remarked upon by Cody and Brown (1969) and Ficken et al. (1974). Other species' songs could help to define the limits of efficacy in acoustical communication. Popp and colleagues demonstrated that as numbers of singing individuals increased, each one sang less often and placed its song in the first quarter of the other species' intersong interval more often than predicted.

The means of communication when density and physical environment vary warrant closer attention. The denser the packing of individuals, the greater the opportunity should be for effective visual communication; however, some environments that provide the resources that permit dense packing might also include vegetation so dense that it would mask easy visual communication. The opportunity for tradeoffs thus exists, but it is unlikely that communication and environment have a simple linear relation.

Other Characteristics of Vocalizations

Less attention has been paid to the nonsong vocal repertoires of warblers than to their songs, and few attempts have been made to catalog them in a species. Lein (1980) found that members of a local Ovenbird population gave 13 nonsong vocalizations during the breeding season, with almost no overlap between males and females. At least 11 vocalizations happened during male-female interactions; only one vocalization was made by both sexes, and even that showed sex-specific differences in sonograms. Thus, these vocalizations figure into the intersexual communication of Ovenbirds. Lein suggests that this broad range of vocalizations is related to the small role of visual signals in the dull-colored monomorphic birds.

Lein contrasts the Ovenbird pattern with that of the brightly colored and dimorphic Chestnut-sided Warbler. These warblers give at least six nonsong vocalizations in male-female interactions, of which three are identical in the two sexes. Ficken reported similarly extensive call sharing. Her adult redstarts gave six nonsong vocalizations, all by both sexes, although two were given more often by females and one more often by males. Nolan (1979) reported ten nonsong vocalizations in Prairie Warblers, five given by both males and females, two heard only from males, and three not identified by sex.

Lein noted that nestling and fledgling Ovenbirds give several vocalizations when begging or responding to alarm, but did not indicate whether any of them develop into the vocalizations of the adults that he recorded. Ficken (1962) found that the "screech" notes of fledgling American Redstarts change into the "snarl" notes of the adults, and at least one of the five juvenile calls given by Prairie Warblers changes into an adult call (Nolan 1978).

These studies do not permit detailed consideration of the function of the vocalizations, but we can infer that calls are given in certain contexts, including both male-male (and, less frequently, female-female) interactions and male-female interactions in which either sexual or aggressive tendencies prevail. Some were usually given when birds were close to each other, at which times singing was likely to be curtailed. Ficken and Ficken (1962a) presented a general synopsis of warbler call notes that identified seven types. Although this catalog is incomplete in terms of the number of vocalizations that both Lein and Nolan reported, the Fickens recorded some types from several species. As they pointed out, an adequate library of vocalizations requires sonographic records accompanied by field observations.

Some species intersperse chip notes between songs, possibly an indication of the level of motivation. Black-throated Green Warblers often utter series of metallic "double-chip" notes between Unaccented Ending Songs, the number probably correlated with the abundance and distance of conspecific singers, or how recently they have encountered another male (Morse 1967b). Double-chips are never given between Accented Ending Songs. These calls bear a remarkable similarity to the beg notes of fledglings, as do analogous calls of Prairie Warblers (Nolan 1978). Other spruce-woods species, especially Blackburnian Warblers, use chip notes in similar contexts, but they do not form as prominent a part of their repertoire as those of Black-throated Green Warblers (Morse 1967b), and they can be distinguished interspecifically, at least by the human ear. The call notes of Yellow-rumped Warblers are also similar to the begging calls of their fledglings. In this series the calls are especially harsh and loud; Allen (in Bent 1953) referred to them as "explosive." It is interesting to relate them to Morton's (1977) hypothesis that harsh, low-frequency calls indicate aggressiveness. Adult Yellow-rumped Warblers are socially subordinate to adult Magnolia and Black-throated Green warblers (Morse 1976a), species whose calls are not loud and harsh. Juvenile call notes fit the motivation-structural rules. My studies of warblers suffer from not including sonographic work, yet they refer to distinct patterns that warrant such analysis for the purpose of learning about the evolution of these vocalizations.

Warblers mob potential predators or predatorlike objects during the breeding season, as well as at other times, using the vocalizations noted above. Ficken et al. (1967) reported that Black-and white Warblers, Pine Warblers, Yellow-throated Warblers, and Ovenbirds joined in a group that mobbed a Chuck-will's-widow *(Caprimulgus carolinensis),* a large goatsucker (Caprimulgidae) that somewhat resembles an owl and that is reputed to swallow small birds whole (Bent 1940). This pattern is most often directed at hawks, owls, crows, and jays, however.

Bill snapping is a nonvocal auditory display given by several warblers (Ficken and Ficken 1962a). These displays are especially frequent in species that clap their beaks under other circumstances. Sherry (1979) noted that broad beaks facilitate the production of conspicuous snaps. Both true flycatchers (Muscicapidae) and New World tyrant flycatchers (Tyrannidae) snap their bills in aggressive situations, especially attacks. The American Redstart, a flycatching species, gives this display frequently.

Visual Displays

Vocalizations constitute only part of a warbler's displays, although they doubtlessly play a dominant role in long-distance communication and in closer-range communication among dull monomorphic ground dwellers such as Ovenbirds (Lein 1980). Visual displays (Figure 2.7) complement or replace vocal displays at closer range, and most are produced in aggressive, sexual, or combined aggressive-sexual contexts.

Courtship by male warblers involves many movements: fluffing plumage, including crown raising, and wing motion (raising, spreading, vibration) are all common in *Dendroica* and accentuate their markings. Several flight movements are used during courtship, and they resemble moves made in male-male encounters. Examples are gliding and "moth" flight (swallow, rapid wingbeats) (Bond 1937; Ficken and Ficken 1962a). Considerable similarities among species and even genera suggest that these characters are evolutionarily conservative. Since selection should be strong for specific distinctness, especially in groups with many closely related sympatric species, the striking coloration found in different species could influence discrimination more than courtship behavior does (Ficken and Ficken 1962a). Little direct evidence exists to support the role of color in isolating mechanisms, however (see Chapter 8).

A few warblers perform courtship feeding (Lack 1940; Nolan 1958). Not only is this behavior a form of display to the female, but it may provide the female with supplemental resources, or supply her with resources more quickly than she would otherwise be able to gather them, thus hastening egg laying (Royama 1966).

Flight with stiffened wing beats, often punctuated by glides, is an agonistic display performed by several species. The display is especially well developed by American Redstarts. They combine these movements in a circling display by which an individual approaches an opponent and then returns to its perch. Two males may even alternate these displays (Ficken and Ficken 1962a). Other species have less exaggerated displays of a similar nature. For example, Chestnut-sided Warblers circle toward an opponent but do not perform exaggerated wing motions simultaneously. Redstarts and Chestnut-sided Warblers sometimes hold their wings out from the body in an exaggerated position subsequent to prolonged encounters (Figure 2.7).

Ficken and Ficken have reported male "soliciting" displays by Golden-winged Warblers following defeat in boundary contests. These resemble the displays of females soliciting copulation: the birds pose with quivering wings, raised tail and crown, and lowered breast. Mayfield (1960) reported a similar display in the Kirtland's Warbler, which, however, did not incorporate crown raising but included gaping. Nolan observed this behavior by a male intruder during a territorial encounter, in the midst of a territory occupied by an unmated bird, and in response to the sudden appearance of a female approaching its nest with food. All of these examples could have been submissive displays, though the context remains unclear. Nolan noted the similarities between this behavior and begging for food: in male-female encounters, some females were carrying food. Soliciting postures of females are extremely conservative (Ficken and Ficken 1962a).

Head-forward displays, often accompanied by gaping, may occur during violations of individual distance. Other threat displays include tail spreading by Yellow-rumped Warblers and redstarts, species with conspicuous tail markings (Figure 2.7). Prairie Warblers, having less conspicuous tail markings, rarely give this display (Nolan 1978). Tail and wing flicks are given by Common Yellowthroats defending territorial boundaries (Stewart 1953). Under similar circumstances Mourning Warblers flick their wings rapidly outward and fan their tails (Cox 1960). Ovenbirds in this situation tilt their tail upward, droop and spread their wings slightly outward, and alternately raise and lower their feet as if kneading the substrate (Freeman 1950). Warblers in territorial encounters frequently sleek their plumage and crouch on their perch, thereby decreasing their apparent size. Male Prairie Warblers perform many of these displays (Nolan 1978), suggesting that these behaviors are widespread among warblers, although they are documented in only a few species.

When alarmed, birds of several species with distinct crown markings raise their crown features and flick their wings and tail. Redstarts under similar circumstances pivot with their tail spread (Ficken and Ficken 1962a).

The visual displays of warblers deserve detailed attention. Results to date suggest considerable similarities among species, even those of different genera. Major differences in presentation can be correlated more with plumage patterns than with phylogeny, although rare displays in well-studied species (e.g., Prairie Warbler) suggest that species' repertoires are large and that a major difference among species lies in the frequency with which a display is performed, rather than the ability to perform it.

Display and Community Structure

It is rapidly becoming clear that bird vocalization is a complex phenomenon and that birds can send more sophisticated messages than has been appreciated until quite recently. Perhaps previous neglect was due to researchers' not having treated the subject in an ecological and evolutionary context (Morton 1986). Characteristics of vocalizations seem to be shaped both by conspecifics and by other animals; they are influenced by environmental variables as well as by the physical attributes and limitations of the vocalizers themselves. They probably carry a combination of honest and deceptive information about the producer, and the producer may have considerable ability to change its vocalizations in ways that permit adaptation to social contexts and environmental conditions. The type and extent of development of these abilities is partly governed by the life-style and environment of the performer, but little effort has thus far been exerted to analyze a population for all possible vocal abilities; some skills are likely to be mutually exclusive, some are not exclusive but unlikely to be highly developed simultaneously, and some are likely to be synergistic in effects.

All this having been said, the results and hypotheses generated thus far do not give a clear sense of what causes variation in song number and complexity. Some species have a single song, but often with considerable variation within populations; others have two songs or two groups of songs; and still others have large repertoires of dozens to hundreds of songs.

Progress on the acoustical qualities of vocalizations has proceeded further than that for repertoire size, although the two factors are not completely separate. The acoustical factors, though complex, may be simpler than those relating to the size of the repertoire, which involves the social milieu of the birds. If so, it is pertinent to view the question of song and other vocalizations from a community viewpoint. This task has not been undertaken and would involve studying interactions among species as well as other variables. Members of high-diversity warbler communities may have smaller repertoires than low-diversity communities, but this does not account for why repertoire sizes vary so much among members of a single community. Differences in repertoire size per se seem unrelated to physical acoustical factors. Phylogenetic factors do appear to affect the size of the song repertoire, however: many *Dendroica* have two songs, *Oporornis* and *Seiurus* have one, and the chat has many. Species with small repertoires may use other methods of song modification or variation to increase their communication skills.

Most attention has been paid to songs, but many species possess an extensive repertoire of calls as well, which are bound to affect song repertoires. Data on suboscines suggest that their "simple" vocalizations play a major role in communication. Given the rich repertoire of calls recorded from songbirds, including warblers, it seems unlikely that they will differ greatly, although calls may not play as major a role in songbird communication as they do in suboscine communication.

Vocalizations are employed along with visual displays, and the extent of vocal development appears to depend on the potential for developing displays in vocal and visual modes. Few studies combine simultaneous analyses of vocal and visual communication. The two kinds of display should interact in a predictable way as long-range and short-range signals, although the balance between the two depends on the environment for transmitting both types of signals, as well as the phylogenetic attributes of the individuals in question (whether there is the potential for conspicuous plumage, vocalizations with certain characteristics, large or small numbers of song types, and so on).

Vocalizations of parulines are usually not as complex as those of the sylviine warblers (Simms 1985). Visual communication is probably less developed in sylviines, perhaps converging in the direction of the North American wrens, largely brown-and-white brush-dwelling birds with complex vocal repertoires (see Kroodsma 1977). This difference between paruline and sylviine warblers could be related to the differences in community structure between the two groups. Groups with many similar species, such as the parulines, experience pressure to produce a simple repertoire to minimize uncertainty (Wiley and Richards 1982). In keeping with this prediction, warblers of western North American communities may have more variable repertoires than do their eastern counterparts.

8 Plumage

Plumage serves a host of functions, the most basic being locomotion and insulation. Life-sustaining functions such as these determine feather characteristics, which in turn affect a bird's ability to be an efficient flycatcher, to maneuver in foliage, and to undertake long-distance flight. In addition to its influence on the properties of flight and endothermy, plumage is likely to have an effect on other traits, too, such as predator avoidance and communication.

The color and structure of feathers are important clues to the functions they serve. Melanin-bearing pigments considerably strengthen feathers, for example, and they often predominate in the areas of a bird's plumage subject to the greatest wear (Burtt 1986).

Variation in Color Patterns

Warblers display an impressive variety of color patterns. Many colors of arboreal species are conspicuous against the birds' usual background, at least to the human eye, which is believed to perceive color patterns in the same range as do birds (Hailman 1967, 1979). Thus, warbler colors most likely function in communication. Species-distinctive colors could serve as isolating mechanisms, perhaps in concert with species-specific song patterns. Many warblers live in environments that are suited to extensive visual communication. In contrast to arboreal species, terrestrial warblers, such as waterthrushes and Ovenbirds, are brown, dull white or dull yellow; they resemble terrestrial species of other taxonomic groups, such as many sparrows and thrushes, instead of their arboreal counterparts (Burtt 1986). These species frequent areas with poor visibility, due to the obstruction of heavy vegetation and to low ambient light.

Relating more subtle differences—such as rump patches, crown patches, facial, head, and breast coloration—to microhabitat is a more formidable task. Color could serve several purposes, many of which are so difficult to separate that the task of assigning function to color has been a conten-

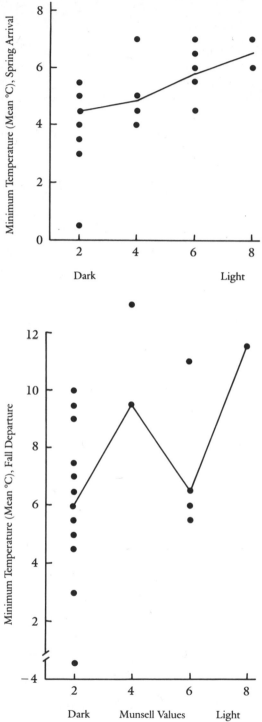

tious one. Burtt's extensive study revealed a wide range of warbler colors that are not randomly distributed over the body. Frequencies of color differ between males and females: males more often have black and females more often have brown or yellow-green feathers. Further, if colors between sexes do not differ, the yellows, oranges, reds, and chestnuts of females are less saturated (less monochromatic) than those of males; this observation concurs with the common impression that they are duller than males. Age and seasonal differences parallel intersexual differences: juvenile and winter plumages of males are more similar to female plumage than to their own summer plumage. Thus, color may play a sexual role in courtship or in male-male territorial interactions during the breeding season, or both. Age and seasonal differences also point to the importance of color in the birds' breeding habitats.

Other hypotheses to explain color variation involve thermal considerations. Burtt argued that the color of legs and mandibles is a factor in determining a bird's thermal regime, since 10–56 percent of a resting bird's heat is lost through its uninsulated legs, and 5–10 percent is lost through the mandibles. Accordingly, species with dark legs, which absorb more radiation, should migrate north earlier in spring, migrate south later in fall, and winter in areas colder than ones frequented by light-legged species. These predictions were sustained for warblers (Figure 8.1); further, there was no correlation between color and size, which could have confounded the results.

Mandibles, which release a modest 5–10 percent of the body's heat, do not show such correlations. Birds with light-colored upper mandibles almost never forage in the open, suggesting to Burtt that the glare of reflection adversely affects visual acuity. Reflectance from dark-colored mandibles is much lower, and birds with dark upper mandibles predominate among species foraging in the open; correspondingly, upper mandibles are dark more often than are lower ones. But these results do not distinguish between glare-reduction and countershading (Endler 1987); that is, dark upper mandibles and light lower ones enhance crypticity.

Figure 8.1 (facing page) Leg color (Munsell values) of different species of warbler vs. mean minimum temperature at spring arrival of the species *(top)* in Wisconsin and vs. mean minimum temperature at fall departure of the species *(bottom)* from Wisconsin. Munsell values are points on a scale from 0 to 10, on which white has a value of 10 and black has a value of 0; hues (reds, purples, etc.) are not represented in Munsell values. Solid lines denote mean temperatures at different Munsell values. (Modified from Burtt 1986; used by permission of the American Ornithologists' Union.)

Burtt (1984) tested the reflectance hypothesis by painting white the dark upper mandibles of Willow Flycatchers. The white-billed birds confined their foraging to shaded areas, as predicted by the reflectance hypothesis.

Other, feathered, areas about the eye—the eyebrow stripe, eyeline, and eye ring—do not relate to a bird's foraging areas. Reflectance is not as high from feathers as from shiny mandibles, and the adaptive significance of the color of these feathers remains unclear. This is just one example of the complex nature of warbler coloration and the difficulty of addressing alternative hypotheses.

Melanin-containing plumage resists abrasion more effectively than do feathers containing other pigments (Burtt 1986), so melanin-based colors (black, brown, yellow-green) should predominate in situations or locations resulting in heavy wear. In support of this prediction, abrasion-resistant colors predominate in dorsal parts, which encounter greatest wear from airborne particles in flight and from contact with vegetation. In particular, the major flight feathers of the wings and tail usually contain these pigments. Although bold flash patterns may be on wing and tail, they are confined to areas of low turbulence and minimal flapping motion—the wing bars and tail spots. In a similar way, feathers that rub against each other, as in the shoulder region, have a predominance of melanin-based colors.

The tendency for melanic feathers to predominate on warblers' backs could also be associated with thermoregulatory factors, ultraviolet radiation, and crypticity (substrate-matching and countershading). Separating these hypotheses, none of which are exclusive, is the most serious question to be asked in further studies of color patterns.

Compromises in coloration due to conflicting selection pressures must arise in some situations. The most conspicuous flash colors in warbler habitats are white and yellow. Flash patterns, generally tail spots or wing bars, are most common in species that occupy habitats with medium to strong light and that have relevant behavioral repertoires—for example, flying frequently. Burtt had difficulty accounting for facial and throat patterns and says little about them, but his approach can be applied to spruce-woods warblers, and I shall do so.

Coloration and Distribution of Spruce-Woods Warblers

Four morphologically similar spruce-woods *Dendroica* warblers (Magnolia, Yellow-rumped, Black-throated Green, and Blackburnian) have quite different breeding plumages, as do several other sympatric congeners. [This statement is based on the assumption that color perception is similar in warblers and humans, which is a legitimate generality, though the basis for color vision is totally different in birds and mammals (see Hailman 1979).] In combination with warblers' distinctive songs, the plumage patterns may form effective territorial and reproductive isolating mechanisms. Hybrids within *Dendroica* are rare and have not been recorded among these four species (Chapter 13). Although potentially functioning as isolating mechanisms, these species' plumages demonstrate trends associated with ecological differences (Burtt and Gatz 1982).

Earlier (Chapter 3), I described the foraging of these species. The males, most conspicuous on their stationary singing perches, are located above their major foraging areas. They also sing while foraging at lower elevations, and the height varies with the species (Morse 1968). These spruce-woods warblers can be separated into species that sing territorial songs from a stationary, conspicuous position and species that sing territorial songs while on the move. Of the stationary group, both species are confined to coniferous forests; of the mobile category, one species is confined to coniferous forests and the other ranges widely in mixed coniferous-deciduous forests.

Black-throated Green and Blackburnian warblers, species that sing while stationary in an exposed position (Morse 1967b), have bright plumages that are especially conspicuous when viewed head on, from below, and from the side: the Black-throated Green Warbler has yellow cheeks, crown, and neck (dorsally) and the Blackburnian Warbler has an orange head and throat (Figure 8.2). "Fidgeting" movements, usually when call notes alternate with songs, enhance the effect of the bright colors. When they "fidget" they flick their wings and tail and move their feet in place.

Yellow-rumped and Magnolia warblers, species that do much of their singing while foraging, possess "flash" patterns rather than bright, face-on plumages (Figure 8.2). These patterns, which are much more conspicuous when the bird is in motion than at rest, consist of a yellow rump and flanks in the Yellow-rumped Warbler and white shoulder and tail markings in the Magnolia Warbler.

The Yellow-rumped Warbler's foraging and other motions result in the rump patch being strikingly displayed. In the dimly lighted understory of spruce forest, an area used more frequently by Yellow-rumped Warblers than by any others, this flashing pattern is particularly conspicuous to the human eye. Yellow-rumped Warblers hawk for insects more frequently

Figure 8.2 Contrasts in light and dark anterior coloration of spruce-woods warblers. *From upper left, clockwise:* Magnolia Warbler, Yellow-rumped Warbler, Blackburnian Warbler, Black-throated Green Warbler. (Illustrations by Brian Regal.)

than do other spruce-forest warblers (MacArthur 1958; Morse 1968), often from the lower dead branches of the understory. The foraging movement therefore simultaneously has considerable display value (see Burtt 1986). The most conspicuous patterns of the Magnolia Warblers are their white subterminal tail band and wing patches, marks that are especially conspicuous when the bird is foraging. The conclusion that foraging movements are a form of display assumes that warblers receiving the signals are highly sensitive to movement, which is probable in light of physiological (Granit 1955) and anatomical (Pumphrey 1961) information on a wide range of birds. Black-throated Green and Blackburnian warblers have white wing bars, a trait held in common with most other *Dendroica*. The wing bars are not as large as the flash patterns of Magnolia and Yellow-rumped warblers and, to the human eye at least, not as conspicuous against their typical background as the flash patterns of Magnolia and Yellow-rumped warblers are in their environment.

Because of the way birds perceive color (see Hailman 1977), white and yellow should be visual signals at low light levels, such as under forest canopy. (This would not prevent white and yellow from being visible at higher light levels too.) If selection occurred primarily for the conspicuousness of flash patterns, maximal selection for white should be in areas of minimal illumination, because of its relative brightness. If the understory (both deciduous and coniferous) used by the Magnolia Warbler is the least well illuminated habitat used by the four species on their breeding grounds, and that used by the Yellow-rumped Warbler is next in level of illumination, my prediction is sustained.

Additional factors may reinforce the selective pressure of illumination. In the Magnolia Warbler's accustomed habitat it frequently appears in a background of deciduous leaves. By the middle of the breeding season insect infestations often result in leaves with large yellowish blotches, which might make a yellow patch cryptic rather than conspicuous. White flash patterns can assume a much greater display value here than yellow ones.

White wings and tail markings are most conspicuous on the Magnolia Warbler. All four species flick their wings and tail rapidly in several contexts, and these patches probably are close-range signaling devices; in the Magnolia Warbler they can facilitate longer-range signaling as well.

In the Yellow-rumped Warbler's accustomed area in the coniferous understory, flecks of sunlight frequently dapple the ground and foliage on clear days. Although this light may have a slightly yellowish hue to the human eye, it contrasts with the yellow markings of this species.

Considering the types of areas frequented by Yellow-rumped and Magnolia warblers and the favorable reflectance and sensitivity values for either white or yellow, background color may be even more important to the two species than moderate differences in the brightness of the two colors. The face-on patterns of stationary singing species, the Black-throated Green and Blackburnian warblers, may be associated with their frequent presence in higher-visibility sites. The Blackburnian Warbler, the species most often in the open because it inhabits the higher reaches of the tree, possesses a color pattern (orange) that is spectrally not as bright as the face-on yellow of the Black-throated Green Warbler, the species more frequently in denser foliage below the treetops.

Although all these species retain the same general color patterns throughout the year, they are usually brightest in the breeding season (Dwight 1900; Ridgway 1902). At other times of the year male plumages are more similar to those of females and juveniles than when males are in breeding plumage. Therefore, selection for the bright patterns probably takes place during breeding, rather than at other seasons.

These predictions are an ad hoc attempt to explain the most conspicuous plumage patterns and related behavior of this bird community. Although these explanations closely fit the observed conditions, there may be legitimate alternative hypotheses; the proposals offered here could be tested but have not been. I acknowledge this deficiency but have presented interpretations primarily as a stimulus to exploration in this neglected area. It would be naive to pass off the riot of warbler color patterns as merely a random assemblage.

Color Patterns as Isolating Mechanisms

The lack of hybridization among similar warbler species suggests that species-specific plumage patterns and/or vocalizations are effective isolating mechanisms. As Hamilton and Barth (1962) have pointed out, however, traditional explanations of color patterns as barriers to interspecific pairing and hybridization present an incomplete picture.

Further, in spite of the arguments for warbler (or other passerine) coloration acting as an isolating mechanism (see Rohwer et al. 1980), very little direct information exists. In one experiment testing male plumage as an isolating mechanism, male Common Yellowthroats who had their distinctive black facial masks painted out (so they resembled females) attracted females as quickly as did unpainted males. Yet they had to fight more often to defend their territories than did other males (Lewis 1972), implying that plumage patterns are more important in agonistic interactions than in intersexual ones, although sometimes agonistic interactions associated with plumage differences affect reproductive success.

Effective territorial display is of extreme importance in dense populations of spruce-woods warblers. Here we would predict selection for well-marked differences between species, with the direction of coloration dictated by ecological considerations. Selection for visual conspicuousness may be even more important to these species for territorial display than for isolating mechanisms. In such a scenario, selection for isolating mechanisms should occur largely for marked differences between species, and selection for territorial display should favor the most quickly discerned characteristics.

Coloration of Males and Females

Although we do not know whether ancestral warblers were sexually dichromatic, the prevalence of monomorphic tropical warblers, presumed to be the ancestors of the parulines, suggests that monomorphism is a primitive paruline condition. Tropical warblers as a group are monomorphic when it comes to color, but the different species may be either bright or dull; bright species occupy well-lighted places, comparable in this regard to bright-colored temperate-zone species. Tropical species are territorial throughout the year, and females as well as males defend their territories (Skutch 1954; Elliott 1969). Plumage monomorphism is associated with permanent pair bonding (Hamilton and Barth 1962). The pattern of monomorphic, bright-colored species is also widespread among tanagers (Thraupinae), a closely related subfamily (e.g., Meyer de Schaunsee 1966).

Male and female plumages of most temperate-zone warblers are moderately dimorphic, and one can readily distinguish breeding males and females of most species. Males have brighter (more saturated) coloration than do females, the intensity of color being related to territorial display and defense. The modest sexual dichromatism of most warblers suggests there is little pressure to develop fundamentally different coloration patterns.

The Black-throated Blue Warbler is a notable exception to this rule: males are slate blue, dark blue, and white; females dull brown and dull white. Ecological and behavioral differences (Black 1975) are comparable to those between males and females of less dimorphic species (Morse 1968). Decreased saturation in female coloration should, however, diminish conspicuousness, which is of considerable importance because of the female's role in incubation and early care of nestlings. Nevertheless, if inconspicuousness were the main factor in plumage color, any pattern that decreased conspicuousness (as the Black-throated Blue Warbler female plumage does) should be favored, and the rarity of this trait in such a closely related group is thus enigmatic. It would be useful to compare the incubation patterns and behavior of Black-throated Blue Warblers with those of species that do not have such a distinct dimorphism.

The sexes of several dull-colored temperate-zone species are extremely similar. Ovenbirds and waterthrushes are primarily terrestrial, and they frequent cover so dense they have little opportunity to acquire visual information from moderate distances; in addition, illumination is often low.

Seasonal Changes in Plumage

Migratory warblers molt and regrow feathers before and at the end of the breeding season. Colors of bright-plumaged males are usually duller during fall and winter than in the breeding season, more like that of females, who also become duller at this time. Many warblers molt head and throat plumage, where bright colors are located, in both fall and spring (Dwight 1900). Moynihan (1960, 1962) proposed that dull coloration decreases the bearer's aggressive character and enhances participation in mixed-species flocks outside of the breeding season. Although color changes may minimize hostile interactions, warblers migrating in fall are nevertheless highly aggressive at times (Morse 1970a).

The late-summer molt resembles that of other brightly colored migratory species, such as tanagers and buntings. Since all of these species molt into their breeding plumage while still on the wintering grounds, hypothetical advantages such as inconspicuousness will not obtain for the trip north. Molting probably must take place before heading north, however: it is an energy-intensive activity that cannot be undertaken in migration, nor can it be commenced after arrival on the breeding grounds because of the pressure to establish a territory as soon as possible. It is the most energy-intensive activity undertaken on wintering grounds (Greenberg 1984b). Relations between migrants and residents before and after they molt into their bright plumage have not been compared. Migrants should increase their aggressive behavior after molting, but rising gonadal titers might independently incite aggression.

Moynihan's hypothesis predicts that migratory species wintering in bright plumage are more likely to hold territories on wintering grounds than those molting into a much duller winter plumage. A survey of warblers wintering in Veracruz, Mexico, by Rappole and Warner (1980) did not demonstrate such a relation, but it was not explicitly concerned with the matter.

I assume that sexual selection and species recognition are significant determinants of bright coloration patterns, and that predation and social facilitation (Moynihan 1960, 1962) provide counterselection. These factors probably dictate between-season color shifts of many parulines (Hamilton and Barth 1962). Baker and Parker (1979) attempted to explain bird coloration as mainly an antipredatory phenomenon, with sexual selection and other factors of secondary importance. Especially significant to them is the notion that conspicuous coloration signals predators that capture is improbable—and would require so much energy that a catch is unprofitable. This notion, reminiscent of Smythe's (1970) pursuit-invitation hypothesis, is based on studies of numerous western Palaearctic species, but not parulines. It predicts that birds should be brightest when juveniles are most prevalent (and hence vulnerable to predators). In fact, parulines (and a variety of western Palaearctic passerines as well) molt into bright plumage before spring migration, when juveniles are scarce, and begin to molt out of this bright plumage when juveniles are most abundant. Bright color patterns are attained shortly before pair formation; the timing fits the predictions for sexual selection but not the predictions for predator avoidance (see Anderson 1983). Thus, my assumptions about plumage change appear to be reasonable, although the subject needs testing.

Immature Plumage

Young warblers of most temperate species retain their fall plumage through the end of their first winter. Two exceptions are the American Redstart and the Olive Warbler (only questionably a paruline), whose males retain a resemblance to female plumage, similar to the first fall plumage, through their second summer. I will concentrate on the American Redstart, since this aspect of its biology has been extensively researched. In other passerine birds, delayed acquisition of adult plumage is related to male aggression and polygyny or to a surplus of male birds, both of which make the number of females a limiting resource (Lyon and Montgomerie 1986). No differences are known between the breeding systems of redstarts and warblers with conventional color patterns. Delayed plumage maturation suggests that young males will be at a disadvantage when competing with older males for mates; thus, individuals attempting to breed in their second year will have low success. Immature plumage might lower the probability of aggression against its wearer, either because juveniles are less conspicuous than adults (the crypticity hypothesis) (Proctor-Gray and Holmes 1981), or because their plumage signals a different status to other males (Lyon and Montgomerie 1986). This prediction also holds if early reproduction is linked to increased mortality.

Ficken and Ficken (1967) attempted to explain the delay in attaining an adult male plumage in American Redstarts on this basis. However, these young birds are physiologically capable of breeding, and some of them do so successfully (Morse 1973; Proctor-Gray and Holmes 1981). Proctor-Gray and Holmes found that the reproductive success of young birds that secured a mate was comparable to that of older ones, casting doubt on the Fickens' explanation.

Rohwer and colleagues (1980) have suggested that juvenile males mimic females in such a way that they gain access to suitable habitats and thereby acquire mates. In many areas, however, second-year and older redstart males often occupy different habitats (Ficken and Ficken 1967). Further, young birds seem to be recognized as males, in spite of their plumage differences, by older males (Morse 1973), and indeed juvenile males do not act like females. It is difficult to envision young male redstarts routinely using this technique to attain territories, as Lyon and Montgomerie (1986) pointed out. Certainly the second-year males that I have observed did not have much success in usurping space when they invaded areas dominated by adult males, which were aggressive and incessantly chased them. Therefore, the hypothesis that juvenile males benefit from mimicking females also has difficulties when applied to redstarts.

Proctor-Gray and Holmes proposed that immature plumage can be selected in the distinct habitats that second-year males occupy, and that these birds or their nests undergo less predation pressure than do older males. In their study, the only four nest predations involved adult-plumaged males that fed nestlings. The high density of redstarts in some habitats could inhibit the success of young birds in the presence of adult males. Large populations, combined with the minimal success of young parents from several passerine species, including Prairie Warblers (e.g., Nolan 1978), could reduce the advantage of selection for adult plumage below that of selection for lower-density species. The data are not adequate enough for an authoritative interpretation of delayed male adult plumage, however. No studies test whether second-year males nesting in sites atypical of older birds shift habitats in the following year. Neither is the mortality of breeding and nonbreeding second-year redstarts known (Proctor-Gray and Holmes 1981).

Recognizing the difficulty of using a single explanation to predict delayed maturation of adult male plumage, Studd and Robertson (1985b) proposed a multifactorial approach. They argued that low intrinsic rates of population increase, high adult survival, high risk or cost of breeding, and low reproductive rate as subadults all select for a delay in breeding. Some of these features fit parulines well (Roberts 1971). Still, this syndrome does not establish why only the American Redstart, among the plethora of eastern warblers, has developed this pattern. There is no reason to suggest that it is longer-lived than other warblers, although data are inadequate to state this with confidence. Its maximum known lifespan (Klimkiewicz et al. 1983) does not exceed that of other retrapped warblers. Moreover, Studd and Robertson used wing-length to predict longevity, rationalizing that birds of large species live longer than do those of small species (Western and Ssemakula 1982). The redstart is one of the smaller warblers, and if this character were of major importance certain spruce-woods species should develop delayed plumage rather than the redstart.

The Studd and Robertson hypothesis suggests that those species experiencing the strongest competition for breeding should be characterized by delayed maturation: competition in this regard could involve problems of acquiring a quality territory—or females. Redstarts are high-density breeders in many areas, and the tendency of young birds to use different territories is well known (Ficken and Ficken 1967; Morse 1973); however, redstart population densities do not exceed those of several other warblers (see Morse 1976b; Erskine 1977). Therefore, the mystery remains about why they, alone among eastern warblers, are slow to acquire mature color patterns.

Other Intrasexual Variation

Although the redstart's age-related dimorphism is unusual among warblers, intrasexual plumage variation is widespread in the breeding strategies of other warbler species. For example, Studd and Roberston (1985a, c) found considerable variation in reddish-brown streaking on the yellow breasts of male Yellow Warblers. The streaking was unrelated to age but was associated with status. Those with the heaviest brown streaking were the most aggressive (Figure 8.3A) and held what were potentially the highest-quality territories. Males with lighter breast streaking were less successful in holding the most sought-after territories, but they did not have lower success at nesting than did heavily streaked birds (Figure 8.3B). Equivalence in nesting success appeared to be a consequence of how males allocated their effort: heavily streaked birds spent much more time and energy in defense (Figure 8.3C), and perhaps as a consequence expended significantly less effort in rearing their young than lightly

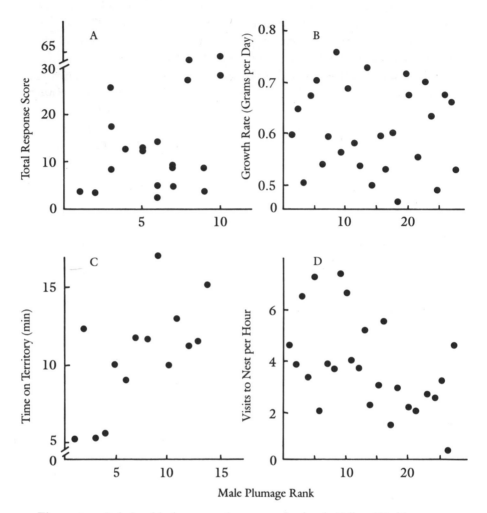

Figure 8.3 Relationship between plumage rank of male Yellow Warblers, based on amount of brown streaking on the breast (high-ranked birds had the most brown), and various parameters: (A) total response scores (based on number of vocalizations and movements) to stuffed mounted Yellow Warblers with 20%, 28%, and 42% of the breast covered by brown streaking; (B) growth rate in terms of change in mass per unit time; (C) time spent on territory; and (D) number of visits to nest per hour. (A and C modified from Studd and Robertson 1985a; B and D modified from Studd and Robertson 1985c.)

streaked males (Figure 8.3D); lightly streaked males spent less time in defense and more in caring for their young. The result was that offspring production on low- and high-quality territories was equal. The tradeoff should help to maintain variation in streaking within the population.

Studd and Robertson's results provide an explanation for variation in plumage patterns of male Yellow Warblers. Combined with information on the breeding systems of redstarts, the satellite-male systems of polymorphic male Ruffs *(Philomachus pugnax),* a communally displaying sandpiper (Rhijn 1973), and others, their results provide insight into relations among social systems, morphology, and environmental conditions. Studd and Robertson's studies are incomplete on at least two points, though: they have not controlled for the role of vocal display, which they acknowledge; and they have not provided independent criteria for establishing territorial quality.

Although such a badge-signaling system had not previously been reported in a territorial system, it may not be rare. Rappole (1983) reported wide age-independent variation in necklace patterns of male (and female) Canada Warblers (Figure 8.4), and variation in facial patterns of Hooded Warblers has long been known (Bent 1953; Rappole and Warner 1980). Rappole (1983) also noted that other warblers exhibit similar levels of variability in plumage patterns; they include Wilson's, Parula, Bay-breasted, Chestnut-sided, and Kentucky warblers and Yellow-breasted Chats.

Plumage as a Means of Discrimination

Burtt's analysis makes it clear that selective pressures may be as much physiological as they are behavioral, and one might argue that physiological considerations such as abrasion, thermoregulation, and ultraviolet shielding constrain color patterns. Thus, we see a tendency for dark dorsa and flight feathers of the wings and tail and for colors whose signal function is confined to certain areas of the body. Nevertheless, the brightness of warbler plumage establishes the potential for behavior-related variation. Coloration differences between males and females, summer and winter birds, adults and immatures, and permanent residents and migrants—and even the range in individual variation within any one of these categories—emphasize that color patterns respond to ecological and social constraints, and that highly visible plumage is influenced by biological as well as physical factors.

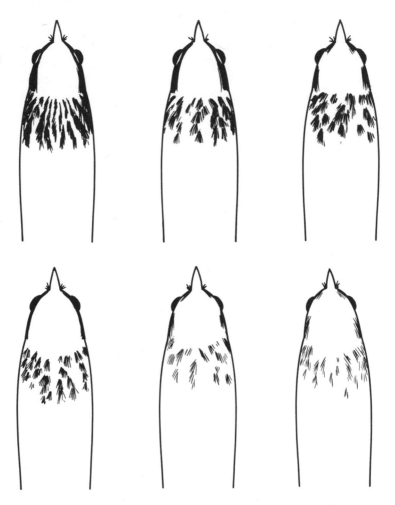

Figure 8.4 Variations in breast striping of adult male Canada Warblers. (Illustrations by Brian Regal.)

Analysis of warbler coloration remains virtually virgin territory, despite considerable interest in it. Surprisingly little progress has been made in evaluating the role of conspicuous, species-distinct coloration in reproductive isolation, although that function is traditionally assigned to it. Given the difficulty of testing this in the field—that is, of testing females directly—this predicament is not surprising. Yet the sparse information from birds in general casts doubt on whether color patterns by themselves identify conspecifics. Vocalizations and visual displays (apart from, or in addition to, color patterns) could, alternatively, play the major roles in interspecific discrimination.

The vocal repertoires of many warbler species, which include different vocalizations sung in different contexts (such as the presence of a female), suggest that singing is central to species discrimination. Given the frequent muting or cessation of vocalizations near other individuals, visual characters may come into play at this point, although this is uncertain. Males direct elaborate behavioral maneuvers to both males and females in their proximity. Nevertheless, closely related species exhibit considerable similarity in their displays, which means either that subtle nuances operate in distinguishing displays between species, or that combinations of displays and color patterns operate as units (although Lewis's Yellowthroat experiments would not support either of these possibilities). Alternatively, though vocalizations decrease in amplitude and often in frequency at close range, they may continue to play the dominating role.

Such hypotheses, if sustained, leave open the question of why highly species-specific plumage patterns arise if they do not function in species discrimination. They could act in male-male discrimination: males respond more forcefully to other males than to females. The level of recognition between conspecifics is still unknown, but interactions between male Golden-winged Warblers and Blue-winged Warblers and hybrids with varying plumages and song patterns suggest that vocalizations play a more important role than plumage patterns in species discrimination.

Although the female's response to male patterns is of fundamental importance, the male faces an analogous task of responding only to females of appropriate species so it does not waste energy on a fruitless heterospecific mating. This problem should not be as detrimental to males as to females, for males have opportunities to mate more than once in a season. Although warblers are facultatively polygynous, however, only a modest percentage of them take more than one mate, and the success of second matings is low. Thus, although the average cost of a heterospecific mating should not be as great for a male as for a female, the proper choice of a mate is no casual matter. To the best of my knowledge, little attention has been accorded to male discrimination of females. If males have to discriminate between females of different species, they must use visual (and probably vocal) cues that probably are not as species-distinct as those that identify males. Yet there are few records of hybridization among most species-pairs that are most vulnerable in this regard. If this is partly a consequence of male discrimination, it behooves us to ask what characters males choose and whether they are able to discriminate with the precision of females.

It is important to work out the relations between visual and vocal communication and to consider how each contributes to both male and female discrimination. Color, song, and physical display all appear to play a part. The basis for the difference in color patterns between North American parulines and their European sylviine equivalents, most of which have not evolved bright color patterns and dichromatism, also remains to be explained.

It is difficult to approach these questions by alternative hypothesis testing, for several reasons: viable hypotheses are usually not exclusive; two-tiered hypotheses (those with physical-physiological and behavioral elements) may imply cause-effect relations yet to be worked out; and the species themselves have independent yet potentially intertwined evolutionary histories, about which we know virtually nothing. Burtt's effort to address the matter by assembling background physical information and then proceeding with hypothesis testing and quantitative observations is an important step toward a more rigorous analysis.

9 Migration

Most temperate-zone paruline warblers migrate great distances, some-times over several thousand kilometers, in their trips back and forth be-tween breeding and wintering grounds. In the winter, the vast majority of warblers retreat south of the United States, or at least to the southern United States, except for the Yellow-rumped Warbler, which can be found as far north as southern Canada. The yearly change of habitat has a marked effect on warblers, both in how they adapt for sustained move-ment and in how they choose wintering and breeding grounds. We might expect extreme specializations (other than for prolonged flight) to be uncommon in highly migratory species because of uncertainties encoun-tered en route and the vicissitudes of matching breeding and wintering habitats. Yet we know little about how migratory abilities affect resource exploitation. The problem is bound to be complex, for species differ wide-ly in how they specialize during winter and summer. We can be sure that major stresses arise: unpredictable food supplies, competition for re-sources with residents and other migrants, predators, inclement weather, and difficulties of orientation (Moore and Simm 1986). In addition, dur-ing flight a bird has to accommodate a metabolic output that may reach ten times its standard metabolic rate (Pennycuick 1975).

Information on the effects of these stresses would add much to our understanding of warbler dynamics. To date, though, only a handful of papers has focused on the issue (e.g., Parnell 1969; Rappole and Warner 1976; Morse 1979; Keast 1980a). One of my purposes here is to draw attention to this gap in knowledge and to stimulate interest in remedying it. My main purpose, however, is to sketch the extent and nature of the birds' movements, the mechanisms for accomplishing them, and the eco-logical problems encountered, including physical constraints, resource ex-ploitation, and interactions with other individuals.

The dates of migration and the paths taken by migratory warblers are well known, though exceptions remain. The mechanisms responsible for performing these movements, such as orientation and navigation, are still controversial; but I do not intend to discuss them as ends in their own right, for detailed treatments can be found elsewhere (e.g., Emlen 1975 or Gauthreaux 1980). Even so, the migratory period plays such a central role in the life cycles of warblers that I shall provide a brief background and introduction to the subject. Although much of the work in the New World involves Neotropical migrants as a whole, a general approach is appropriate since the migrants are dominated by warblers, both in terms of individuals and numbers of species (Nisbet 1970; Keast 1980b). Nisbet's results suggest that in many years warblers make up 70 percent or more of the Neotropical migrants. My intent is to relate the information presented here to the function and evolution of migration, for these aspects of migration have received inadequate attention despite the interest in mechanisms. Questions of demography and sex, the role of competition in timing and seasonal distribution, and the relation between present and past conditions all deserve attention (Gauthreaux 1979).

Routes Taken

Most species of warblers migrate along a broad front over much of their route—a consequence of their wide longitudinal distribution during the breeding season. Particular attention has been paid to migration along the eastern seaboard of North America, where migration is heavy during the fall. From radar observations it is clear that most migrants (warblers and others) in this area move in a southwesterly direction (Drury and Keith 1962). Yet some individuals overshoot the Atlantic coast (which runs in a northeast-southwest direction), perhaps because they are nocturnal migrants, and find themselves over water at dawn. These birds often land on ships (Scholander 1955; McClintock et al. 1978). Many deaths probably occur as the birds exhaust their resources over water. Nevertheless, they can fly for considerable distances, as attested to by the frequency of sightings in Bermuda (Bradlee et al. 1931) and by the rash of sightings in western Europe, a source of constant interest and attention by British birders reporting in *British Birds* (e.g., Alexander and Fitter 1955; Nisbet 1963). Most birds arrive there unaided because of strong west or southwest winds (Elkins 1979). A comparable number of European vagrants is not found along the eastern coast of North America, which we would expect if transatlantic vessels were carrying those birds across the ocean. Thirteen species of parulines have so far been recorded in the British Isles alone, and the number continues to grow (Simms 1985). Robbins (1980b) compiled a list of the next likely species to be found as vagrants in Europe, some of which subsequently appeared. Among the most frequently reported species in the British Isles are Blackpoll Warblers, Black-and-white Warblers, and Yellow-rumped Warblers. In a single fall, at least 11 or 12 Blackpoll Warblers arrived in the British Isles (Simms 1985). Such birds have not served as the nuclei for new colonizations, although a Yellow-rumped Warbler has wintered in England (Simms 1985).

Most birds that overshoot land probably return to the eastern coast of North America, as implied by mass movements toward shore on mornings following heavy migrations (Baird and Nisbet 1960; Murray 1976). Several workers (see Ralph 1981) have pointed out that an especially high percentage of the birds drifting offshore along both Atlantic and Pacific coasts are immatures and that many of them are off course.

Although east-coast migrants generally parallel the continental coastline, no evidence indicates that they follow it during the night; instead, they probably adjust to landmarks at dawn (C. J. Ralph, pers. comm., 1987). Many eventually move through Florida and into the Greater Antilles, a wintering area of several North American species. This route requires flight over a substantial amount of water, though birds remaining near or on course will not cross waters approaching the breadth of the Gulf of Mexico. Still, given the migrants' lack of precision along the northeast coast in the fall, one wonders how many are lost over this large expanse of water. Reports of West Indian–wintering species in Central America and vice versa (Lack and Lack 1972; Lack 1976) may represent displaced birds that managed to make landfall. Do these geographically distinct wintering records imply that birds initially reaching the wrong area were faithful to these sites in subsequent years?

In addition to stragglers that find themselves accidentally over water, another group regularly migrates over the water off the east coast of North America (e.g., Williams and Williams 1978), a route one-third shorter than a mainland-and-island route but which requires a prodigious nonstop flight. The major warbler species—and perhaps the only passerine species—to take the ocean route is the Blackpoll Warbler. Only this passerine species is known beyond reasonable doubt to use this route (Nisbet 1970; Ralph 1981), although McClintock et al. (1978) named other candidates: the Cape May Warbler, the Yellow-rumped Warbler, and the American Redstart. Following the arrival of strong high-pressure fronts in the fall along coastal New England and the Maritime Provinces of Canada, these species take off and fly nonstop to the northern coast of South

America (Figure 9.1). They take advantage of strong tail winds as far as Bermuda, and then move southward until they pick up the southeast trade winds blowing toward the coast of South America. They fly at high altitudes that vary considerably from one part of the journey to another. The birds cross Bermuda at about 2,000 meters (6,500 feet), yet by the time they reach Antigua they may be as high as 6,500 meters (21,000 feet); they average an altitude of 3,000 meters (Williams 1985). Then they begin to descend to an altitude of about 300 meters (1,000 feet) when they reach Trinidad. This remarkable story has been learned from radar surveys at weather stations, from portable radar on oceanographic vessels along the migration route, and from ground observations. It explains the rarity of the Blackpoll Warbler in the southeastern United States in the fall, although numerous Blackpolls, among the lightest on record, appear at Island Beach, New Jersey, following heavy migration. These probably are birds returning to nearest land after finding themselves over water in the morning; Murray (1965, 1979) interpreted their presence on shore to mean that Blackpoll Warblers are not long-distance, overwater migrants. However, the conditions of Murray's birds, the usual scarcity of this species south of his study area in the fall, and radar information suggest that these birds may merely be poorly prepared or incapable migrants that have turned back, while the majority have continued southward over the ocean. The vast majority of birds in Murray's samples are immatures making their first migratory flight.

The wind is not as favorable on the return trip, however. For this remarkable journey to be retraced in the spring, Blackpoll Warblers must migrate in a more conventional way—across the Caribbean via the Antilles. If they meet adverse weather patterns during long overwater hauls, they are vulnerable to high mortality. Bent (1953) reported that northward-migrating Blackpoll Warblers landing on boats or islands near their destination were emaciated, and many subsequently died. The weakened state of these birds reveals how much the fall migrants depend on favorable wind directions during their much longer nonstop movements.

Eastern warblers have three major routes to and from the tropics: the route over the Antilles, in itself a major wintering area for warblers; a nonstop flight between the Yucatán Peninsula and the Gulf Coast of the southern United States; and the western shore of the Gulf of Mexico. Birds making nonstop spring flights from the Yucatán Peninsula may simply be responding to a geographic puzzle: after progressive northward movements they have run out of land. This does not happen to birds

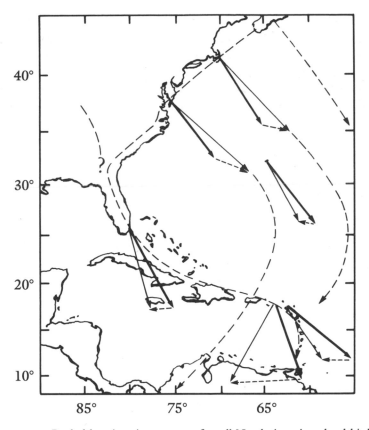

Figure 9.1 Probable migration routes of small North American land birds to the Caribbean islands and South America (dashed lines). The lines forming triangles indicate the relation of the wind to the heading and the track of the birds. The broken lines of the triangles show the direction of the wind (with relative wind speed indicated by the length of the line), the darker solid line represents the average heading of the birds, and the lighter solid line shows their average track. The birds continually exhibit a southeasterly heading, but the trade winds create a drift that turns them toward the southwest as they approach their destination. (Modified from Williams and Williams 1978; copyright © 1978 by Scientific American, Inc. All rights reserved.)

moving in a northerly direction to the west of this location in their northward spring movement, or to southward Gulf migrants in the fall. But these hazardous overwater movements imply that the dangers of more circuitous routes over land are formidable, especially ones that expose land-based birds to unpredictable conditions. Birds stopping where they have scant opportunity to obtain food (Rappole and Warner 1976) jeopardize their health if they remain. Additionally, land-based migrants may face habitats strikingly different from those on either their wintering or breeding grounds.

The route across the Gulf of Mexico is almost certainly the most common route of eastern parulines, even though it requires an overwater flight of nearly 1,000 kilometers. Large numbers of several species take it both during the fall and spring (Lowery and Newman 1966; Buskirk 1980). Though northward-bound, cross-gulf migrants take off from the northern end of the Yucatán Peninsula in the spring, many start below its northern tip (Rogers et al. 1982; Moore and Kerlinger 1987). Some cross-gulf migrants continue to fly northward after they arrive on the United States Gulf Coast (Lowery 1945). This extra distance serves them well because in some places extensive marshes run north of the coastline for more than 50 kilometers.

Migrants take flight in both spring and fall in response to local weather patterns. With light tail winds they can make the crossing in 20 hours, and strong tail winds shorten the crossing to as little as 12 hours (Buskirk 1980). But weather fronts far in the distance at takeoff sometimes expose them to inclement conditions before they complete their journey. To be assured of an uneventful crossing a bird would have to predict weather conditions up to 1,000 kilometers away, 12–24 hours in advance (Buskirk 1980). Weather patterns are not always stable enough for them to make such predictions. Especially serious are strong head winds, which prolong a flight as long as 30 hours and increase the migrant's vulnerability as a consequence of extra energy demands.

The greatest number of migrants cross the Gulf of Mexico when weather is likely to be favorable (mid-April and early October), yet not all species cross then, and the prime periods are by no means times when good weather is assured. Migrants arriving on the Yucatán coast through the third week of September were primarily warblers (Buskirk 1980).

Many birds appear to cross the Gulf with energy to spare (Odum et al. 1961), and their fat deposits are adequate for them to fly several hundred kilometers more, which accounts for their ability to continue inland (Lowery 1945). This excess energy serves two functions: to ensure that the bird finds satisfactory areas to forage after its crossing, and to help it survive if it gets caught in a headwind while over water. Since so much of the northern Gulf Coast consists of unsatisfactory habitats, the premium for being able to move onward should be high. Even outlying islands in the marshes are likely to be saturated by migrants (Lowery 1974). The farther away from the shore the birds move, the lower their density should be and, perhaps, the greater the probability that they will encounter favorable habitats.

The amount of fat found in new arrivals (Odum et al. 1961) is consistent with the observation that, after migrants cross the Gulf, they move farther northward the night after the crossing. Only when birds arrived late from over the Gulf (delayed by headwinds over the water) did they appear to be nearly exhausted of fat reserves, and only then did a high percentage of them remain more than one day when flying conditions were favorable (Gauthreaux 1971).

Displaced Migrants

A substantial number of eastern warblers are regularly found in California during the fall migration. DeSante (1973, 1983; also see Diamond 1982) suggested that these birds fail to distinguish between 0-degree and 180-degree bearings (thus moving north when they should move south, and vice versa) or suffer from mirror-image misorientation (confusing left with right). McLaren (1981), following DeSante, also proposed this explanation to account for the appearance of the western Black-throated Gray, Townsend's, and Hermit warblers on Sable Island, off the coast of Nova Scotia.

DeSante attempted to test this proposal by running Blackpoll Warblers (eastern vagrants) caught on the Farallon Islands (43 kilometers west of San Francisco) in orientation-cage experiments, which measure the compass bearing of the birds' movements. Although the results suggested that mirror-image misorientation may play a role in vagrancy, additional evidence is needed to evaluate this novel hypothesis.

It seems likely that most coastal birds that lose their way would fly until they exhausted their resources, fall to the water, and drown. This is consistent with the great number (83 percent) of first-year birds reported in offshore flights (also see Ralph 1981). But it is of interest to note that several vagrant birds banded in California have returned to the same site as many as three years in a row, and thus are probably wintering in the area (Diamond 1982). This pattern would provide the basis for establishing new wintering grounds for species and could partly account for the origin of otherwise improbable wintering grounds. The Townsend's Warbler and the eastern race of the Yellow-rumped Warbler have disjunct west-coast wintering areas that are well isolated from their major wintering areas and that lie west of their migration route. "Misorientation" might account for other disjunct wintering grounds and for many persistent vagrant species, including western warblers and other species that habitually turn up along the east coast in the fall (e.g., McLaren 1981; Diamond 1982). Since the breeding grounds of some warblers, such as the American Redstart, have been and still are expanding northwesterly, it is likely that navigational discrimination is still being sorted out (Diamond 1982). Redstarts and Blackpoll Warblers initially migrate eastward in the fall from their most westerly nesting areas (Bent 1953). With rapid habitat change occurring, particularly on the tropical wintering grounds of these species, pressure for major shifts in migratory pathways may be strong (DeSante 1983).

Migration and Orientation

There has been much disagreement on the reason for, and significance of, the different ratios of adults and first-year birds in diverse areas on fall migration routes, as measured by the extent of skull ossification (Norris 1961). Proportions of young birds captured in the fall along the eastern and western coasts of North America are higher than those captured inland: in coastal situations they fall in the range of 85–95 percent, as opposed to 65–70 percent inland (reviewed in Ralph 1981). The inland ratio is what one would expect if adults reared broods of four (Hall 1981).

Although the high ratios of juveniles in coastal areas may be an artifact of adults migrating earlier before migration studies begin, or of adults and juveniles having different routes, the inland ratios do not support such arguments for at least the vast majority of species (Hall 1981; Ralph 1981). Murray (1966) has argued that the difference lies in the behavior of adult and juvenile birds finding themselves over the ocean at daybreak and subsequently flying back to land. Murray believes that adults simply move farther inland upon return than do juveniles. Few data support this hypothesis, and kills from tall buildings and television towers suggest that during the height of migration, percentages of adults and juveniles found along the coast and inland differ. For instance, kills at a skyscraper in coastal Boston produce a coastal ratio, whereas those from a television tower 60 kilometers inland give an inland ratio (Baird, in Ralph 1981).

Ralph has proposed that two hypotheses are consistent with the data now available. Juveniles that overfly could learn from mistakes and not occupy the edge of the seashore in subsequent years. Yet many misorienting birds find themselves so far at sea by dawn that it would be impossible to get back to land, and so selection would be strong. There is no independent evidence that the birds learn under such circumstances. Alternatively, migrants that do not closely approach the edge of the shore may be selected, so one need assume no further selective factor to account for the commonly reported differential distribution.

The Blackpoll Warbler, the major passerine migrant over the western Atlantic, has adult/juvenile ratios along the coast similar to those of most species inland. What prompts its long overwater flight in the first place is not clear. Weight-loss studies on Blackpoll Warblers show that they have unusually low metabolic costs during migration (Hussell and Lambert 1980). Close congeners—Bay-breasted Warblers, for instance—are morphologically similar but do not adopt this migratory route. Blackpoll Warblers do deposit greater amounts of fat before embarking than do closely related coastal migrants (Nisbet et al. 1963). A few other species, including Bay-breasted and Tennessee warblers, also have a high adult/juvenile ratio along the northeastern coast (Ralph 1981), so the story is not a simple one. Ralph suggested that the latter species may have adopted a different orientation pattern, migrating along the fringe of the coast in that way. In this instance, birds leaving the northeastern coast might intersect land around Virginia (Richardson 1978). The Blackpoll Warbler, however, is the only one of these species that is common on Bermuda and also has a high adult/juvenile ratio there.

Dunn and Nol (1980), studying fall warbler migration across Lake Erie from Long Point, Ontario, a long peninsula jutting into the lake, found that juvenile birds were more prone to return to land after starting across the lake than were adults. This behavioral difference might account for the high proportion of juveniles captured at one banding station along the shore and the high proportion of adults at another station.

Foraging and Habitat Selection by Migrants

Migratory warblers face formidable problems in adapting to the rapidly changing panorama of habitats that they find as they move between winter and summer nesting areas. For example, fall-migrating spruce-woods warblers that breed in coniferous forests sometimes have to find food in areas as variable and different as palmetto-scrub or savanna-like grassland. Cape May Warblers, birds of the spruce forests during the summer, often migrate in large numbers along the outer coast of Maine, an area characterized by treeless islands covered with rocks and savanna-like grasslands. On some occasions Cape May Warblers are the commonest birds on these islands, foraging in the lush grass (Morse 1980c), a situation for which mice might seem better prepared! Although insects may be abundant, it would not be surprising if the birds gained little if any mass. The contingencies of migration may limit the degree of specialization to such preferred sites as conifers.

The habitat selection of warblers during migration is inadequately known. Parnell (1969) demonstrated that several spring migrants in North Carolina exploit sites resembling those of their breeding grounds more than would be predicted by chance. Most species that were habitat generalists or habitat specialists on migration had similar patterns on their breeding grounds. But the two exclusively conifer-nesting species, Yellow-rumped and Blackpoll warblers, did not exhibit this pattern, prompting Parnell to suggest that in lieu of favored sites they did not respond strongly to secondary sites. Parnell concluded, however, that niche choice—that is, the selection of certain areas within a habitat—might be of greater importance to those migrants than habitat choice. Overlap between species was higher during migration than during the breeding season, probably because more species were present (Power 1972).

I compared the foraging Blackpoll Warblers during spring migration in Maryland with patterns during northward migration in Maine and on breeding grounds in the mountains of New Hampshire (Morse 1979). The structure of all of the habitats differed markedly. The Maryland site was a mature deciduous forest with trees 25 meters or more in height; the site in Maine consisted of tall Red and White spruces of similar height; the breeding area in New Hampshire had greatly stunted trees, sometimes less than 3 meters in height, although more frequently 5–8 meters. This alpine woodland is typical of the Blackpoll Warbler's breeding grounds on most of its nesting range across northern North America, although it nests in taller forests in many areas. In each location the Blackpoll Warblers foraged in the middle of the crowns of the trees, rather than in the tips of the vegetation, although they foraged more in the tips of the vegetation in Maryland, in part because of the differences in deciduous and coniferous vegetation. In addition, this difference may have been associated with prey insects feeding in the unfurling leaves. Foraging patterns varied more in location and kinds of prey taken than in the birds' movements in catching prey (see MacArthur and Pianka 1966).

Lawrence (pers. comm., 1983) undertook a broad study of the foraging patterns of warblers migrating through Rhode Island and eastern Massachusetts in the spring. He also found that species foraging in the treetops on their breeding grounds tended to work high in the trees and those frequenting low vegetation on their breeding grounds tended to work low in the vegetation. Overlap was broad, and individuals of most species foraged in areas with emerging insects, which correlated with the unfurling of leaves.

In studies of spring-migrant warblers in Illinois, Graber and Graber (1983) censused both the amount of food available and the rate of food capture by individuals, primarily Tennessee, Palm, and Blackpoll warblers. They found clear differences in the birds' foraging patterns depending on the number of available insects. When prey was abundant, movements were moderate and the birds captured many prey—virtually all lepidopteran larvae, by far the most abundant resource. The calculated intakes, with estimated energy gains (Figure 9.2), were high enough (estimated to be 14.9–19.0 kilocalories per bird per day) for the warblers to realize a substantial gain (estimated expenses being slightly over the 11.0–13.0 kilocalories required for daily maintenance). This success contrasted with food intake in the low-prey area, where the birds foraged extremely rapidly early in the day but were unsuccessful in finding enough prey to permit a gain (estimated intake of 7.2 kilocalories per bird per day). By midmorning these birds had ceased foraging, and few were to be seen on the following mornings, suggesting that they had moved to other areas. Graber and Graber emphasized the importance to these birds of finding one or more areas of high prey availability along the migratory route. Type of prey is significant also: the lepidopteran larvae, in addition to being abundant, have a high water content, allowing the birds to replenish water lost in migratory flight (Hart and Berger 1972; Berger and Hart 1974). Graber and Graber's study was not accompanied by observations on the foraging niche patterns of these birds.

Laursen's (1978) studies of sylviine warblers on spring migration in Denmark are relevant to our study of paruline warblers in North America. Two variables were found to correlate with stomach contents of these sylviines (and presumably their foraging): food abundance (strongly affected by the spring climatic patterns) and the number of birds present.

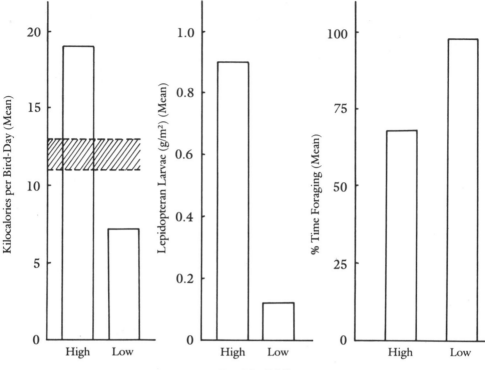

Figure 9.2 Foraging patterns of spring migrating warblers in Illinois during periods of high and low food availability. *Left*, energy intake; *center*, density of principal food items (larvae); *right*, percent of time spent foraging. Cross-hatched area in left-hand figure denotes estimated daily caloric requirements. (Modified from Graber and Graber 1983.)

Where food was abundant, prey differed little among the warbler species; when many birds were present, however, the sylviine warblers clearly segregated the foods that they took. Laursen assumed that the stomach contents sampled reflected a tendency to exploit hunting sites differently as a consequence of how many individuals were present; unfortunately, he did not make foraging observations to test this hypothesis.

Moore and Simm (1986) found that Yellow-rumped Warblers about to begin migration foraged in a risk-prone way; that is while fattening they chose variable rewards in preference to constant ones of the same average abundance (Figure 9.3). In the process, they adjusted their feeding behavior to consume more food items in a foraging bout, to handle items more rapidly, and to select more profitable items than previously. Upon attaining maximum body mass, they foraged in a risk-averse way, choosing predictable rewards in preference to variable ones of the same mean abundance. Under natural circumstances, "safe" foraging habitats might quickly cause them to seek alternative resources or sites if they perceived resources to be inadequate. If migrants are exposed to inadequate conditions, however, taking a chance is clearly the least of evils, their only opportunity to avoid starvation.

In fall migration warblers exhibit a strong tendency to join mixed-species resident flocks led by chickadees and titmice (Morse 1970a; Keast 1980a). Participation in these groups will draw warblers into closer contact and possibly increase the conflict for food. However, they may be led to the most favorable feeding sites by following the residents, at the same

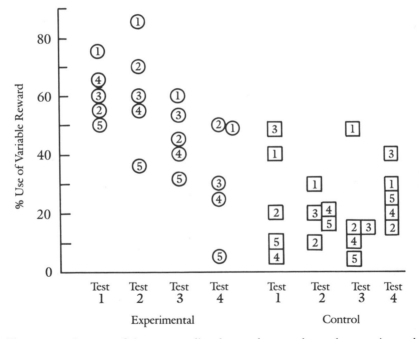

Figure 9.3 Percent of times unpredicted rewards were chosen by experimental (circles) and control (squares) Yellow-rumped Warblers during the fattening period (tests 1 and 2) and after fattening (tests 3 and 4). Experimental birds on a 16:8 hour daylength, control birds on a 10:14 hour daylength. Responses of five experimental and five control birds numbered according to individual. (Modified from Moore and Simm 1986.)

time perhaps benefiting from the predator-alarm responses of others (Morse 1970a, 1977a). Warblers make up the majority of migrant species joining flocks at this season, and they may also constitute a substantial part of the flocks, sometimes almost as many as the residents. Migrants differ conspicuously from residents in that no single species is likely to assume an abundance comparable to that of individual resident species. These warblers appear to be casual members, continually joining or dropping out, rather than central parts of the flock (Morse 1970a). Participation in different habitats and over time also differ. In my studies a greater number of warblers joined flocks in mixed spruce-birch forests than in adjacent spruce forests, even though many were the birds that earlier bred in spruce forests (Magnolia, Black-throated Green, Yellow-rumped, and Blackburnian warblers). The basis for this shift appeared to be the high frequency of lepidopteran larvae, mostly leaf rollers, on the birches.

Although joining the flocks concentrated the birds, different species foraged in different strata and thus were dispersed within the site, even though none may have experienced difficulty in obtaining adequate food while foraging. These differences in foraging sites suggest that the species have innate preferences for sites, even in the absence of feeding contingencies. Their choices resembled those of the breeding period, when the flexibility of some species allowed them to exploit areas normally frequented by their dominants even though they retained distinctive foraging patterns.

Hutto (1985b) compared the habitat selection of migrants in Arizona in spring and fall. Fifteen of the 26 species were warblers, and in some of the habitat types they made up an even larger proportion than this. He found a close correspondence between habitat-exploitation patterns and the distribution of food in the habitats, as measured by sticky traps. Several species differed in their exploitation patterns between spring and fall. Hutto concluded that birds forage where they are most efficient unless interspecific competitive interactions force them to modify their first choice.

Territorial and Aggressive Behavior during Migration

Birds on migration are generally not believed to be territorial or very aggressive. Even so, Rapppole and Warner (1976) found that Northern Waterthrushes *(Seiurus noveboracensis)* were strongly territorial on spring

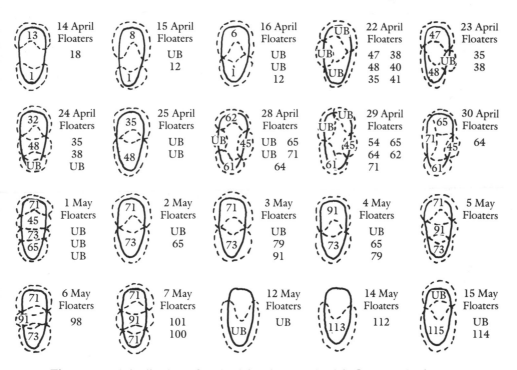

Figure 9.4 Distribution of territorial and nonterritorial (floaters) Northern Waterthrushes around a temporary pond in southern Texas during spring migration. Dashed lines denote territorial boundaries. Numbers refer to specific birds; UB, unbanded birds. (Modified from Rappole and Warner 1976.)

stopover sites around the edges of small ponds in southern Texas, a behavior that prevented some individuals from finding satisfactory foraging areas. Territorial waterthrushes often remained for several days at a site in the study area (Figure 9.4), putting on enough fat to sustain them during migration. The site was one of only a few suitable stopovers in several hundred kilometers along this part of the coastal area. Individuals unable to obtain a territory quickly moved on.

Since no behavioral study of migrant warblers other than Rappole and Warner's has used color-marked birds, it is difficult to determine whether other individuals remain for considerable periods of time. All we can do is make general observations that numbers of a species often grow in a staging area prior to the advent of good weather that triggers migration, or that an unusual species may be seen in an area for several days before

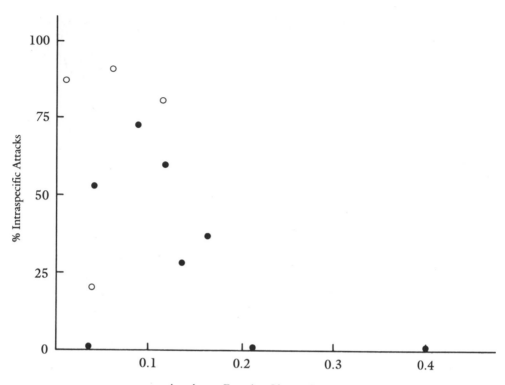

Figure 9.5 Frequency of hostile interactions—fights, chases, and supplanting attacks—of warblers (black circles) and other species (white circles) during late summer, expressed in interactions per foraging observation (attempt to capture a prey item). (Modified from Morse 1970a.)

disappearing. Moore and Kerlinger (1987) noted that warblers might remain from 1 to 7 days in one place after crossing the Gulf in the spring, and they proposed that some of them also temporarily defended territories where they gained weight. In contrast to Rappole and Warner's waterthrushes, Moore and Kerlinger's warblers showed little aggression. Murray (1976) noted that few of the birds marked in large banding projects undertaken along coastal regions in the fall are captured a second time, which suggests a mobile pool. Yet stays of 2–3 days might profit from territory formation. Given the territorial behavior in Rappole and Warner's waterthrushes, it is clear that this phenomenon deserves more attention. Unfortunately, heavy movement can frustrate studies with marked birds, because a great deal of effort is required to capture and handle a large number of individuals, only to have most of them quickly leave.

One often sees aggressive behavior among warblers associated with mixed-species flocks of chickadees and other birds in late summer. Aggression is over twice as frequent among warblers as in resident members of these groups (Figure 9.5), and almost one-third of the hostile interactions in those flocks involve warblers. Some interactions are initiated by birds still on their breeding territories that attempt to drive off members of the visiting flock or other permanent residents; a majority are interspecific. Black-and-white Warblers often are aggressive at this time. Some of them—at least the ones observed in Louisiana in September—are truly migratory birds, and perhaps are even temporarily territorial.

The proportions of interspecific and intraspecific interactions of warblers in flocks differ strikingly from those of most resident species in being primarily interspecific (2:1)—partly a consequence of the numerous warbler species present. Nevertheless, intraspecific interactions occur more frequently than predicted by chance. Since the various warbler species continue to differ in their foraging habits during migration, one would expect conspecifics to be in closer contact with each other than with other warblers.

Little detailed work has been done on display patterns of migrating warblers. Rappole and Warner (1976) reported aggressive visual displays associated with territorial behavior. Since warblers begin to sing as they leave their wintering grounds, however, and continue to do so on migration, these patterns could play a role in determining resource use in migration and behavior upon arrival at their breeding grounds.

Interactions between Migrants and Permanent Residents

Migrants in passage may be forced to interact with unaccustomed species in areas they are not well equipped to exploit. Unless resources are abundant at such times, the opportunities for the migrants may be poor, both for exploiting resources and for aggressive interactions. If resources are common but patchy, however, migrants could profit by following residents to the resources (Morse 1977a).

In turn, by virtue of their sheer numbers, migrants must sometimes exert a strong effect on the food supplies of an area, even if migration is tuned to periods of high food abundance. At times, the temporary population density of areas along migration lanes can exceed the normal density severalfold. Whether the sizes of resident populations in these areas are consequently depressed is not known, even though some geographic areas must differ markedly from others in this regard, solely as a consequence of the numbers of migrants passing through once or twice a year. In the spring, migrations reach their peak at the very time that residents are breeding and their food requirements are at a maximum. Since many migrants feed on insects, it is less likely that they will have as great an effect on residents during fall migration because residents depend on non-mobile arthropods or other food sources in the winter. Still, if migrants managed to depress a resource that would later, in diapause, have been prey for the residents, they could have an important effect on the food supply. One might test this hypothesis by comparing resident population densities in areas of high migrant density (the tips of peninsulas along flyways, for example) with resident densities in otherwise similar habitats (perhaps at the heads of the same peninsulas).

Avoiding Predators

Being unfamiliar with the areas through which they pass, migrants can be especially vulnerable to predators (Errington 1946). Since they may be in poorer physical shape than residents, it is difficult to determine how much the vulnerability of transients results from being in an unfamiliar site and how much it is a consequence of substandard condition.

Two Old World falcons, Eleonora's *(Falco eleonorae)* and Sooty *(F. concolor)* falcons, have unusual autumn nesting times that coincide with the periods when vast numbers of migrants (on whom they prey) traverse their nesting areas about the Mediterranean and Sahara, respectively (Walter 1979; Clapham 1964). No similar adaptations are known for any American raptors, but it would be profitable to investigate more subtle modifications of predator behavior in response to migrants. For example, numbers of migrating Sharp-shinned Hawks often peak along with those of their most important prey species, although causality has not been established.

Some predators subsist mainly on migrants in season, but per capita mortality of prey species is probably low. Moving together, migrant prey are so great in number that they are less vulnerable than are widely scattered birds. This predator-satiation effect has been amply reviewed in relation to populations of other animals; it probably is important in co-ordinating the timing of migration, although physical factors and food availability must play major (and probably often greater) roles as well.

Climatic Stresses on Migrants

Physical factors sometimes result in heavy mortality of warblers during the migratory period. I will consider two problems in detail: the appearance of inclement weather while birds are migrating over large stretches of open water, and cold weather during migration.

Overwater migration is hazardous, and ample evidence of catastrophic mortality verifies this. Many birds are killed during flights over the most well traveled overwater route in the Western Hemisphere, the Gulf of Mexico (see Morse 1980c). On occasion windrows of drowned migrants—mostly warblers—wash up at such places as Galveston Island, Texas, if the prevailing tides and winds push the carcasses onto shore. Presumably only individuals that fall into the sea near the coast wind up on the shore. Strong onshore winds that exhaust small birds would wash them away from land; others might sink or be eaten either by fishes or by pelagic birds, such as gulls, before reaching shore. Thus most catastrophes probably go unnoticed.

Warblers sometimes land in exhausted and emaciated condition on oil rigs and ships off the Gulf Coast (e.g., Lowery 1946); reluctant to leave, they subsequently die there (Gauthreaux 1971). Similar observations have been made along the north coast of the Yucatán (Paynter 1953). Such birds often contain little or no visible body fat and are seriously desiccated (Odum et al. 1961). Rogers and Odum (1966) argued that energy depletion is a more serious problem than water loss; some individuals have expended considerable amounts of their pectoral musculature. Deaths from exhaustion, less spectacular than the catastrophes caused by violent weather, occur while most individuals are flying successfully to the mainland. The number of birds that die in this way is unknown, but the loss of the birds that are unable to fly the last 50 or fewer kilometers to the mainland could be large. We are probably seeing the cutting edge of natural selection here. In contrast, mortality might be nonselective when migrants encounter severe storms over large bodies of water, except to the degree that selection occurs for crossing these water bodies when inclement weather is minimal (see Buskirk 1980). But it would be profitable to compare the body condition of storm-downed birds and those picked up from the decks of offshore oil rigs and ships.

That birds fly over the Gulf suggests that greater dangers await migrants taking the shore route to and from the Neotropics. The circumgulf route is much longer for birds flying between southern Mexico and eastern North America. Birds going from the Yucatán coast to the southern Louisiana coast travel less than one-third the distance by flying across the Gulf instead of following the shore through Mexico and Texas: the most direct route is under 1,000 kilometers, the least direct route over 3,000 kilometers. The distances saved obviously change with the site of origin, but in perhaps the most important comparison, from central Guatemala directly south of the Yucatán, an area through which large numbers of migrants pass, the difference between routes is still twofold (Figure 9.6). By making such measures, one can calculate where the greatest number of trans-gulf and circum-gulf migrants should nest. As one gets farther north and farther west, the difference in distance between the two routes is less, but taken in terms of migration distance alone, the advantages of the overwater route are substantial for birds nesting in much of eastern and central North America. A direct test of the patterns in Figure 9.6 would be whether they coincided with the birds' migration routes. Such a test, strictly applied, would be impossible in the absence of information on the breeding and wintering areas of individuals, but migratory routes of species and distinguishable races with well-defined breeding and wintering areas may permit a partial assessment of this question. Counts of migrants in southern Texas and along the south-central Louisiana coast should provide useful initial information for such an analysis.

Rappole and Warner's (1976) study of waterthrushes highlights the problems by migrants taking the coastal route, which are compounded by the number of other individuals present. The same patterns hold for several tree-dwelling warblers that Rappole and Warner observed in less detail.

The nonstop migration from northeastern North America to northern South America performed by the Blackpoll Warbler is an even more spectacular example of the pains taken to avoid the overland route. Blackpoll Warblers may be able to remain aloft longer than species that habitually cross the Gulf of Mexico, which in turn may be able to remain aloft longer than birds taking the circum-gulf route. Williams and Williams (1978) estimated that Blackpoll Warblers are in continuous flight as long as 88 hours in their passage from eastern North America to South America; in comparison, 25–40 hours are needed to fly over the Gulf of Mexico in

Figure 9.6 Minimum distances, in kilometers, between selected points for trans-gulf and circum-gulf migrants.

poor conditions (Buskirk 1980). The Blackpoll Warblers' southern movement is mostly passive, gliding with prevailing winds (Richardson 1974). If headwinds are blowing, it is unlikely that the birds can remain aloft long enough to complete the crossing.

What evolutionary lines of warblers are preadapted for extended ocean crossings, and how much are the capabilities subject to modification? If long-distance oceanic migrants and overland migrants were randomly distributed among taxonomic lines (genera, species groups, populations), one might conclude that the capabilities could be easily attained, whereas segregation of the trait of long-distance flight to subgroups would indicate that preadaptations of phylogenetic constraints are more important. This issue could have an impact far beyond the act of migration, for it could affect where a species winters or summers. The wintering grounds of Blackpoll Warblers are in areas where few if any other northern migratory warblers penetrate, through much of Amazonia and occasionally even south into subtropical Brazil (Sick 1971). Although they move into an abundant and diverse avifauna, Blackpoll Warblers could, by occupying a different geographic area, exploit niches that would be contested by congeners in other areas. This migratory act could partly account for the abundance of the species, one of the commonest breeding birds in northern Canada.

Migrating birds, the greatest number of which are warblers, are sometimes caught in the air over bodies of fresh water, forced down, and drowned. In May 1974, for instance, observers reported thousands of warblers washed up on the shores of Lake Manitoba following a storm (Houston and Shadick 1974). In May 1976, in one of the largest warbler disasters recorded, as many as 200,000 warblers, thrushes, and jays washed up on the shores of Lake Huron, having been caught aloft in a spring storm (Jansson 1976). Henshaw (1881) reported a similar event on Lake Michigan after an autumn storm. These "wrecks" take place over water bodies much smaller than the large seas regularly traversed by migrating birds, and the distances involved presumably fall well below the maximum flight range of many migrants. Much has been written about the ability to migrate when good flying weather will be guaranteed at least for a sizable part of the flight (Williams and Williams 1978; Buskirk 1980). Are birds' abilities to predict weather near the northern end of their migration inferior to their abilities in other situations, or are weather conditions less predictable there?

Untimely cold snaps are a second major climatic factor affecting migrants. Snowfalls or unseasonably cold weather, often accompanied by heavy rainfall, curtail insect activity. These conditions may occur in both spring and fall, but since most warblers migrate well before cold weather sets in, spring is the time of greatest vulnerability for many. Keast (1980a) noted the foraging behavior of 10 species of warblers that had advanced into cold wet weather before trees had come into leaf. Foliage-gleaning species hunted on the ground, in low shrubbery, and on bare trunks along with Black-and-white Warblers. Tramer and Tramer (1977) reported the response of foliage-gleaning warbler species in Ohio to record-setting cold in late September. These birds, including Tennessee, Magnolia, Bay-breasted, and Blackpoll warblers and Ovenbirds, fed on berries (honeysuckle, yew) until they exhausted the supply. Then they proceeded to creep through the grass on lawns, apparently after small arthropods on the undersides of grass blades. American Redstarts even plucked grass seeds, behavior that Bent (1953) listed for a few warbler species, probably under similar circumstances. In these conditions the birds probably lost considerable mass.

Many bird-watchers in the northern United States or southern Canada have observed spring disasters, which can destroy large numbers of insectivorous migrant birds (e.g., Finch 1975) and severely depress breeding bird density (Griscom 1941; Zumeta and Holmes 1978). At such times birds become so tame they can be captured by hand, either because they are emaciated or intent on searching for insect prey (Tramer and Tramer 1977; Keast 1980a).

Diversity and Timing

The dates warblers begin migrating correlate with the distances they must migrate to their breeding grounds. In New England, the Yellow-rumped, Palm, and Pine warblers are early spring migrants, and also the only species that winter in large numbers within the eastern United States. This pattern also holds within species that show enough geographic isolation to permit testing. In Arizona locally breeding Yellow Warblers were the first to arrive, followed by individuals of races breeding immediately to the north. The Alaskan race was the last to arrive, two months after the breeding residents, and the most southerly winterer (Phillips 1951). In contrast, considerable overlap can occur among races of this species in the fall. Raveling and Warner (1978) retrieved prairie- and tundra-nesting birds on the same night from television tower kills in Minnesota. The birds may have commenced their migrations as much as 3,200 kilometers apart.

Many Yellow-rumped Warblers are also extremely late fall migrants (Murray 1979; Keast 1980a). They are the predominant warbler species at the end of the migratory season, which is consistent with the short distance that they have to traverse. Early movements happen as well, which probably consist of the tropical wintering segments of this species.

In addition to the distances to be traveled, feeding habits affect migration. The ability of Yellow-rumped Warblers to live on vegetable matter allows them to move when most others cannot. For instance, Whittle (1922) reported massive northward movements of this species along the South Carolina coast in late February, in spite of subnormal temperatures.

These birds fed on large stands of Wax Myrtle *(Myrica cerifera)* along the coast; their more insectivorous relatives have no such assured food supply at this time.

Species wintering in the northern Neotropics and summering in the northern United States and southern Canada overlap the most in their migratory period (Keast 1980a). They account for the great temporary diversity seen in migration, during which an observer can record as many as 20–25 species within a limited area during a single day. Most of these individuals have traveled roughly similar distances and have approximately the same distance to go. This abundance of species and individuals is consistent with the distance-time hypothesis, that migration date is a function of distance traveled.

Many fewer species are to be found in later groups of migrants, those wintering far to the south and often breeding far to the north. The Blackpoll Warbler is one of these, with the most southerly wintering range of any species. Among other late spring migrants are Bay-breasted and Tennessee warblers, which also breed only at northerly latitudes—though usually not as far north as Blackpoll Warblers—and some of the *Oporornis* warblers, such as Mourning and Connecticut warblers, which also have northerly breeding areas and southerly wintering areas (Keast 1980a).

A factor tempering the distance-time hypothesis is the time when breeding areas become habitable. Black-throated Blue and Black-throated Green warblers, though primarily northern nesters, have disjunct populations, accorded subspecific rank, in the lowlands of the southeastern United States, as well as extensions of the nominate race running down the Appalachians to latitudes that the coastal races inhabit. Although the wintering areas of the coastal disjuncts are not clearly demarcated, they do not differ markedly in latitude from the regions occupied by at least some individuals of the nominate race (AOU 1957). Yet, coastal races may arrive back on their breeding grounds ten days to three weeks before their higher-elevation conspecifics (Bent 1953), which will not experience favorable conditions until later in the season because of the difference in altitude. In these instances, conditions anticipated on breeding grounds must play a role in setting arrival times and initiating migration. This probably accounts for why northerly-wintering Yellow-rumped Warblers do not return to their breeding areas earlier than they do, and why early-migrant species have more variable arrival dates than do later migrants (Weydemeyer 1973; Slagsvold 1976).

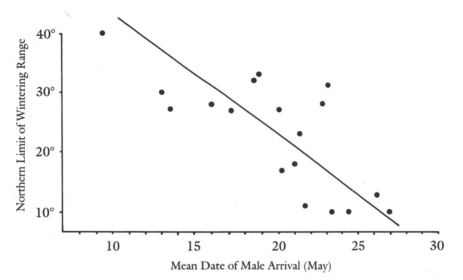

Figure 9.7 Relationship between mean arrival date of migrating male warblers in southern Ontario and the northern limit of their wintering ranges. (Modified from Francis and Cooke 1986.)

In several warblers with western and eastern races, the western races breed at lower latitudes than the eastern ones, often at high elevations. Investigation of their migratory periods might provide further insight into the question of migratory distances and times. Two possibilities for this analysis would be Orange-crowned and Wilson's warblers.

Information on the fine-tuning of arrival times comes from migration dates of spring migrant warblers on the north shore of Lake Ontario in Canada. Francis and Cooke (1986) found a strong correlation between arrival dates of male migrants and the latitude of their wintering ranges (Figure 9.7). Not only did males of a species arrive before females, and older birds before young ones, but the earliest species to migrate had the greatest male-female differential in arrival time. Further, large males of a species usually preceded smaller ones, but although females had a similar pattern, males returned sooner than females of similar size.

These results do not explain why more southerly wintering species do not commence migration sooner, which could accommodate for these differences; nor does it explain how mechanisms between males and females differ to ensure that females arrive later. It is sometimes assumed that females commence migration later, but there is no evidence for such a difference, although Nisbet and Medway (1972) have shown that females of a sylviine warbler, the Eastern Great Reed Warbler (*Acrocephalus orientalis*), commence their northern journey from Malaysia later than do males wintering in the same area. Rogers and Odum (1966) have also determined that females of at least one Neotropical migrant, the Orchard Oriole (*Icterus spurius*), accumulate fat later in spring than do males. One thus assumes that departure dates differ among species and individuals in a way that accommodates their hardiness and distance to travel. Clearly Yellow-rumped Warblers can withstand more severe climatic conditions than can species normally migrating later (e.g., Chestnut-sided Warbler, Black-throated Blue Warbler), but whether midperiod migrants are hardier than late migrants (Blackpoll Warbler, Mourning Warbler) is unclear.

Juveniles from different geographic areas may have to commence migration at different ages, which has implications for their ability to cope with stresses. This is bound to be of importance for species that occupy a wide latitudinal range, such as American Redstarts and Blackpoll Warblers. I am unaware of considerations of this issue in relation to paruline warblers, but it has been explored for the European sylviine, the Willow Warbler. Following the suggestion that Scandinavian birds would have to commence migration earlier than do German ones (Gwinner et al. 1972), Högstedt and Persson (1982) compared the times to first migration in Willow Warblers from northern Sweden, southern Sweden, and southern Germany; their evidence did not support this argument. Still, "leapfrog" migrations (Cox 1968), in which the most northerly members of migratory populations move the farthest, should provoke situations in which northern young, typically later hatched, will commence movements relatively sooner than their more southerly counterparts. In some species, the earliest movements are by the most northerly members of these populations. In Yellow-rumped Warblers, for instance, the early migrants are the ones that winter south of the United States (Murray 1979). This factor could determine the extent to which far northern areas can be colonized for breeding.

Population Dynamics and Migratory Attributes

The formidable costs of migration speak to the magnitude of advantages arising from a migratory life-style. The distinguishing feature of the migratory season is the large role played by density-independent factors. Given the possibility of high mortality during migratory periods, the question arises about the degree to which density-independent factors affect a species' attributes. Predictability of environmental conditions is bound to be minimal in relation to birds' ability to respond to it, for example. Many individuals embark on long nonstop journeys that easily transcend the ranges of single weather patterns, and birds are extremely vulnerable while flying over water or other inhospitable areas. Compounding this problem, the energy demands of long flights are prodigious, especially for overwater passages, which cannot be quickly aborted. As a result, physiological deprivation is not unusual—but it is not the norm, either. Although disproportionately large numbers of individuals take these long hops when the weather is likely to be moderate, complete safety is by no means assured.

High mortality during migration may favor strong selection, especially in response to environmental factors, although not to the total exclusion of density-independent factors. Density becomes important when resources, or resource spaces, are severely limited, as reported by Rappole and Warner. If resources are scarce at the time of rigorous environmental challenges, most migrants could be severely affected, but some individuals are likely to tolerate such conditions better than others. At this time differences in the abilities of individuals to forage efficiently could exert stronger selection than at most other times.

If a species maintains a strong migratory habit, it may find certain lifestyles closed to it; for instance, the long, pointed wings associated with long-distance migrants may restrain their ability to evolve highly maneuverable wings (e.g., Winkler and Leisler 1985). Perhaps this is why the frequency and diversity of aerial specialists is much higher among tropical residents than among migrants. Selective factors related to successful migration may preadapt individuals to range expansion and the formation of new populations, as suggested by the history of off-course migrants. With breeding and wintering ranges subject to rapid change, the ability to find opportunities in uncertain circumstances may account for rapid colonization of new areas as well as for the completion of long migratory journeys.

10   The Wintering Season

Until recently virtually all detailed ecological and behavioral studies on wood warblers had been conducted on the breeding grounds. Fortunately, that is now rectified, although our knowledge of warbler behavior on their wintering grounds still lags far behind our understanding of the breeding season.

The wintering season may be a key to warbler biology, and some argue that it is the most critical period of the life cycle (e.g., Bennett 1980). Migrants are integral members of the communities they occupy during the winter, not merely waifs temporarily displaced from their northern homes—a formerly prevalent attitude (see Griscom and Sprunt 1957). This is an important, but not surprising, conclusion, for many birds spend over half of the year on their wintering grounds (Schwartz 1980) and over twice as much time there as on their breeding grounds.

Wintering Range and Habitat

Since most temperate-zone warblers depend on insects for sustenance, their geographic range changes or contracts during the winter, when insect prey is scarce. A majority of species retreat to tropical or subtropical latitudes. The wintering grounds of most temperate-zone warblers overlap their summer breeding areas little if at all, and other warblers vary markedly in the habitats or altitudes that they inhabit in different seasons.

Chipley (1980) and Keast (1980b) have noted two patterns of winter distribution among warblers: some species occupy small geographic areas at high densities, others occupy much larger areas at low densities. Members of large genera contribute a disproportionately high number of species to the first group. These birds may have difficulties in coexisting with other species as a result of their ecological similarities. The second group has a disproportionately large number of representatives from monotypic genera. Many of the latter, like the Black-and-white Warbler, have foraging patterns that differ from the leaf- and twig-gleaning habits found in such large genera as *Dendroica* and *Vermivora*. Local resources may not support birds with an unusual foraging habit in high densities.

Males and females of a species may have different wintering ranges and habitats (Rappole and Warner 1980). In Veracruz male Hooded Warblers outnumbered females 8:1 in primary forest, but females slightly outnumbered males in secondary forest. Migrants in passage had a roughly 50:50 sex ratio. Lynch et al. (1985) reported a similar segregation of sexes on the Yucatán Peninsula, and Morton et al. (1987) demonstrated experimentally that males and females actively chose the different habitats. It is well known that males and females of several temperate-zone wintering species of various taxonomic groups have nonidentical wintering ranges (e.g., Ketterson and Nolan 1976), so it would not be surprising to learn that this pattern was common among wintering warblers as well.

More than one geographic race of a species may winter in the same area. Several species, including Yellow and Wilson's warblers, were represented by more than one race on Rappole and Warner's (1980) study sites in southern Veracruz. Eastern and western races of some species winter together, but races of other species may be geographically separated. Separation may stimulate the differentiation of races; for unless their different races segregate by habitat or niche when wintering in the same area, the species that share wintering grounds would diverge only as a result of events on the breeding grounds (and possibly in migration). Those that occupy disjunct wintering ranges may undergo differential selection in both (or all three) areas (Salomonsen 1955).

Habitats Exploited

Willis (1966) and several Old World ornithologists (see Morse 1971b) have tended to assume that wintering warblers and other Neotropical migrants frequent highlands, edge, and second-growth habitats and that they avoid lowland primary forest. Further, migrants have been assumed to exploit temporary resources. This pattern fits some Neotropical species, but others are denizens of lowland forest. Although the density of migrants in lowland forests or even undisturbed upland moist forests is lower than the density of migrants in "marginal" habitats (Tramer and Kemp 1980), certain species, such as Swainson's, Worm-eating, and Kentucky warblers, concentrate in undisturbed moist lowland forest (Karr 1976; Willis 1980). The lowland forest in the New World tropics has been extensive enough that even if wintering densities of migrants were low, most migratory species must have wintered there (Pearson 1980; Terborgh 1980).

Conventional wisdom held that migrants' heavy use of second-growth, edge, and highland sites was due to lower densities of permanent residents in these habitats. This argument was attractive for second-growth and edge habitats, since they were purportedly only infrequently occupied by permanent residents, which would therefore not become specialized to these habitats. Migrants frequently do not give way to permanent residents, however. Also, permanent residents usually do not take up the migrants' niche space when migrants go north (Hutto 1980; Rappole and Warner 1980). These observations force us to reassess the role of migrant species in the avifauna of their wintering grounds.

Densities of winter and permanent residents vary according to the habitats chosen by oscine and suboscine birds (Stiles 1980). Tramer (1974) noted that dry forests, where migrants are abundant, are characterized by a high ratio of resident oscines to resident suboscines. Migrants, composed primarily of oscines, often are much more common in dry forests than in moist ones (Waide 1980). Large parts of the northern Neotropics are, or were, composed of dry forests, which may account for the migrants' high relative abundance there.

Migrant species of two types are likely to use young second-growth areas. Savanna and scrub dwellers, of which Yellow-rumped Warblers are an example, arrive in the first year or two after destruction of the forest. Later, mature forest species come down to use the developing canopy, sometimes when it is no more than one or two meters high (Rappole et al. 1983). Stiles (1980) noted that 70–95 percent of the canopy-dwelling species in Costa Rica exploited this habitat. By contrast, brush-frequenting species such as Orange-crowned Warblers, Common Yellowthroats, and Yellow-breasted Chats establish territories in young second-growth forests. Some species, including Magnolia and Wilson's warblers, only move through this habitat, and Rappole et al. (1983) concluded that resources were inadequate to support them on a longer-term basis. Species of the forest floor or its lower reaches do not, however, find conditions in young second-growth forests that resemble those of old-growth forests, and thus usually do not occupy them (Rappole et al. 1983), although Greenberg (pers. comm., 1987) has found that Ovenbirds, Kentucky Warblers, and Wood Thrushes *(Hylocichla mustelina)* exploit young second growth in Panama.

Even though many Yellow-rumped Warblers winter in temperate regions, their summer and winter habitats differ. They are primarily confined to coniferous breeding sites, but they occupy a wide range of habitats during the winter, most of them at least partly deciduous.

Movements during the Winter

Migratory species often move between fixed breeding and wintering grounds, and that migration is confined to a limited period. Records of site tenacity for many species, including warblers, at their wintering sites support this perception, and this pattern might hold for species that move to stable wintering areas like warm tropics not subject to severe fluctuations in moisture. Rappole et al. (1983) listed examples of winter site fidelity for 26 species of warblers, some of which returned to the same site for several years. To test the homing abilities of one of these species, Stewart and Conner (1980) found that Palm Warblers displaced from their wintering sites returned to those sites (from distances of 10–22 kilometers).

In the fall other warblers move shorter distances southward into areas that periodically test their survival ability. Terrill and Ohmart (1984) found that the wintering ranges of western Yellow-rumped (Audubon's) Warblers differed from year to year in Arizona and adjacent Mexico; the birds retreated farther south in years of severe weather. Thus they could exploit the most northerly sites and yet rapidly depart if inclement conditions set in. These birds retain a nocturnal migratory disposition—a restlessness for migration, known as Zugunruhe—well into winter, which corresponds with their tendency to move southward in adverse weather conditions and a correlated difficulty of finding insect prey. Fruits apparently are not important winter foods for this population, as they are for Yellow-rumped Warblers elsewhere.

Extreme changes in warbler density during the winter at low latitudes indicate that major population shifts also take place there. For instance, Russell (1980) found that migratory species (especially Blackpoll Warblers) regularly inhabited arid thornscrub forest in Colombia until the leaves dropped from the trees in November. Willis (1980) reported that migratory species left Barro Colorado Island in October, when the habitat became unsuitable. He proposed that residents and other migrants, presumably affected by deteriorating conditions, attacked these birds and forced them to move on. Incursions of migrants at times long past those of normal fall migration may be part of the same phenomenon. Pearson (1980) noted that in December migratory species moved into the forests of lowland Peru, possibly from the highlands, because a highland resident, the tanager *Tersina viridis,* arrived at the same time as the wintering species. Conversely, Johnson (1980) recorded a rapid increase of migrant density in the uplands of Colombia during late winter, which he associated with a decline in conditions at lower elevations.

Many movements are much shorter. Morton (1980) found that most migrant species in Panama withdrew from the driest areas in late December. Those species, including Bay-breasted and Tennessee warblers, changed from eating insects to eating fruit during the winter. When fruit ripened at the dry sites in March, these warblers moved back to them. Tennessee Warblers, also nectar feeders at times, moved from the Pacific coast of Panama in early December to the Atlantic coast in mid-January, following the blooming of the vine *Combretum fructicosum.* In contrast, Chestnut-sided Warblers in the same locality remained rigidly territorial (Greenberg 1984b).

Wintering warblers vary in their ability to shift habitats and seek new regions. Habitually territorial and insectivorous species, such as the Kentucky Warbler, did not exhibit regional movements (Morton 1980). These species were confined to habitats unlikely to dry out, such as riverine locations, freshwater swamps, moist lowland forests, and mangrove swamps. In other places that migrants remain stationary during winter, temporary resources do not figure prominently into their economy either (Emlen 1980; Rappole and Warner 1980). Emlen reported that fruit and nectar were limited in the pine forests of the Bahamas and Florida, but warblers exploited them when they were available. Omnivorous foliage gleaners such as Bay-breasted Warblers were the most flexible, both in habitat shifts and regional movements.

To sum up, wintering warblers vary widely in character from sedentary to opportunistic. The most northerly winterers move as necessary, but it is becoming clear that many migrants and residents relocate with rainy and dry seasons in strongly seasonal tropical areas. Even within low-latitude areas of modest wet-dry seasonal change, species differ in their tendencies to move or remain as conditions change.

Foraging and Resources

In general, migratory warblers forage in the same way during winter as they do on their summer breeding grounds, though differences in vegetation inevitably dictate some change (Eaton 1953; MacArthur 1958).

Coniferous-nesting species find no conifers south of Nicaragua or montane fir forests south of Guatemala. Black-throated Green Warblers that I observed below the timber line (3,050 meters) in Costa Rica concentrated their activities in the needlelike leaves of Elfin Oaks *(Quercus costaricensis)*, probably the most coniferlike tree in the region (Morse 1980c). These birds foraged primarily in the canopy, well inside the tips of the vegetation, which were monopolized by resident Flame-throated Warblers *(Parula guttaralis)*, a species that foraged much like Northern Parula Warblers and Golden-crowned Kinglets at their summer sites. Rabenold (1978, 1980) reached similar conclusions about Black-throated Green Warblers in nearby western Panama's oak forests at about 2,000 meters, although these birds used different parts of the habitat more than breeding individuals do in Maine and North Carolina. Rabenold reported that their foraging flights were confined to short distances in Panama (less than 1 meter), implying that they do not search for discontinuously rich patches, as they do on their breeding grounds. The foraging differences could be a consequence of either foliage structure or distribution of insect prey.

Hutto (1981a) studied four species of warblers on their Wyoming breeding grounds and their western Mexico wintering grounds. Of six foraging variables, the birds separated from each other most in their absolute and relative foraging heights. Relative height remained constant between seasons, but absolute height varied widely because the height of the two habitats differed. Other variables, such as resources taken and methods used, remained constant between seasons, consistent with the idea that foraging techniques are more difficult to change than locations (MacArthur and Pianka 1966).

The difference in summer and winter foraging of the Yellow-rumped Warbler is linked to winter frugivory, primarily on Bayberry (Wilz and Giampa 1978) and Wax Myrtle. Like other frugivores, they often form flocks, a social system linked to exploiting patchy, temporarily abundant resources (reviewed by Morse 1977a), though they sometimes glean for insects. A combination of sociality and vagility characterizes Yellow-rumped Warblers wintering south through Central America (Emlen 1980; Rappole and Warner 1980). In Panama they sometimes even forage in grasslands (Karr 1971).

Many migratory warblers exhibit a wider range of exploitation patterns—in both foliage categories and height—on their wintering grounds than on their breeding grounds; however, the degree of flexibility differs from species to species. For instance, Black-and-white Warblers forage on large limbs or trunks (Tramer and Kemp 1980) much as they do in North America, whereas other species have intermediate levels of change. The Black-throated Green Warbler, considered a highly stereotyped species on its breeding grounds (Morse 1968; Rabenold 1980), has intermediate flexibility. American Redstarts are quite flexible on both wintering and breeding grounds, shifting within a few hours to totally different patterns of exploitation (Holmes et al. 1978; Bennett 1980).

With the exception of wintering Yellow-rumped Warblers, warblers have traditionally been considered obligate insectivores, yet as noted earlier, results from tropical wintering areas prove that this is a gross oversimplification. Morton (1980) reported striking changes in food items taken by several warbler species in Panama over the winter, from insects during the first (moist) part of the winter, to fruit during the last (dry) part of the winter. Species accomplished this by shifting habitats, as well as parts of the habitat exploited. Morton emphasized the role of fruit or nectar in the yearly cycle of some warblers, and how the importance of this alternative food source differs among species. Few species pass up abundant, easily procured resources, even if they are not adept at handling them (Rappole et al. 1983). Tennessee and Bay-breasted warblers were primarily frugivorous during the dry season: their stomachs were more than two-thirds filled with fruit. Stomachs of Chestnut-sided Warblers had over one-quarter fruit by volume; Prothonotary Warblers, 10 percent; and Golden-winged Warblers, no fruit (Morton 1980). Fruit procurement of Chestnut-sided Warblers varies among individuals during the winter because they are highly territorial and select the sites well before fruiting time (Greenberg 1984b). Greenberg (pers. comm., 1987) has found a strong phylogenetic component to frugivory—widespread among *Dendroica* and *Vermivora* and rare or absent in *Oporornis, Helmitheros, Mniotilta,* and *Wilsonia*.

Morton reported that Tennessee Warblers feed heavily on nectar during the first part of the dry season; indeed, they may be significant pollinators of *Combretum fructicosum*. Their face and throat can be so discolored with the red pollen of *Combretum* that they easily pass for another species. Morton suggested that this "war paint" even aids the defense of territorial sites near these flowers.

Working in Colombia, largely on Blackburnian Warblers, Chipley (1977, 1980) drew different conclusions from Morton. At the beginning and end of the winter, when densities were low, Chipley discovered that the birds exploited their habitat as they did on their northern breeding areas (foraging primarily in the tops of the trees); during the middle of the winter, when their numbers increased, they foraged at all heights, even on the ground. Chipley attributed the change to intraspecific competition in the absence of other similar species. Seldom, at least at low latitudes, do single species of Neotropical migrants become numerous enough to exhibit as strong an effect as the one Chipley reported (see Keast and Morton 1980). Although his study area was in Colombia, beyond the range of the highest migrant densities, migrants made up 50 percent of this community, and Blackburnian Warblers made up 57 percent of the migrants. Chipley hypothesized that the inconsistency of Blackburnian Warbler foraging patterns reported from other tropical areas during the winter (Slud 1960; Buskirk 1972) may be due to densities of conspecifics and potential competitors.

Other warblers take fruit while in the temperate zone, although little is available until late summer or fall. For instance, Cape May Warblers feed so heavily on fleshy fruits during fall migration that they damage grape crops in the eastern United States (summarized by Bent 1953).

Exploitation Systems

Greenberg (1984b) has compared the exploitation systems of two wintering sympatric species: Chestnut-sided and Bay-breasted warblers at Barro Colorado Island in Panama. These two species have strikingly different foraging, sociality, and movement patterns during their time on the wintering grounds.

Chestnut-sided Warblers are highly stereotyped in their foraging heights, concentrate on foliage rather than branches and twigs, and even specialize on undersides of leaves. Further, they concentrate on tree species with planar leaf distributions. In comparison, Bay-breasted Warblers are much less stereotyped in their choice of height and substrate. Although they do not use leaves as often as Chestnut-sided Warblers, they forage over both the top and bottom sides of them. The two species differ in foraging motions because of their feeding sites: the Bay-breasted performs more routine gleaning of the whole branch and the Chestnut-sided more rapid leaping because they feed primarily on the leaves. Bay-breasted Warblers are opportunistic in their feeding; at times they forage on lawns or along roads in response to a flush of insects—actions that Chestnut-sided Warblers do not perform. Both warblers take large amounts of fruit, but this tendency is far more variable in Chestnut-sided Warblers because of their territoriality, as opposed to the much wider ranging Bay-breasted Warblers, which seek out fruit trees over larger areas.

Investigating these two species on their breeding grounds in Maine, Greenberg (1984a) found that their foraging patterns reflected those of the wintering grounds. The Bay-breasted Warbler is a coniferous forager in the summertime, however, and consequently has to exploit its substrate differently from its winter regime of broadleaf foliage in Panama. Greenberg separated warblers into two exploitation types. One, characterized by the Chestnut-sided Warbler, defines the small, highly stereotyped bird that remains within its wintering territory. This type includes other species that are broadleaf (or dense coniferous) foragers on their breeding grounds, such as Yellow, Black-throated Green, and Magnolia warblers. The second type, like the Bay-breasted Warbler, is composed of large, primarily coniferous summer foragers that have more flexibility in foraging and site performance. Yellow-rumped, Cape May, and, provisionally, Blackpoll warblers are in this group.

Although these similarities may hold for the species Greenberg studied on breeding and wintering grounds, the relations of the genus as a whole are more complex. For example, although the Magnolia Warbler has an affinity to dense coniferous habitats on many of its breeding grounds (Morse 1968; Hall 1984), it seems questionable whether its habitat's cover is denser than that of Bay-breasted or Blackpoll warblers. The Black-throated Green Warbler, another coniferous nester in many areas, is sometimes territorial on its wintering grounds (see Bent 1953). Its summer coniferous haunts are probably not denser than those of Bay-breasted and Blackpoll warblers. That some Black-throated Green Warbler populations exploit deciduous foliage extensively (e.g., Holmes 1986) raises the question of whether populations vary in exploitation patterns and also provides a possible test of Greenberg's hypothesis.

If Chestnut-sided Warblers are especially efficient foragers and find particularly favorable microhabitats for exploitation, this could account for the differences Greenberg found between them and Bay-breasted Warblers. He may have uncovered a cost arising from the need to shift foraging substrates during the year. As a roving budworm specialist on its breeding grounds, the Bay-breasted Warbler's tendencies toward vagrancy may be unusually well developed.

Some migratory warblers respond to habitat features that are uncommon or absent in the temperate zone. A prime example is the use of dead leaves or clumps of dead leaves hanging on trees, which likely enhances diversity of bird species in tropical forests (Remsen and Parker 1984). Although resident antwrens and wrens exploit this substrate, migratory warblers use the sites as well (see Greenberg 1987a). Warblers that forage in dead leaves on their wintering grounds include Golden-winged Warblers, Prothonotary Warblers, and Worm-eating Warblers (Morton 1980; Greenberg 1987b), of which the Worm-eating Warbler is the most specialized (Greenberg, pers. comm., 1987).

Dead-leaf clusters are important foraging sites for Worm-eating Warblers. In the laboratory these birds exhibit an inordinate curiosity about the clusters, although by the time they reach their wintering grounds this preference diminishes. Yet they normally encounter clusters; and given the numerous prey likely to be hiding in them, they quickly form a preference for foraging in these sites (Greenberg 1987a). Among the species that Greenberg (1984a) studied, Worm-eating Warblers were unusual in retaining the ability to respond to novel factors. Other species were flexible in their learning only up to 6–8 weeks of age. The Worm-eating Warbler's response could be a consequence of the unavailability of dead-leaf foraging sites during the usual critical period for learning, which would be spent on or near its natal ground. How such differences evolved is unclear, but similar ontogenetic studies of unrelated dead-leaf specialists, such as tropical species that exploit these sites, would be instructive. The Worm-eating Warbler is unusual in its obsession to manipulate substrates, behavior that occupied over 50 percent of the time in the laboratory and that was not extinguished in the absence of rewards. The flexibility could be due to this species' regular shift between live- and dead-leaf foraging on breeding and wintering grounds (Greenberg 1987b).

Territorial and Aggressive Behavior

Until recently, little attention has been paid to the possibility that migratory birds such as warblers hold winter territories, although certain behavioral characteristics of these birds suggest that this trait is widespread. Time and again only one individual of a warbler species is found in a mixed-species foraging flock, a situation resembling that of resident flock members in the tropics, most of which are territorial. Migrants join the flocks only when these groups cross their (the migrants') accustomed sites. Other wintering warblers are completely solitary, shunning both flock members and conspecifics (Bent 1953).

Recent work confirms such impressions of territoriality among migrants and reveals that some individuals remain on a site for as many as 6–7 months (Schwartz 1964, 1980), are aggressive toward intruding conspecifics (Rappole and Warner 1980) and sometimes to individuals of other ecologically similar species (Tramer and Kemp 1980), and reoccupy sites in subsequent winters (Loftin et al. 1966; Diamond and Smith 1973). Schwartz (1980) came to the conclusion that previous territory holders enjoy an advantage in claiming their old sites.

Wandering birds ("floaters") occur in at least some wintering populations, and when a resident is removed one of these birds quickly occupies its place. Of all these studies, however, only Rappole and Warner (1980) and Morton et al. (1987) have tested this notion by removing territorial birds. Rappole and Warner also confirmed that territorial birds only follow a flock while it is in their territory.

The probability that a migrant holds a territory differs with season and locality as well as species. Tramer and Kemp (1980) found that among migratory warblers only the Tennessee Warbler held territories in the highland forests of Costa Rica, although several others are territorial in some lowland areas (Morton 1980); the difference between the two areas in the extent of territorial behavior is probably due to differences in resource availability. Changes in resources appear to account for changes in territoriality of Bay-breasted and Tennessee warblers over the winter (Morton 1980). Bay-breasted Warblers, for instance, arrived on their wintering grounds when insects were few and competitors were many. Under these circumstances they were solitary and territorial, but when nectar and fruit became abundant in midwinter they gave up their territories and foraged in flocks. At this time food was probably superabundant and also concentrated in a small area that would be hard to defend. These birds varied in sociality over distances no greater than a few hundred meters, and they differed over the winter at a single site. In contrast, Greenberg (1984b) reported that Chestnut-sided Warblers were invariably territorial on their wintering grounds, even when Bay-breasted Warblers at the same sites had given up territories.

Both sexes may hold territories, which supports the resource-limitation argument (Rappole et al. 1983), and they may use the same territorial displays (Rappole and Warner 1980). Obligately territorial species retained the same plumage patterns throughout the breeding and nonbreeding seasons and tended to be monomorphic, although some species, including the Chestnut-sided Warbler, do not fit this pattern (Morton 1980). Both Black-and-white Warblers and Ovenbirds occasionally sang, although not as frequently as Least Flycatchers, Yellow-bellied Flycatchers *(Empidonax flaviventris),* and White-eyed Vireos *(Vireo griseus)*. Rappole and Warner (1980) associated these songs with territorial defense. The warblers usually gave chip notes, however. Rappole and Warner suspected, but presented no evidence, that both males and females of the singing species sang.

Aggression rates of many wintering warblers are low (Chipley 1977; Post 1978), considerably below those on their breeding grounds and in the north during late summer (Morse 1970a). In some species overt aggression is almost lacking at this season (Tramer and Kemp 1980). Nevertheless, hostile interactions were observed on rare occasions, especially when a wandering individual violated a territorial boundary. Rappole and Warner (1980) used live decoys of some species (in net cages) to increase the frequency of these interactions enough to permit quantitative study. This technique worked well with the Hooded Warbler, but in some species it was not successful. It was unclear why these stakeouts did not work for species that were aggressive toward live intruders. Most observations involved intraspecific aggression (Rappole et al. 1983).

Contrasted with this behavior is that of Cape May and Palm warblers, which aggressively defend localized point sources when century plants *(Agave braceana)* come into bloom in the Bahamas (Emlen 1973). These plants are widely spaced and offer abundant sources of nectar and insects—contested by several species of wintering warblers as well as by resident hummingbirds and Bananaquits. One Cape May Warbler spent 90 percent of its time over nearly an hour guarding the plant and attacking other birds, and only 10 percent feeding, which Emlen suggested might be maladaptive, especially since insect stores in the vicinity were high. Without information on food intake and energy expenditure it is impossible to determine whether these birds maintained a positive energy budget. Emlen suggested that the birds encounter point sources so infrequently that they do not respond appropriately. Since they were not marked, it was not possible to determine detailed time budgets; however, nearly a bird a minute approached the century plant that Emlen observed. Kale (1967) noted a migrant Cape May Warbler in the Dry Tortugas (probably the same bird) that spent three days repeatedly attacking and repelling warblers of several species that attempted to visit a century plant inflorescence. Hence, such behavior may be important at times.

Pine Warblers wintering in the southern United States are also highly aggressive (Morse 1967c, 1970a). They attack not only conspecifics but also ecologically convergent Brown-headed Nuthatches *(Sitta pusilla)* (Morse 1967c) and Yellow-throated Warblers (Gaddis 1983) when they are on the same wintering grounds. It would be worthwhile to determine whether Pine Warblers have a despotic system like that of the Harris' Sparrow *(Zonotrichia querula)*, whose success depends on suppressing other individuals as much as possible (Rohwer 1975). During winter, at least, they vary in (yellow) color enough to make a badge-signaling system feasible.

Participation in Flocks

Flocking is a widespread activity of warblers outside the breeding season. Flocks vary greatly in size and composition, and the tendency of warblers to flock varies with habitat (Tramer and Kemp 1980). Both temperate- and tropical-wintering species may participate in these groups, although some warblers show little predisposition to join them (Stiles 1980). Nonparticipants include species that are permanently territorial (waterthrushes, Ovenbirds, Hooded Warblers), although other species that hold territories in their winter habitats temporarily join flocks that move across their territories. The flocks are composed of a single or many species, migrants alone or migrants and residents mixed. Species may be intraspecifically solitary or gregarious.

In the temperate zone Yellow-rumped Warblers often form unispecific flocks at the northern edge of their winter range, which may be partly a consequence of their dependence on patches of Bayberries (Hausman 1927). Farther south, Yellow-rumped Warblers often participate in the mixed-species flocks that form about chickadees and titmice, but the more Yellow-rumped Warblers there are, the greater the probability that they will split from other flock members and move on their own as single-species flocks. Other warblers in the same areas, such as the Pine Warbler, may be gregarious in flocks, although often aggressive toward the other members of the flock. In contrast, Orange-crowned Warblers occur one to a flock and are probably territorial (Morse 1967c, 1970a).

In tropical areas Tennessee Warblers are at times gregarious and form one-species flocks when hunting insects (Tramer and Kemp 1980), but they are aggressive and territorial when feeding on floral nectar (Morton 1980). This difference is probably due to the spatial patterning and consequent defendability of the resources. Other species, such as Bay-breasted Warblers, are sometimes intraspecifically social and participate in flocks led by resident species such as antwrens (Hespenheide 1980; Greenberg 1984b); at other times they are territorial (Morton 1980).

By joining a flock migrants may both improve feeding opportunities and reduce vulnerability to predation. Hespenheide (1980) noted that most migrants joined flocks when insects were scarcest, which is consistent with the observation that flocking enhances foraging efficiency under extreme conditions, perhaps whenever efficiency falls below that realized by defending a territory. Participating in flocks with its conspecifics probably improves the probability that a Yellow-rumped Warbler will find berries (see Ward and Zahavi 1973). Bay-breasted Warblers were similarly social when feeding on fruit (Morton 1980), but they also participated in antwren flocks (Hespenheide 1980). By contrast, Powell (1980) reported that species vulnerable to predation are most likely to join flocks. Small hawks, residents and wintering migrants, are common in tropical forests, although we have no more information on the impact of these predators in the tropics than in temperate zones.

The advantage of temporary participation by territorial birds in flocks is hard to explain since they presumably know their sites well and would not greatly benefit in either finding food or avoiding predators. Even so, perhaps flocking does enhance a bird's ability to hunt and evade. Study of flocking behavior might provide a fresh perspective on variety in sociality and use of space.

Interactions between Wintering and Resident Birds

The evidence that substantial numbers of migrants occupy primary forest, that more permanent residents occupy second-growth forest than was once thought, and that migrants are not always subordinate to residents when they meet indicates that relations between migrants and residents are more complex than was initially believed. In some areas, such as the Colombian highlands (Johnson 1980), residents and migrants separate as Willis (1966) noted; in other areas they do not. Although residents appeared to affect the migrants' feeding patterns, the converse was not true. Elsewhere in Colombia, Hilty (1980) noted that migrants were almost always subordinate to residents at fruiting trees and that they, too, probably had minimal impact on the residents.

Resident and migrant birds on Barro Colorado Island in Panama may segregate. Since migrant small-insect feeders there are larger than residents, Willis (1980) suggested that they might enjoy a competitive edge.

There is little evidence, however, that these residents or migrants shift their foraging patterns in response to each other (Willis 1980; Tramer and Kemp 1980): interactions are far less frequent there than over ant swarms (Rappole et al. 1983). It is possible that migrants and year-round residents compete only when transient as well as wintering individuals are present, resulting in the highest densities (see Emlen 1980). Residents and migrants may then cause other migrants to move to less saturated areas (Willis 1980), but evidence is lacking. Negligible aggressive behavior between migrants and permanent residents is what one would expect if they pursue distinct life-styles.

Direct correlations between numbers of residents and migrants in some areas, such as the Bahamas (Emlen 1980) or Campeche (Waide 1980), also suggest that population densities are influenced by something other than competition. In several instances, residents show no sign of filling niches left vacant after the migrants leave (e.g., Rappole and Warner 1980).

Residents sometimes alter their feeding patterns after migrants depart, however. Miller (1963) proposed that migratory species might depress the insect food supply in the Colombian highlands enough to shift the residents' breeding seasons. On Jamaica, the resident Mangrove Warbler is confined to mangroves when migrants are present but expands into surrounding habitats when the migrants move on (Lack and Lack 1972). This and other West Indian insectivorous birds breed immediately after the migrants leave. The timing could be coincidental, but it is consistent with the hypothesis that species adjust their breeding seasons to the departure of the migrants.

Resident and migratory West Indian parulines may truncate each other's distributions in some areas. On the Greater Antilles, which support numerous migrants, the resident warblers are species of the edge or ecotone. The residents are not so confined in the Lesser Antilles, which support few migrants. Since the number of resident species per habitat is similar in both the Greater Antilles and Lesser Antilles, this effect is not a mere artifact of differences in diversity. Faaborg et al. (1984) also noted that if these West Indian resident species cannot breed because of drought, their numbers decline and the numbers of winter migrants subsequently increase, the migrants filling sites normally taken by residents. Cape May and Prothonotary warblers wintered in their Puerto Rican study areas only in drought years (Faaborg et al. 1984).

A transect study by Stiles (1980) in Costa Rica from sea level to 3,500 meters emphasized the importance of considering oscine and suboscine distributional patterns over altitudinal or geographic gradients for understanding resident-migrant relations. Oscines are more common in uplands than in lowland forests. In the uplands both migrants and residents belong to temperate-zone families, but that is not the case in lowland wet forest faunas. With altitude, resident and migrant oscines decrease in the same proportion, indicating that they constitute a single integrated fauna. No forest-interior migrants occurred over 1,500 meters in Stiles's transect, and by this height many of the suboscines had dropped out. Stiles speculated that substitution of oscines for suboscines at higher altitudes has resulted in resident oscines exploiting forest-interior habitats of higher elevations. Permanent-resident oscines may be more similar to migrants in resource-exploitation patterns than are suboscines of lower altitudes, and as a result they perhaps exclude migrants from this upland habitat.

Resident and migrant warblers sort out mostly by habitat and feeding zone, but some geographic separation also occurs. In the West Indies residents concentrate in thickets and undergrowth, and migrants are largely foliage gleaners; in Central America resident warblers concentrate in the undergrowth and the air column, while migrants are predominantly foliage gleaners. Bill and body size play little role in the ecological separation of resident and migrant species. Although this broad pattern holds in most instances, the separation between migrants and residents is not complete—migrants include some brush and ground feeders as well as aerial foragers.

Broad patterns of allopatry are also to be seen between migrant warblers and ecologically similar residents. For instance, American Redstarts are often separated geographically or by habitat from the behaviorally similar tropical redstarts (*Myioborus* spp.), and resident Yellow (Mangrove and Golden) Warblers occupy mangroves when migrant Yellow Warblers are present in other habitats. On a larger scale, Keast (1972, 1980b) found an inverse relation between the abundance of parulines and their ecological counterparts, the small tyrannids, throughout most of Central America. Keast (1980c) further argued that these tyrannids decrease the opportunities of warblers in low-latitude tropics (see Chapter 15). Greenberg (pers. comm., 1987) noted that small tanagers, antwrens, and antbirds may be as strong ecological counterparts to parulines as are the small tyrannids.

Migrants form more of the bird fauna in the northern than in the southern Neotropics. Percentages of migrants decline as one moves southward (Keast 1980c; Terborgh 1980) (Figure 10.1), and their impor-

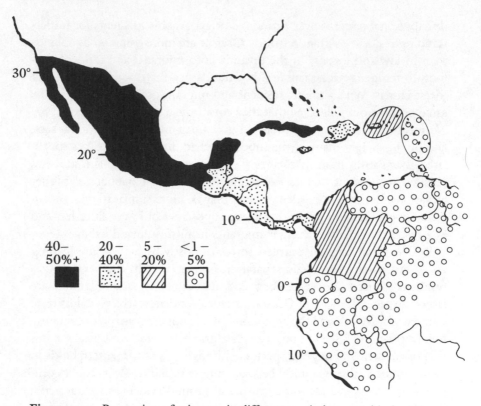

Figure 10.1 Proportion of migrants in different tropical geographical areas during the winter. (Data from Terborgh 1980.)

tance must change for this reason alone. In the more southerly parts of the migrants' winter ranges, as in Peru (Pearson 1980) and Colombia (Hilty 1980; Johnson 1980; Orejuela et al. 1980), migrants concentrate in second growth and in uplands, as was reported for migrants in general in earlier papers. These patterns differ considerably from Rappole and Warner's (1980) and Waide's (1980) conclusion that one cannot generalize to say that migrants concentrate in highlands, use seasonally abundant food, or capitalize on patchy resources more than residents. The reason for this difference is obscure, though Terborgh (1980) tentatively related it to the food available at different altitudes.

Given that migrants penetrate tropical communities, the question of why residents do not usually exhibit ecological release in their absence is an enigma. Hutto (1980) cited the "mental conservatism" (stereotypy) of resident tropical species, which has been noted by several workers, such as Terborgh and Weske (1969) and Lack (1971). If stereotypy is high in tropical resident species, what is the basis for it?

Stiles (1980) argued that the timing of the departure and return of the migrants precludes invasion of their niches by residents. Migrants are absent when resources are most abundant, so resident young that are just learning to forage for themselves will confront numerous returning migrants, both birds in passage and winter residents, if they invade the migrants' niches. Stiles noted that migrants depart Costa Rica around the beginning of the rainy season, when resources for insectivores and small frugivores will increase, becoming more than adequate for reproduction without any expansion of their foraging repertoires. The return of migrants when young are learning to forage for themselves may cause them to specialize on the activities they can perform best, or differently from the migrants, which is also what their parents do. Stiles pointed out that numbers of migrants during fall migration greatly exceed those that remain for winter. Thus, residents would be exposed to an especially high level of training at an early point in their lives, which would exaggerate any conservatism.

This is an attractive hypothesis, especially since it is one in which the migrants play a continuing part. It is consistent with the results of Klopfer's (1965, 1967) laboratory experiments that tropical residents, even oscines, are more stereotyped than their temperate-zone counterparts, and that much of this difference is a consequence of early learning. The hypothesis is also consistent with information on resource availability in Costa Rica (Buskirk and Buskirk 1976; Janzen 1980), but it is without direct corroboration. Greenberg (pers. comm., 1987) noted that some tropical residents, such as antwrens, are more specialized than the Bananaquits and tanagers studied by Klopfer, and that early learning plays less of a role in their stereotyped behavioral patterns. Many antwrens mature and disperse long before the first migrants arrive in late summer.

To evaluate the effect of migrants on residents' behavioral ontogeny we need data on the behavior of color-banded resident juveniles immediately before the return of migrants, during the period of maximum influx of migrants, and subsequent to that period. The hypothesis is most easily tested where migrant influx is high.

Alternatively, residents may fail to exploit prime wintering habitats because large insects are scarce there, even though these sites contain an abundance of small insects (Greenberg 1986). Residents seek large prey in rearing their young. Comparisons of insect size distributions in key habitats would be the first step in evaluating this hypothesis, which could be mediated through high levels of stereotypy.

Rappole and Warner (1980) argued that residents attempting to expand into migrants' places in the migrants' absence would have to compete in two niches upon the migrants' return—their own plus that of the migrants. Yet migrants are also forced to exploit a second niche—the other being their breeding grounds. Given the difficulties that migrants are likely to experience during the rest of the year, both on their breeding grounds and in migration, it is difficult to imagine that they would be better adapted for this task than year-round tropical residents that might simply be attempting to exploit two adjacent niches. If the residents expanded their niche during their greatest resource demands—during the breeding season, for example, as in the Mangrove Warblers studied by Lack and Lack (1972) and the resident community studied by Chipley (1980) in Colombia—they might be exploiting an important opportunity only at times of major need. Rappole and Warner's (1980) explanation is an *a posteriori* attempt to explain a counterintuitive phenomenon. Of course, this only heightens the mystery!

Hutto (1980) suggested that nest-predation pressures account for the failure of residents to invade habitats in western Mexico used primarily by migrants during the winter. Spacing out to avoid predation could contribute to limiting resident densities but does not totally account for the drastic decrease from winter to spring of 64 to 2 individuals per hectare that he reported. Since insect prey abundance increased during the period, the state of the food supply does not explain this remarkable anomaly.

Even if Central America is a region of plenty while migrants are gone, such an abundance of resources is not the rule throughout the tropical wintering grounds. For instance, this pattern does not hold in Peru (Pearson 1980), where the dry and rainy season are mirror images of the seasons in Central America. From the preceding arguments one can predict that the reproductive periods of species resident in Peru will be timed differently in relation to the migrants. The periods when migrants are absent would not be a time of superabundant food, so either the resident species should move into the migrants' niches after their departure or reproduction should occur while migrants are present. If permanent residents breed when migrants are present, this could help to explain the low density of migrants in South American forests (unless resources are superabundant).

In interpreting differences in responses of resident birds following the departure of migrants, it is important to separate instances of potential competitive release into those in which the residents gain access to major new resources not present when the migrants are present, and those in which resources do not change. Buskirk and Buskirk (1976) and Janzen (1980) referred to flushes of insects that become available shortly after migrants leave. Although no one has attempted to correlate insect gradations with shifts in the distribution of permanent residents, they may be examples of shifts that are independent of competitive release. Waide (1981) reported shifts in the foraging behavior of dry-forest residents in Campeche, Mexico, that coincide with the departure and return of migratory species; however, in only one instance did shifts measurably affect niche overlap between the two components, thereby refuting the argument that these shifts are affected by migrants.

Predation Avoidance on the Wintering Grounds

Flocking should be especially advantageous for avoiding predators for any individual that has just settled in an area (Stiles 1980), because it may not know where hiding sites lie or what the local complement of predators may be. Maintaining a territory or a home range, though, should confer the same advantage once an area has been learned. By holding a territory more than one winter, the period of vulnerability might be minimized. Since some migrants occupy the same wintering areas in successive years (Diamond and Smith 1973; Loftin 1977) the benefit of avoiding predators in this way might be substantial.

Yet, no quantitative data exist to demonstrate that new migrant inhabitants and residents experience different risks from predators. Small hawks (*Accipiter, Micrastur,* etc.) are considered major predators of small birds in Neotropical forests, although snakes are more often the cause of mortality than they are at temperate latitudes (Skutch 1976).

Data on the ratio of migrants to residents taken by predators, such as might be gathered on plucking posts, or nests, would clarify whether small hawks are a great danger to small passerine birds in the Neotropics. Ratios taken from the beginning to the end of migrants' yearly tenure on their wintering grounds would illuminate whether increasing familiarity with an area and changing tendencies to participate in flocks affect predation levels. Such data would be especially valuable if taken in conjunction with censuses of bird populations.

The Effect of Physical Factors on Wintering Populations

Low temperature and related factors have a major impact on birds wintering at high latitudes (Lack 1954; Fretwell 1972), and for that reason probably strongly influence the population dynamics of species like the Yellow-rumped Warbler. Temperature may be significant at lower latitudes, also, and a second physical factor, moisture, may figure in there as well. A third factor, distance to closest wintering area, further affects migrant density on wintering grounds.

Species wintering at high latitudes are constantly subject to the effects of unpredictable winter climate (Morse 1980a). Yellow-rumped Warblers that winter in areas of occasional high snows may even find much of their food covered at times. In spite of the importance of such phenomena in setting the northern limits of wintering populations, little of a precise nature is known about the effect of severe winter weather on these birds. Weather-related population decreases could signify either local mortality or dispersal, and Terrill and Ohmart's (1984) data on western Yellow-rumped Warblers suggest that facultative dispersal may occur under such circumstances.

Pine Warblers winter throughout the southern pine forests in the United States and thus have a more northerly winter distribution than most of the other parulines, except the Yellow-rumped Warbler. They feed heavily on insects during the winter, although they take other food, such as pine seeds and fruits (Bent 1953; Morse 1970a), if available. They are vulnerable to unseasonably cold winter weather, however, and suffered heavy mortality during the unusually cold winters of 1976 and 1977 (Robbins et al. 1986). The loss was corroborated by lowered breeding densities in subsequent years. This disaster indicates clearly the possible costs of northerly wintering.

In addition to species that winter at high latitudes, other warblers are exposed to severe weather because they fail to move southward when their conspecifics do, only to succumb to normal winter conditions or to survive at feeders that provide exotic food supplies (mealworms, citrus fruit, and the like). These stragglers are unlikely to be of evolutionary significance to the populations of which they are a part. Normally they would be culled from the population as a result of their abnormal behavior. Occasionally, however, this type of behavior could be of major importance. Given the tropical origin of warblers, and the return to the tropics of most species in the winter season, it is possible that northerly-wintering populations of Yellow-rumped Warblers are descended from individuals that had this type of abnormal wintering habit. Diamond (1982) has suggested that some western-wintering populations of eastern Yellow-rumped Warblers may be the result of abnormal migration.

Although temperature has frequently been mentioned as an important variable affecting the suitability of northern areas as wintering grounds, few references discuss its effect on migrants wintering at lower latitudes. The oft-reported tendency for migrants to frequent highlands holds up in the southerly parts of the wintering range (Costa Rica, Panama, and South America), but not in higher-latitude tropics and subtropics, where no correlation occurs in moist areas (Terborgh 1980). In fact, in western Mexico, Hutto (1980) reported many fewer migrants in the uplands than in the lowlands, which led Terborgh (1980) to propose that winter temperatures fall low enough in the uplands to depress insect activity.

The Neotropical region contains areas arid enough to exert a physical effect on individuals living there, and not surprisingly these regions usually have few migrants wintering in them (Russell 1980). A more important factor is the fluctuation of wet and dry seasons, which dictates the availability of both insects (Wolda 1978; Hespenheide 1980) and fruit (Morton 1980).

Even if the regular interposition of wet and dry seasons is the most important moisture-related variable facing migrants in the Neotropics, unusual periods of drought exert an effect, especially in marginally acceptable areas. Orejuela et al. (1980) reported that during an unusually severe dry season in Colombia some wintering migrants did not put on the fat that they normally would accumulate prior to their northward migration. If such stressful conditions were widespread they would have a severe effect on entire populations of these birds. Some species initially in the study area may have avoided this problem by leaving the site before conditions became severe.

The Whitethroat *(Sylvia communis),* an Old World sylviine warbler, probably provides the best example of the impact of unusual drought on the wintering ground. Whitethroats breed throughout most of Europe and winter in the vast arid Sahel region, south of the Sahara Desert of Africa. During a period of severe drought in the Sahel during the 1960s and 1970s, birds on the breeding grounds declined precipitously (Winstanley et al. 1974), only rebounding somewhat after the drought temporarily ameliorated (Batten and Marchant 1977). Climate-mediated population fluctuations of this sort are not rare on wintering grounds, but often they can be documented only when they fortuitously happen in the midst of ongoing studies.

Unusual wetness can produce strong effects too. The winter of 1980–81 was the wettest on record (57 years of data) in the Panama Canal area, with substantial rains falling during the usual dry season. The rain inhibited the activity of the pollinators of fruiting trees, with the result that fruit production was low and late. Frugivorous migratory species, including Bay-breasted Warblers, Catbirds, and Wood Thrushes, that normally entered the area in large numbers never appeared (Martin and Karr 1986).

A third factor logically included under this heading is the effect of distance on the numbers of individuals that winter in an area. Although selective pressure should favor individuals that move far enough to escape severe weather, additional travel only increases energetic costs and dangers (Terborgh and Faaborg 1980). Thus, it is not surprising that the proportions of wintering migrants drop from 50 percent or more of the avifauna in Florida, some parts of Mexico, and the Greater Antilles to less than one to a few percent in equatorial South America (Terborgh 1980) (Figure 10.1). Combined with the tremendous distances required to reach the equatorial zone is a diversity of resident species, some of which are ecologically similar to the migrants (Keast 1980b). Equatorial South America has the richest avifauna in the world (Pearson 1980) and large numbers of putative warbler equivalents (Keast 1980c). Terborgh and Faaborg (1980) reported a comparable decline of migrants through the West Indies. Although the Greater Antilles, nearest to the northern mainland source, have high proportions of migrants, their number declines to one percent or less on the most distant of the Lesser Antilles. Even though the Lesser Antillean islands have fewer species than the Greater Antillean islands, within-habitat population densities and species numbers on the Lesser Antilles do not differ significantly from those on the Greater Antilles. The distance effect appears valid, then, although it is perhaps exaggerated by the small land areas of the Lesser Antilles (Terborgh and Faaborg 1980).

Diversity and Partitioning of Wintering Warblers

Many species that breed together are allopatric on their wintering grounds; however, MacArthur (1958) reported that the number of winter overlaps (half or more of wintering ranges in common) of spruce-woods warblers and other northeastern breeding species were about what one would predict by chance. He thus concluded that the winter community was randomly composed in relation to the summer community. Chipley (1980) noted that MacArthur's assignments of species to allopatric or sympatric categories were conservative and concluded that the frequency of sympatry was lower than would be predicted by chance (Figure 10.2). In fact, each of MacArthur's five study species has a different center of winter abundance (Hutto 1985a). Greenberg (1986) has also noted that 75,000-square-kilometer blocks in the eastern United States routinely encompass 8–11 breeding *Dendroica* species. A similar measure of the winter grounds

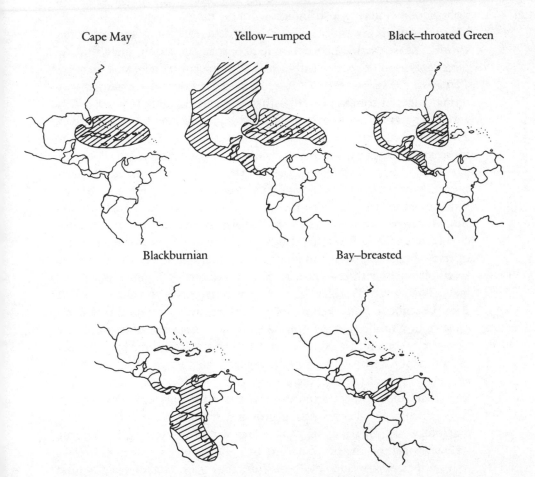

Figure 10.2 Distribution of wintering spruce-woods warblers (cross-hatched areas). (Data from AOU 1983.)

yields no more than 5–7 *Dendroica,* despite their small wintering areas. If sympatry is low during the winter, these species may be avoiding each other (or have done so historically), which suggests that competition played an important role in setting their ranges in the past (Keast 1980b,c). This distributional pattern does not establish the role of competition but does satisfy the demands of a null model, which many distributions attributed to competition fail to do (Connor and Simberloff 1979, 1984). In that the criteria of null models have been criticized as overly conservative (Gilpin and Diamond 1984a,b), rejection of the null hypothesis (Chipley 1980) would be an especially strong statement about the nonrandom wintering distributions of these birds.

Local wintering warbler faunas are not randomly composed taxonomic entities (Keast 1980b). In addition to allopatric separation, there is altitudinal separation of species within several of the largest genera: *Dendroica, Vermivora, Oporornis, and Wilsonia.* These species often have narrow wintering ranges in comparison to warbler species belonging to small genera and having few migratory equivalents (e.g., Black-and-white Warbler, American Redstart, waterthrushes) (Chipley 1980). Two exceptions to intrageneric allopatry perhaps explain why this relation is still not more prevalent. When the similar Blue-winged and Golden-winged warblers, normally allopatric on their wintering grounds, come together, they respond more strongly to each other than to resident species (Morton 1980). Black-throated Green and Townsend's warblers, two other similar species, coexist in the Costa Rican highlands. The Townsend's Warbler is the only interspecifically aggressive migrant species there and may have been responsible for how the two species partition resources (Tramer and Kemp 1982). Townsend's Warblers have only recently begun to winter in Costa Rica (Greenberg, pers. comm., 1987). Further, the two species frequently separate by altitude (Keast 1980b). Segregation of species among habitats, however, is usually less extensive on wintering than on breeding grounds (Keast 1980b; Terborgh and Faaborg 1980), and although a few species are habitat specialists (e.g., the Prothonotary Warbler and Common Yellowthroat), most are opportunists in winter. Thus, although migrant warblers separate on their wintering grounds, segregation occurs at a geographical scale. In contrast, habitat and within-habitat segregation are well developed on the breeding grounds (Keast 1980b).

Nevertheless, considerable ecological segregation takes place within habitats among migrant warblers that winter on Hispaniola (Terborgh and Faaborg 1980). Highland pine-forest and lowland rainforest communities both consist of seven warbler species, with no overlap of species in the two areas. This diversity exceeds that of the resident insectivorous birds. Neither of these two assemblages represents an integrated complex on the breeding grounds; in fact, both lowland and highland groups contain northern and southern (North American) breeders (Terborgh and Faaborg 1980). For this season, Terborgh and Faaborg maintain that the members have been selectively screened. According to this argument, members from a large pool of species succeed because their features complement those of other members. This argument is consistent with the observation that more species are occasionally found in these regions than are represented in the assemblages. For instance, Lack and Lack (1972) noted that in addition to the 18 warblers observed wintering in Jamaica, as many as 12 others were regular but uncommon. They concluded that there was no shortage of participants and that the remaining 12 species were ecologically excluded. Six of these species winter commonly on the nearest Central American mainland, Honduras. And Puerto Rico, a similar-sized West Indian island, also has 18 wintering species of warblers (Post 1978). Terborgh and Faaborg (1980) noted that less common species in Hispaniola mainly have wintering ranges in Central America, and also that small numbers of several common West Indian–wintering species are found in Central America, though not common there. These species may assume a greater role if the status of common wintering species changes.

Hutto (1980) reported a group of wintering migrant warblers and other species in western Mexico that does not differ in composition among habitats. Sympatry of these species is higher in Mexico than on western breeding grounds, in contrast to what was reported for other areas by Chipley, Keast, and others (Figure 10.3). The difference could be due to the limited geographic area in which western warblers winter (mainly west of the mountains) and to their low species diversity and within-habitat sympatry on the breeding grounds, rather than to selective screening on the wintering grounds. Further comparison of the western wintering community with others would be profitable.

Although many species of warblers are less specific in habitat and microhabitat selection on their wintering grounds, they may differ in their exploitation patterns, as Greenberg (1984b) noted in his study of Chestnut-sided and Bay-breasted warblers. Rappole and Warner (1980) provided two illustrations of this point. Although intraspecifically territorial, Kentucky Warblers and Ovenbirds are not interspecifically territorial, even though both feed on or near the forest floor. Kentucky Warblers

Figure 10.3 Summer and winter overlap of eastern and western warblers. Geographic centers of distribution are depicted for breeding (black circles) and wintering (white circles) populations of western species and breeding (black triangles) and wintering (white triangles) populations of eastern species. Species with ranges depicted in the Gulf of Mexico and Caribbean Sea have wintering populations both in Mexico and Central America and in the West Indies. (Modified from Hutto 1985a.)

feed mainly by gleaning insects from undersides of live leaves above the forest floor, while Ovenbirds invariably take their prey from the substrate, which Kentucky Warblers do in less than 10 percent of their foraging maneuvers. In a similar manner, Magnolia and Wilson's warblers, intra-specifically intolerant gleaners of the upper, outer tree canopy, are tolerant of each other's presence. Magnolia Warblers search below the horizontal plane on undersides of overhanging leaves; Wilson's above the plane, capturing most of their prey in flight. The behavioral basis for selective tolerance deserves attention, for it could enhance community diversity.

Implicit in this discussion is that interactions are more probable among migrants than between them and residents (Schwartz 1980; Willis 1980). If winter communities are composed by selective screening, what role does such screening play in forming communities on the breeding grounds, especially given the importance of the wintering season to a bird's biology? Selective screening would favor specialization more than the formation of random assemblages, although not as much as would be expected if an entire complex summered and wintered together. But even if inter-specific interactions are of prime importance, responses to conditions unlike those on breeding grounds could have a compensating aspect: they may select for characteristics conferring high survival value during migration.

Factors That Determine Wintering Distributions

Migratory warblers are a highly successful group on their wintering grounds, not "second-class citizens" that merely insinuate themselves about the tropical residents as openings occur. Residents' frequent failure to fill the migrants' niches when they leave for the north indicates that migrants play an unique role in their communities. In Central America and the West Indies, where wintering warblers have been most studied, migrants leave for the north during or just preceding a time of plenty, such that resident populations experience minimal resource pressure and therefore cannot expand their areas of activity in a major way, even though they usually breed at this time. If migrants return as conditions are becoming more difficult, their spaces are not likely to be taken and so they have no difficulty in reclaiming them. Arriving when resident young are establishing themselves, migrants can prevent the young from exploiting their niches. Alternatively, insects that become available in winter may not be satisfactory for breeding.

Although this seasonal pattern holds over much of Central America and the West Indies, climate differs with latitude and does not fit the pattern for the northern Neotropics in parts of tropical South America. Therefore, it is important to evaluate the status of migrants and residents in the latter areas as well. One might predict that migrants would have greater difficulty establishing themselves to the south than where the season of plenty coincides with the absence of migrants. That factor by itself could contribute to the decrease in migrant populations wintering farther south, even though increased distance and suboscine diversity (Keast 1972) could account for there being fewer migrants to the south. Smaller seasonal fluctuations in productivity in low-latitude areas of South America could decrease opportunities for temporary residents, and perhaps affect their ability to coexist with year-round residents. More information on relations between residents and migrants in those latitudes would aid our understanding of the generality of relationships reported to the north. Another key part of the picture will be dispersal patterns of migrants within the tropics, as they relate to changes in productivity during their stay. Several workers have documented such movements, but they need more systematic evaluation.

The impact of migrants on other migrants presents a related question. Even if migrants do not overlap with residents in their activities, they still might interact with each other, which implies limits to diversification. Many species with similar foraging patterns are allopatric during winter, even if sympatric during the breeding season; this suggests the potential for interactions if the species were together. Species in large genera have restricted ranges and high densities, and those with few close relatives have broad wintering ranges and low densities, a pattern consistent with present or past exclusion of high-density species by each other. The selective screening that Terborgh and Faaborg (1980) saw as an influence on migrant warbler species' abundance in the West Indies may be central in the determination of the abundance and diversity of the birds elsewhere as well. This possibility would be consistent with Lack and Lack's (1972) arguments that pools of migrant species can easily saturate habitats and that ecological constraints determine which and how many species coexist within the community.

Why don't warblers winter farther north and avoid long migrations? A limited trend in that direction has occurred, but there are few signs that birds breeding at high latitudes follow the trend. Being mainly gleaners of mobile insects, warblers would have to alter their foraging in winter if they remained too far north. Feeding on fruit is an alternative, as is clear from studies in low-latitude wintering grounds. Wintertime fruit is scarce at high latitudes, however, and Yellow-rumped Warblers may have fully exploited such resources already. Other occasional fruits are heavily exploited by nomadic cardueline finches during northern winters, and therefore are not readily available to warblers.

11   Rare Species

European colonization of North America has probably affected population sizes of most native birds, but it has been especially marked in some warbler species. Certain species are also rare as a consequence of extreme habitat specialization, and presumably their numbers were small prior to human disturbance. This rarity may be the hallmark of species well on their way to natural extinction, perhaps as a consequence of the receding of Pleistocene glaciation and changes in climate and vegetation. Species that are common today but were rare in the recent past may also be extreme specialists, the only difference being that human modifications to the environment have greatly increased their access to favorable conditions. In this chapter I present brief case histories of several species that are or formerly were rare. My intent is to elucidate the basis for rarity.

Rare North American warblers can be divided into two groups: local endemics and species with a wide range but low density. The latter can naturally change into the former, but some species also attain high densities within a restricted area. Among the former group I survey Kirtland's and Golden-cheeked warblers. An example of the latter group is the Bachman's Warbler, which occupied damp forests in the southern United States before declining in number over the past century. I will also discuss another species of the same region, Swainson's Warbler, which chooses habitats similar to Bachman's Warbler but has not declined drastically over the same period.

Kirtland's Warbler

The Kirtland's Warbler has probably been decreasing in numbers for hundreds or thousands of years; it may properly be said to be a "relict" or senescent species. Ironically, as a result of its endangered state, we know more about it than most other warblers. Kirtland's Warblers have certain traits characteristic of endangered species: their habitat specifications are circumscribed, at least in the breeding season, and they occupy a limited range within their chosen habitat type. Hatching success of undisturbed eggs is 85 percent (Mayfield 1960); for comparison, Prairie Warblers have a success of 97 percent (Nolan 1978). Kirtland's Warblers have recently experienced nest parasitism from Brown-headed Cowbirds (up to 70–75 percent: Walkinshaw 1983).

The reason for this species' rareness in the first place, which almost certainly preceded the cowbird menace, is puzzling. The bird is confined to Jack Pine forests, though sometimes it nests in Red Pine *(Pinus resinosa)* plantations within its limited breeding range of a few counties in Michigan (northern part of the lower peninsula). It breeds only within the small southern isolate of Jack Pine in Michigan, perhaps 200,000 hectares in area, separated from other stands by Lakes Michigan and Huron. It is not yet clear how or if this locality differs from all others; certainly it has not supported Jack Pines more than 6,000–8,000 years, when the pines followed the receding glacier northward. Within its range Kirtland's Warbler is largely confined to stands of greater than 32 hectares (80 acres); it occupies trees 6–22 years of age in areas that have burned recently, with heights of 1.3–6.0 meters. It is further confined to forests growing on Grayling sand, a particularly porous soil, which enables the warbler to recess its nests in the ground, a trait unusual in this genus. As a result of these and other constraints, no more than 10 percent of the Jack Pine area in Michigan may be satisfactory habitat. Colonies on plots smaller than 32 hectares are usually unsuccessful; more often than not, their inhabitants are unmated males, apparently unable to find or to attract females to these sites. This pattern of females being more discriminating in site choice than males is widespread among passerine birds (Morse 1985). Areas of young pines larger than 32 hectares were formerly characteristic of northern lower Michigan because of the wide sweep of level land, which results in large fires because there are no natural boundaries to stop them. It is unlikely that northern lower Michigan is unique in this regard.

Mayfield (1960) and Walkinshaw (1983) pointed out that an important aspect of a young Jack Pine stand after burning is that no other species are characteristic of it. This habitat is marginal for the relatively few species found with Kirtland's Warblers, suggesting that Kirtland's Warblers have been unable to compete successfully for space in other kinds of habitats. If so, this status is not new, in that their habitat selection has become so finely honed. Mayfield suggested that areas less than 32 hectares may be so small that excessive competition and predation infringes from adjacent areas, making them untenable sites. This prediction historically precedes arguments about the consequence of forest fragmentation on the species diversity of Neotropical migrants in small plots (see Whitcomb et al. 1981; Ambuel and Temple 1982). However, Probst (pers. comm., 1987) knows of no evidence that directly supports Mayfield's explanation.

Within large tracts of young pines the Kirtland's Warblers are colonial, eschewing seemingly similar sites to nest in loose groups within other similar areas, although they possibly differ in subtle ways. One routinely finds clusters of between 2 and 30 pairs separated by substantial distances of similar habitat (Figure 11.1). Once colonized, sites are occupied for several years, although sometimes sites in seemingly satisfactory habitats have been abandoned because of nest failure or chance local extinction.

Historical records, plus regular censuses in recent years, give a sense of trends in this species' population size over the past century (see Mayfield

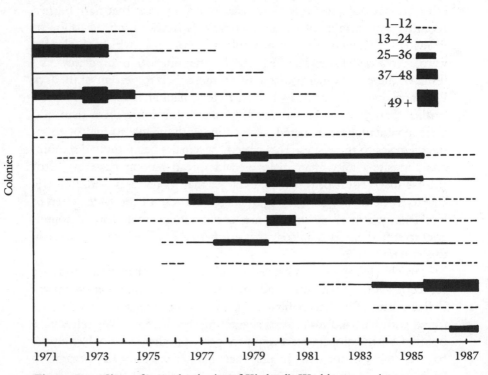

Figure 11.1 Sizes of several colonies of Kirtland's Warblers over time. (Modified from Probst 1986; Probst, pers. comm., 1987.)

1960; Walkinshaw 1983). The vast majority of reports in spring migration, as well as the extralimital records (away from usual breeding and migration areas), were made in the late 1800s; further, most data on this species from its only known wintering grounds, the Bahamas, come from that period also. Mayfield and Walkinshaw take this information to indicate that numbers of Kirtland's Warblers were higher then than now, although even then there were probably no more than a few thousand pairs. Since the breeding grounds were not known at that time, we have no corroboration of migration and winter data. But this period came just after northern lower Michigan was first lumbered. Jack Pine was considered worthless as a timber tree, but adjacent areas of White *(Pinus strobus)* and Red pine were extensively lumbered, leaving heaps of inflammable slash that led to large fires throughout the region. These episodes should have been followed by an increase in satisfactory nesting habitats, exceeding the ones available before or after. All one can say is that the timing of the fires fits the numbers of birds recorded in migration and on wintering grounds.

Numbers almost certainly declined between this period and the first census in 1951 (Ryel 1984). Fewer than 500 pairs were recorded in 1951, and by 1971 the population had shrunk to about 200 pairs. This was a time of heavy nest parasitism by Brown-headed Cowbirds: half to three-quarters of the Kirtland's Warblers' nests were parasitized, a higher frequency than the one recorded earlier. Mayfield and Walkinshaw assumed that Kirtland's Warblers never encountered cowbirds until the late 1800s, and records of the cowbird's range expansion (see Mayfield 1960) establish that its impact was low prior to this time. In the early part of this century such students as Wood (Wood and Frothingham 1905) and Leopold (1924) expressed concern about the cowbird's impact. Kirtland's Warblers have no known defenses against cowbirds, which is not surprising in view of their lack of contact with them until recently. Once the ability to reject eggs evolves, it spreads rapidly through a population (Rothstein 1975b), yet Kirtland's Warbler is by no means unique in lacking this ability. By virtue of its small numbers, however, it is much more vulnerable to extinction than are species with extensive ranges that suffer high parasitism when in contact with cowbirds. For instance, the Red-eyed Vireo suffers just as much parasitism when cowbirds are common, but vireo populations in large forests not visited by cowbirds, such as the Canadian deciduous forests, are free from their depredations. Further, the probability that rejecter behavior will evolve in such a small gene pool is bound to be

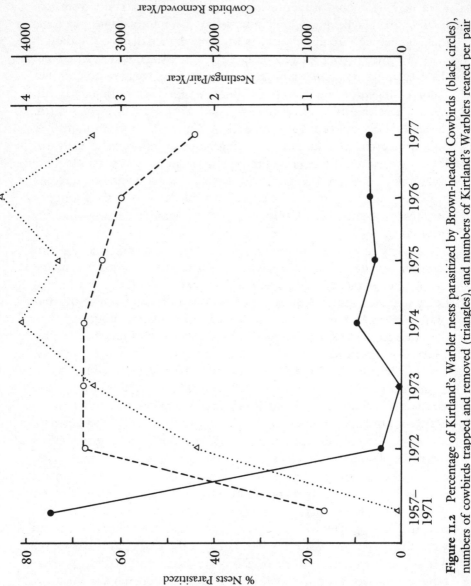

Figure 11.2 Percentage of Kirtland's Warbler nests parasitized by Brown-headed Cowbirds (black circles), numbers of cowbirds trapped and removed (triangles), and numbers of Kirtland's Warblers reared per pair (white circles). Size of yearly sample varied from 26 to 54. (Modified from Walkinshaw 1983.)

low. Thus, the Kirtland's Warbler is unusually vulnerable because of its small population size. In such a situation the entire population could easily become extinct before rejection mechanisms evolved.

For this reason cowbird control was undertaken in the 1970s. The program has reduced parasitism from 75 percent to 6 percent, with a corresponding fourfold increase in the number of fledglings produced (Figure 11.2). One might expect this change to be followed by an increase in population size, but that has not happened; rather, over the past several years the population has stabilized at a low level of roughly 200 pairs. The reasons are unclear. Vigor of the breeding adults is probably not a consequence, for mortality of adults on the breeding grounds is extremely low. Probst (1986) noted, however, that previous estimates of annual productivity and recruitment (Ryel 1981) have not separated postfledging losses from those suffered away from the breeding grounds, or taken into account that some males (at least) are unmated and that many abandon territories. Probst estimated that there may be an 18–35 percent fledgling loss prior to migration, on the basis of Walkinshaw and Faust's (1975) reports that 33 percent of fledglings were not seen again after they left their nests and Nolan's (1978) estimate of an 18 percent loss in Prairie Warblers, a congener that occupies a similar habitat. The period immediately after fledging is one of high mortality in other passerines (Lack 1954). Probst and Hayes (1987) noted that at least 15 percent of all males are unmated, especially ones in marginal habitats (40 percent unpaired), although the number might be partially offset by polygynous matings (Radabaugh 1972). All told, the number of juveniles departing their natal areas could be 40 percent or more lower than the number predicted by calculations based on singing males and fledging success.

Another factor affecting the size of breeding pools is the low level of juvenile returns to their area of birth, a trait common to many passerines. Birds not returning to established colonies have a low probability of finding mates. Although data are difficult to gather, observations of singing males in extralimital areas and of young birds in different colonies suggest that many individuals survive the winter and migration yet fail to join a breeding population. Mayfield (1983) maintained that numbers of extralimital birds have risen following cowbird control, citing Kirtland's Warblers observed as far away as Ontario, Quebec, and Wisconsin, some being banded birds from the Michigan population. So long as they fail to meet females, these birds make no more reproductive contribution than those that die young. There is no information on whether any extralimital individuals eventually find a colony, and it is unrealistic to expect that we will obtain adequate information on this point. One of the Ontario birds sang at the same site at least two years, and a bird on the Upper Peninsula of Michigan probably returned two years also (see Probst 1985).

The problem of locating a breeding colony becomes progressively more serious as breeding ranges decline. During the cowbird-induced decline, the range shrank to one-fourth its former size, and the east-west width shrank to one-third. Perhaps decreasing breeding range is an especially difficult trap for Kirtland's Warblers, since they are a successional species that depends on finding new nesting sites with regularity and is especially inclined to occupy new areas. Although in widely distributed species most vagrants might find mates, the tiny range of Kirtland's Warbler may often preclude it.

The failure of the population to rebound may be due to its being limited elsewhere. This unaccounted mortality probably does not happen on migration lanes, for one would not expect the population to remain as stable as it has if that were so. Two rapid drops, however, one perceived (though not verified) in the late 1920s and early 1930s (Trautman 1979), and one in the winter of 1973–74 (Walkinshaw 1983), may have resulted from hurricanes killing birds either in passage or immediately after their arrival on wintering grounds. Heavy mortality of warblers also may have occurred in the winter of 1970–71 and the spring of 1971, as a result of extreme drought on their Bahamian wintering grounds. The drought took place immediately before the breeding census that recorded the catastrophic 10-year decline (Radabaugh 1974).

The alternative explanation is that wintering grounds are now limiting. Little is known about them, except that the species is apparently confined to the Bahamas. The few workers who have tried to find Kirtland's Warblers recently in the Bahamas have not been very successful. Radabaugh (1974) found one bird in 800 hours of fieldwork on 11 large Bahamian islands, and Emlen (1977) found none in 500 hours of fieldwork on Grand Bahama Island. The older records suggest that they occupy low broadleaf scrub. Whether they would favor young pine is unclear, but pinelands have been extensively cut there. Terborgh (1980) and Rappole et al. (1983) have suggested that this and other species could have dropped to catastrophically low levels as a result of habitat change on their wintering grounds before the birds were known to science. Many of the larger islands were under extensive agriculture from the 1550s, so environmental modifications may have had a major effect on distributions.

Golden-cheeked Warbler

The Golden-cheeked Warbler, another endemic species with a narrow breeding and wintering area, bears similarities and differences to Kirtland's Warbler. It, too, is a narrow habitat specialist, being confined as a breeding species to the "cedar" brakes *(Juniperus ashei)* of central Texas, on and about the Edwards Plateau (Pulich 1976). Pulich estimated its population to be 15,000–17,000 birds. Its distribution is coincident with that of the Ashe Juniper, and even within that range it nests only in certain types of juniper—it uses mature trees that, according to Pulich's estimate, must be at least 50 years of age. Thus, Pulich's picture suggests that Golden-cheeked Warblers differ from Kirtland's Warblers in requiring mature vegetation.

Pulich proposed that the basis for this species' limited breeding distribution is its use of the fibrous bark of Ashe Juniper for its nests, which, in his experience, constitutes the major building material of the Golden-cheeked Warbler. Although two other junipers occupy large areas in Texas, the Redberry Juniper *(J. pinchotii)* in the western part of the range and the Red Cedar *(J. virginiana)* in the northeastern part, Pulich found no indication that the birds use them for materials. This putative nesting requirement has not been tested experimentally, and it seems unlikely to have been a limiting factor traditionally. Areas with stringy Ashe Juniper bark may have other critical traits, such as certain foliage growth patterns.

Golden-cheeked Warblers are not locally limited within the areas they currently occupy; but they exhibit a loose colonial distribution, as noted for Kirtland's Warbler. Pulich reported that the groupings usually contained fewer than six territories, separated from other colonies by considerable amounts of identical habitat, but one aggregation consisted of 21 pairs.

Pulich's assertion that the species is confined to only part of the available habitat must be viewed in the context of Kroll's (1980) analysis of used and unused sites. Kroll reported that sites occupied over a five-year period and sites not occupied during that time differed clearly and consistently in multivariate analyses of vegetation variables, though they were similar to the human eye. Key variables included the presence of Ashe Juniper and Bigelow Oak *(Quercus durandii brevioloba),* distance between trees, density of the two tree species, height of stand, and age of juniper. The birds required oaks for foraging and were most active there, although they concentrated on Ashe Juniper for nesting materials and singing perches. The oaks were especially important when newly hatched nestlings were being fed. Sites occupied by Golden-cheeked Warblers all contained junipers at least 40 years old, an age when the trees slough bark that provides nesting material.

The Golden-cheeked Warbler's range has decreased as a result of a cedar eradication program. Prior to the initiation of this program in 1948 the acreage of all three *Juniperus* species was much greater than it was when Europeans first settled in numbers. Recollections of longtime residents revealed that cedars within the range of Golden-cheeked Warblers grew only in areas of marked relief, such as steep slopes and cliffs of limestone canyons and ravines, and that tall prairie grass penetrated to the edge of the geologic relief (Pulich 1976). Overgrazing and control of fires degraded the land, followed by a massive invasion of the bird-dispersed junipers. Prior to then the prairie lands had been maintained by fire, either natural or set by Native Americans. Vegetational analyses have documented a massive seeding of cedars between 1854 and 1884—the beginnings of the mature cedar brakes. Thus, earlier in this century the warblers may have been more common than they are now or were before that time.

Kroll (1980) emphasized that Golden-cheeked Warblers do not use all parts of the dense, unbroken, old growth—the famed cedar brakes; they occupy only the edges of these impenetrable areas. Thus, Kroll argued, this bird is an ecotonal species, depending on edges or an open mosaic containing mature cedars and scrub oaks. Although Bigelow Oak was a critical resource in Kroll's study area, other species of scrub oak have the same function in other parts of the Golden-cheeked Warbler's breeding range. Pulich's assertion that the birds are loosely colonial should be reanalyzed in this light.

Kroll's analysis, together with Pulich's reconstruction of vegetational history, indicates that the bird's requirements match conditions that were prevalent prior to the arrival of Europeans. Then, the Golden-cheeked Warbler would have been an edge species, confined to linear tracts of habitat in canyon gullies and hilltops, sites that provided opportunities for vegetational diversity.

It is also possible that these birds are confined to areas with few competitors. No other warblers that nest in this habitat overlap heavily, if at all, with Golden-cheeked Warblers in resource use. This pattern resembles that of Kirtland's Warbler, although Kirtland's Warbler is confined to shorter successional periods than is the Golden-cheeked Warbler. Golden-cheeked Warblers might exist indefinitely at sites along the juniper-grassland ecotone.

Golden-cheeked Warblers are also parasitized by Brown-headed Cowbirds. Their situation differs from that of Kirtland's Warblers in that, although heavily parasitized, the Golden-cheeked Warblers have a defense against the cowbird—they desert their nests and undertake a second, more successful breeding effort. Although the cowbird is still laying eggs when warblers are establishing their second clutches, warbler success, with regard to cowbird parasitism, is much higher (65 vs. 25 percent for the first nesting). Pulich attributed the improvement to larger numbers of other species nesting later in the season and consequently being at risk from the cowbird. In contrast to Kirtland's Warblers, Golden-cheeked Warblers may have been in contact with cowbirds for a substantial time. If the area containing most Golden-cheeked Warblers was formerly adjacent to prairie, cowbirds probably ringed their pre-European breeding range, putting many of them at risk. The first workers to collect nests of Golden-cheeked Warblers reported cowbird parasitism. Records are inadequate to determine whether the current rate of parasitism has increased. Pulich's modest sample from the 1960s shows more parasitism than do nests collected by an oologist in the 1930s, but the earlier sample may have concentrated on unparasitized clutches. Pulich's results are from an early nesting in one year and a late nesting in the preceding year, and thus we need more samples.

It is possible that Golden-cheeked Warblers are severely restricted on their wintering range, but even less is known about this part of its life than is known about Kirtland's Warbler. Pulich made two attempts but could not find the species on its wintering grounds. Only 10 specimens have been taken from the wintering area—the region from the mountains of east-central Guatemala to Honduras and adjacent Nicaragua. These warblers are mid- to high-elevation winterers and range from 1,500 to 2,500 meters altitude or more, in pine or pine-oak forests, habitats that are probably declining as a result of increased land use. Golden-cheeked Warblers are also reported from a number of migratory records through Mexico, but Pulich argued that the specimen records do not admit adjacent Mexico as a wintering area. Identification becomes a severe problem on the wintering grounds, since the Golden-cheeked Warbler bears a close resemblance to its sibling species, the Black-throated Green, Hermit, and Townsend's warblers, especially when it is not in breeding plumage; further, these species appear in many areas where Golden-cheeked Warblers have been erroneously (Pulich believes) reported. Pulich noted that several museum specimens he examined were misidentified, instead being Black-throated Green Warblers.

Kroll (1980) studied 12 Golden-cheeked Warblers briefly on their wintering ground in Honduras, a pine-oak forest at an elevation of 1,500 meters, where Monroe (1968) had earlier collected specimens. Except for an open pine overstory, this habitat resembled the breeding sites in having an understory of pine and oak. The birds foraged primarily in the oak understory, a pattern probably mediated by the presence of sibling species, Black-throated Green, Hermit, and Townsend's warblers. The four species appear to partition the foraging environment, with the other three foraging higher. The possibility that this constraint affects the preferences of Golden-cheeked Warblers on their breeding grounds warrants attention. Because the Golden-cheeked Warbler's winter range is overlapped completely by these other species (AOU 1983), they may be major determinants of its resource-exploitation patterns, especially since it spends most of the year there.

Bachman's Warbler

Much less is known about the nesting biology of Bachman's Warbler than of the Kirtland's and Golden-cheeked warblers, mainly because it has become extremely rare since the early 1900s, and has seldom been seen since 1950. What we do know about Bachman's Warbler indicates that it differs from the other two species in nesting over a vast area; its range encompasses scattered parts of the lowland deciduous forests of the southeastern United States. The numbers reported in migration in Florida, south of their known breeding range, were truly impressive at one time. In the 1880s and 1890s at least a few hundred were collected, and thousands more seen (Stevenson 1972). These observations were not merely the consequence of this species being collected with unusual frequency because of its novelty, for Brewster (1891) related its abundance to that of other migrating species. He reported that it was the seventh most common migratory warbler, vireo, or kinglet, less abundant than Northern Parula, Yellow-rumped, Black-and-white, and Yellow-throated warblers, Blue-gray Gnatcatchers (*Polioptila caerulea*), and Ruby-crowned Kinglets (*Regulus calendula*), but more abundant than such common species as Orange-crowned Warblers and Red-eyed, Solitary (*Vireo solitarius*), and Yellow-throated vireos (*V. flavifrons*). If these migration data can be legitimately compared with those of Kirtland's Warbler, which also migrates through the southeast, Bachman's Warbler was immensely more common than Kirtland's Warbler, for which only handfuls of observations have been made on migration (Walkinshaw 1983). The fact that only two observations of Bachman's Warblers were made in Florida between 1949 and 1971 (Stevenson 1972) emphasizes the change in abundance. What is unclear is how much this species is a microhabitat specialist. There is very little information on the conditions at the birds' nest sites, and, not surprisingly, interpretations of these fragmentary data differ (Hooper and Hamel 1977; Hamel 1986).

The species has been considered a bird of the virgin bottomland forests and swamp forests (Stevenson 1972; Shuler 1977); however, a closer analysis of nesting data reveals plants associated with nests that almost invariably are species native to disturbance areas (Hooper and Hamel 1977), so the Bachman's Warbler may have been a second-growth species, rather like its close relatives, the Blue-winged and Golden-winged warblers. If so, this raises the question of why it should have become rare so recently, particularly since the stronghold of this species in South Carolina, I'On Swamp, was extensively cultivated for rice prior to 1865 (Hooper and Hamel 1977). The habitat may now be unsatisfactory, but cutting of old growth elsewhere in the species' range should have provided adequate second-growth areas to prevent catastrophic declines in its numbers.

Limitations on the wintering grounds have often been held responsible for this decline (Terborgh 1974; Rappole et al. 1983). The Bachman's Warbler's wintering area is also limited, confined to western Cuba and the adjacent Isle of Pines. This area is mainly agricultural, a substantial part of it planted for sugarcane. Hence, the species' resources had been decimated here rather than on its breeding grounds. This area has been agricultural since the 1500s, however, and it is simplistic to ascribe the problem solely to the wintering grounds, unless the extent, or the nature, of the agricultural practices have changed radically. Rappole et al. have suggested that this species and the Kirtland's Warbler (which also winters in an area intensively cultivated from the 1500s) suffered a population bottleneck during that time and never recovered, perhaps because of limited genetic variability resulting from it. This argument is more attractive for Kirtland's Warbler than for Bachman's Warbler, given the large numbers of Bachman's Warblers reported in the late 1800s.

Remsen (1986) has added a component by suggesting that this species was a bamboo specialist. Cane *(Arundinaria gigantea)*, a species of bamboo, once was one of the dominant features of the southern lowlands, with canebrakes extending for miles—at times forming understory, and at other times being the dominant element. Remsen's evidence is mostly circumstantial; its attractive aspect is that the demise of the bird parallels the decline of these monumental stands of Cane, a species that has decreased radically because of land clearing, floodwater control, and irrigation schemes. Although still widely distributed, Cane seldom reaches its former abundance. Remsen noted that the range of Cane matches the former breeding areas of Bachman's Warbler, and most reports on nest materials and plants growing in the vicinity of the nest mention Cane. Surprisingly, only Meanley (1972) had previously suggested that Bachman's Warbler might be a Cane specialist. Once Cane is eradicated it requires a long time to reestablish itself because of its unusual reproductive system, which produces seeds only at long intervals. Vegetative reproduction is the sole means of propagation at other times, and this kind of propagation does not result in long-distance dispersal. Remsen also proposed that the Bachman's Warbler's fate might have been tied to the catastrophic biology of the Cane, whose plants die over a wide area following mass flowering. If so, the numbers seen in the 1880s and 1890s could indicate an abundance of Cane prior to a major flowering. Remsen did not present information that would enable testing this idea, but Bachman's Warblers may have been much less common prior to then, with only the two specimens collected (in 1833) up to the 1880s on the breeding grounds (Bent 1953), although Gundlach (1855) reported them wintering in Cuba.

Remsen's argument follows from his experience with bamboo specialists of tropical America. In contrast with those areas, there are no known Cane specialists, unless Bachman's Warbler qualifies. Specializing in Cane might account for the earlier dearth of records of Bachman's Warblers on their breeding grounds, although the data from I'On Swamp, a stronghold of the species, do not follow this pattern. The characteristics of the area, a site of massive disturbance, including rice cultivation, are not consistent with Cane being a dominant feature. It is thus unclear what role Cane plays in Bachman's Warblers' lives, though many nests were built with it or in it.

Swainson's Warbler

The Swainson's Warbler is currently one of the least common North American warblers, but it was probably much rarer in the recent past than it is at present. Its early history resembles that of Bachman's Warbler, even to breeding range and habitat (lowland forests of the southeastern United States). A handful of specimens were collected by Bachman in the early 1830s, and then only a few records were made in the next 50 years. Observations then increased markedly in the 1880s, as did observations of Bachman's Warblers, but Swainson's Warbler did not decrease rapidly subsequent to that period. Indeed, while Bachman's Warbler began to decline precipitously in the 1940s, another population of Swainson's Warblers was discovered in the southern Appalachian Mountains in a totally different habitat. These birds frequented areas with a dense rhododendron understory, in contrast to the Cane- or pepperbush-frequenting birds of the southern and Gulf coastal plains.

Why did Swainson's Warbler prosper while Bachman's Warbler approached extinction? The answer is complicated because so little is known about the breeding biology of Bachman's Warblers. On the other hand, since Bachman's Warblers were in many regards typical *Vermivora* (Ficken and Ficken 1968b), we can gain some sense of their habitat use patterns if we combine information on their congeners with what we know of Bachman's Warbler itself (see Stevenson 1972; Hooper and Hamel 1977).

Swainson's and Bachman's warblers differ in their morphology and foraging habits—the former is a ground forager, and the latter a gleaner—but they occupy the same habitats. Although Remsen proposed that Bachman's Warbler was a canebrake specialist and suggested that its demise was associated with the precipitous decline of this plant, Swainson's Warbler also frequents canebrakes over much of its range (Meanley 1971). It also occurs in low densities at the north end of its range (Dismal Swamp of Virginia and Pocomoke Swamp of Maryland and Virginia) in lowland forests with an abundant understory of Sweet Pepperbush *(Clethra alnifolia)*. Pepperbush's growth habit resembles that of Cane (Meanley 1971). The survival of Swainson's Warbler in the southern coastal plain raises doubt about Remsen's argument that linked the decline of Bachman's Warbler to the disappearance of large canebrakes, for Meanley (1971) emphasized the close affinity of Swainson's Warblers to the same habitat. Although Swainson's Warblers may also have declined as Cane disappeared, the decrease has not devastated their numbers in the way that other factors have affected Bachman's Warblers.

The difference in fortunes of the two warblers could lie in their wintering grounds. Swainson's Warblers also winter on Cuba (as well as Jamaica), but some use the Yucatán Peninsula; Bachman's Warblers have not been reported on the mainland in winter. The mainland wintering population of Swainson's Warblers could have helped to maintain this species, if agriculture on Cuba destroyed the lowland habitat following Western colonization, as proposed by Rappole et al. (1983). The basis for the marked increase in the late 1880s is unclear; previously farmed habitat in Cuba could have become available then. If these birds were also wintering on the Yucatán Peninsula of Mexico, however, why were they rare in the first place? Birds wintering on the Yucatán may have increased at this time, but I am unaware of any changes that might bring an increase in wintering numbers. Rappole et al. (1983) maintained that the Swainson's Warbler's winter area in the Yucatán may be the difference between its success and that of Bachman's Warbler, and Kirtland's Warbler as well. Judging from the stray warblers reported in Mexico and Central America that normally winter in the West Indies, and vice versa, new wintering populations may not infrequently develop (see Lack and Lack 1972). We know that migrants of many species, once established, return to their wintering sites year after year (Keast and Morton 1980). Whether this behavior enhances the ability of others of the same species to colonize the site is unknown, but given that some geographic races of warblers winter in areas discrete from each other (see Ramos and Warner 1980), this relationship is possible.

Swainson's Warblers may have recently moved into a new habitat, the mixed mesophytic forests of the southern Appalachians, characterized by a heavy understory of rhododendron. Structurally, these sites resemble the Cane and Sweet Pepperbush understories that characterize their habitats elsewhere. However, they considerably extend the range and vegetation type, which should minimize the danger that a single catastrophe, such as the demise of Cane, would have a crippling impact on this species. It has generally been assumed that the southern Appalachian population is a new one, having expanded from the coastal plain population; the apparent rapid increase of the species in the Appalachians supports this explanation. Although this bird is difficult to see, it is vocally conspicuous, and a population comparable to its present size would not have remained undiscovered for long. It is not certain that the Appalachian population is a new one, however, for Meanley (1971) reported that a small sample of Appalachian specimens had whiter underparts than those of the coastal plain, whose underparts were distinctly yellow. Thus the Appalachian birds may represent a genetically distinct population that is not of recent origin, one that more likely increased after a severe bottleneck in the same way as the birds multiplied on the coastal plain. In fact, Meanley and Bond (1950) erected a new subspecies, *alta*, for the Appalachian birds, on the basis of the color of underparts. (These birds do not differ systematically in size.) The fifth AOU *Check-list* (1957) did not recognize this subspecies, but this need not mean that there are no differences between this population and the birds of the coastal plain.

Incipient Extinctions:
Kirtland's and Golden-cheeked Warblers

At least on their breeding grounds, both Kirtland's and Golden-cheeked warblers are restricted to habitat islands, the Kirtland's to early-successional Jack Pine forests in upper Michigan and the Golden-cheeked to juniper forests of the Edwards Plateau in central Texas. Both species seem to be well along in the march to extinction. The Golden-cheeked Warbler has higher densities and a larger population than does Kirtland's Warbler. The Golden-cheeked Warbler is a recent scion from the Black-throated Green Warbler superspecies; Mengel (1964) proposed a recent Pleistocene origin. Given its high local population densities, it should be able to increase its population if environmental changes were to expand its habitat. Since some populations of Black-throated Green Warblers, its sister species, occupy primarily deciduous growth, as does a closely related western species, the Black-throated Gray Warbler, occupancy of deciduous growth might be a viable option for the Golden-cheeked Warbler.

The characteristics of the Kirtland's Warbler make it a prime candidate for extinction. In contrast to the Golden-cheeked Warbler, it does not appear to be a recent scion of a closely knit superspecies, its closest relatives being unclear; and the species itself (as a distinct entity) must be much older than the Golden-cheeked Warbler. Apart from its pattern of habitat selection, one of its most characteristic features is its large size—it is the largest *Dendroica* warbler. Probst (pers. comm., 1987) noted that, at 14–15 grams, it falls into the size range of the vireos, and its sluggish foraging habits are more reminiscent of that group than of most *Dendroica*. Possibly it has been further encircled evolutionarily by its similarity to the vireos, another well-established New World group. Probst has suggested, however, that Prairie and Nashville warblers may be more likely competitors of the Kirtland's Warbler—the Prairie Warbler in habitats with a larger deciduous component, and the Nashville in older coniferous areas—than are vireos.

Mengel (1964) has examined the matter of natural extinctions with regard to the isolates produced by Pleistocene glaciation or its associated climatic changes. He noted that if we were to consider current vulnerable species—Kirtland's, Golden-cheeked, Bachman's and Colima *(Vermivora crissalis)* warblers—as incipient extinctions and to postulate similar losses for each Quaternary glacial cycle, 12 paruline extinctions would have occurred in the past million years or so! None of these four species appears to have been pushed largely to their present condition by human intervention. Kirtland's and Golden-cheeked warblers are unusual among parulines in that they occupy small parts of their habitats. If unoccupied parts of the habitats turn out to be unsatisfactory, their requirements are more exacting than those of any other parulines.

Formerly Rare Species That Have Increased

In contrast to Kirtland's and Golden-cheeked warblers, certain species have increased greatly in response to habitat changes wrought by humans over the past few hundred years. The Chestnut-sided Warbler is probably the best example. This species was so rare in Audubon and Wilson's time, not over 150 years ago, that the two observers saw it only once and twice, respectively (Bent 1953), putting it in the same league as the Bachman's and Swainson's warblers. Subsequently, the Chestnut-sided Warbler has become one of the commonest breeding birds in eastern North America, a gain usually attributed to an increase in frequency of young deciduous forest, which it favors. This species appears to be rather specialized in its habitat and foraging repertoire (Greenberg 1984b); therefore, its favored habitat may simply have expanded greatly in size. Prior to the early 1800s it must have been confined to sites of former forest fires, stream-bank areas of catastrophic flooding that are occasionally driven back to early-successional stages, or to successional growth about beaver ponds. Burleigh (1927) also reported finding Chestnut-sided Warblers in high-altitude stunted oak forests in the southern Appalachian Mountains, but no information indicates whether this species was present before the early or mid-1800s. Also, some workers believe that Native Americans burned sizable areas in northeastern North America to maintain second-growth forest for deer browse (Cronon 1983, but see Russell 1983), which would have provided satisfactory breeding habitat. Chestnut-sided Warblers were among the most abundant species in chestnut-sprout areas that followed the great blight of the early part of this century in the eastern United States (Bent 1953).

It is of interest that the increase did not occur quickly after European colonization, for which there are two possible explanations. First, most of the areas, once laboriously cleared, were kept in farmland or pasture, which probably remain in too early a successional state to favor these birds. Only later, after farming areas were abandoned and allowed to revert to forests, or after the use of wood increased, might substantial areas of young deciduous trees become available. The availability of this habitat would have lagged behind the opening of rich prairie lands to the west, which led to an exodus of farmers from the eastern United States. Given the age of forests frequented by the Chestnut-sided Warbler, we would expect a lag of 30–50 years between the abandonment of pasture and the population explosion of this species.

Reports by ornithologists in the 1800s give us a sense of how this species increased. Bent (1953) has summarized much of this information. The most detailed records come from eastern Massachusetts, the result of a long and continuing tradition of field natural history. Brewster (1906) reported that Thomas Nuttall first began to find Chestnut-sided Warblers in the Cambridge, Massachusetts, area around 1830–31, but the species gradually increased in later years. Increases in Chestnut-sided Warblers may have taken place later in the New York City area, in that Chapman (1917) reported the establishment of this species as a breeder in New Jersey toward the end of the nineteenth century, as did Dugmore (1902).

The Chestnut-sided Warbler is by no means the only species whose numbers have soared in concert with the increase of second-growth habitat. Bent (1953) remarked on at least five warblers in this regard: Chestnut-sided, Prairie, Nashville, Golden-winged, and Mourning warblers. Early reports of Nashville Warbler abundance parallel those of Chestnut-sided Warblers. It too began to increase about Cambridge in the late 1830s, according to Samuel Cabot (Brewster 1906). Numbers jumped rapidly, but by the latter part of the century they had decreased in eastern Massachusetts (Brewster 1906; Bent 1953), probably because much of the deciduous forest had reached a maturity not satisfactory for this species. One does not find similar reports on a decline in the abundance of Chestnut-sided Warblers at this time, although the two species often have similar habitats—the representative members of their genera in such areas.

Prairie Warblers are characteristic of earlier-successional stages, old bushy fields being typical sites. Therefore, their increases probably began before that of Chestnut-sided and Nashville warblers; and although the early workers comment on increases of this species, one does not see references to their finding only a handful of Prairie Warblers. This difference could be partly because Prairie Warblers occupy understory vegetation in Pitch Pine *(Pinus rigida)* barrens. The Prairie Warbler was one of the few warbler species to decrease in numbers in the Breeding Bird Survey over the past 20 years. Robbins et al. (1986) attributed the decline to reduction in the area of old fields in early stages of succession.

Changes in the land must have caused other shifts in the populations of warblers that were not so well documented because they did not involve a species over its entire range. For instance, Yellow Warbler populations may also have waxed in eastern North America because of the increased availability of early second-growth habitat resulting from human activities. In addition to its presence around the edges of marshes, this species frequents hedgerows and suburban areas, types of sites not earlier present. Satisfactory habitat should have been nearly as limiting for Yellow Warblers in the northeastern United States prior to European colonization as it was for Chestnut-sided Warblers. A major difference between the two species might lie in the Yellow Warbler using earlier-successional areas than the Chestnut-sided; today one frequently sees a rough zonation between the two species about the edges of abandoned beaver ponds and similar places where succession is well under way (Morse 1966b). The Yellow Warbler does occupy coastal scrub vegetation, which would have provided a refugium, and it may also occupy spruce forests on small outer spruce-clad islands along the northeastern coast; however, these areas, especially the latter, would not provide many nesting sites.

Other species that were not very rare prior to Western colonization, but whose numbers have probably changed significantly, include Magnolia and Canada warblers. Both species currently reach high densities in recently logged coniferous-deciduous forests. Although Magnolia Warblers use a variety of habitats, they attain maximum densities in these mixed forests and dense young spruce growth. In the latter sites, the result of earlier clear-cutting, it is the numerically dominant species, and it reaches among the highest densities of any nesting warbler (see Hall 1984), other than budworm specialists in the midst of outbreaks.

Specialization and the Road to Extinction

It is a popular notion that extinct or endangered species are highly specialized taxa that are vulnerable to even slight changes in environment. Some species discussed in this chapter fit this pattern, and others do not.

The dependency on particular resources, such as highly permeable nesting substrates or the bark of a single tree species, marks a species as a specialist if that resource is not widespread and ubiquitous. Kirtland's and Golden-cheeked warblers are unusual among their group in the specificity of demands, Kirtland's Warbler being one of only two ground-nesting species of *Dendroica* and the Golden-cheeked Warbler being the only member of the *Dendroica virens* superspecies complex to have unusual nesting demands. How did these species come to be restricted to such peculiar resources in the first place? Not only is the answer to this question locked in evolutionary time, but reconstructions of possible scenarios are hard to concoct. Monographers of both species have emphasized that the birds live in habitats with low species diversity and few if any other resident warbler species. Regardless of the basis for this distribution, their segregation may have placed them where resource availability was at best modest but where within-habitat constraints from other species were minimal. In an analogous manner members of island faunas have evolved extreme modifications in the absence of interspecific constraints. That conclusion does not explain why Kirtland's Warblers would adopt ground nesting, however, and it certainly does not account for the Golden-cheeked Warbler's habit of using the bark of a single species of cedar, especially when two congeneric cedars are within reach. Analyses of these traits should be enlightening, for they may provide insight into the forces that hasten changes and then extinction. They represent only one explanation for rarity; as I stressed earlier, problems on the wintering ranges might be paramount.

If the basis for the rarity and confinement of Kirtland's and Golden-cheeked warblers is obscure, the reason for the Bachman's Warbler's isolation is even more so. Bachman's Warblers were, until recently, widely distributed as breeders, in contrast to Kirtland's and Golden-cheeked warblers, and in many ways they are typical *Vermivora*, although we know too little about the species to make that statement with confidence. The Bachman's Warbler is typical in being the sole *Vermivora* species in a major habitat type; yet there is one special factor that deserves attention, and that is the presence of the Northern Parula Warbler. This species is common throughout much or all of Bachman's Warbler's previous breeding range; and where it is common in the northeastern coniferous forest, it is in the near absence of a *Vermivora* species (Morse 1967a). Farther to the west, the Tennessee Warbler becomes common in similar forests (Erskine 1977). It is unclear why Northern Parula and Tennessee warblers occupy different geographic areas, but the important point is that the Northern Parula Warbler often occurs in the absence of the *Vermivora*, the situation that now holds in the southeastern lowland forests. Any explanation of the Bachman's Warbler's decline must account for the Northern Parula Warbler.

There has not yet been found a distinct biological difference between species that increased in numbers and those that decreased. The fact that several currently common species were rare before the mid-1800s suggests that they may have been (or are) specialists, also. If so, the major difference between currently rare and common species is that human-induced changes have made some habitats rare and others common. It is clear from Greenberg's work that Chestnut-sided Warblers are more stereotyped in habitat selection than such conifer-summering species as Bay-breasted Warblers, possibly an evolutionary response to predictable conditions. The only difference between the success of Chestnut-sided and Bachman's warblers is that the habitat of the former has changed in an essential feature. This interpretation is useful in accounting for changes in abundance, but it is not useful with regard to the Chestnut-sided Warbler's origin and persistence as a species. The finding of breeding birds in stunted montane vegetation of the southern Appalachians (Burleigh 1927) could provide a clue to its origin, and further attention to these birds might be enlightening. Their occupancy of this habitat may be a more useful guide in the attempt to sketch their history than is their use of "natural" second-growth areas, such as new growth forming after the die-off of Appalachian chestnut forests (Brooks 1947) or successional growth following the abandonment of old beaver ponds (Morse 1966b). Although the latter areas were probably important to the earlier success of Chestnut-sided Warblers, it is less probable that they were instrumental to the evolution of a species that has such strongly programmed behavioral patterns.

Analysis of Prairie, Nashville, Golden-winged, and Blue-winged warblers for behavioral similarities would help to test these ideas on the evolution of the Chestnut-sided Warbler, as each of them also increased in response to secondary succession of deciduous forests. Studies of these four warblers should proceed on breeding and wintering grounds in order to determine whether their specialization in one area, as with Greenberg's Chestnut-sided Warblers, is matched by specialization in the other area. In light of contentions by some workers that the rare species are limited on their wintering grounds, it is unfortunate that they cannot be studied adequately there, if at all. The argument would nevertheless be plausible if strong stereotypy on the nesting grounds was a good predictor of behavior on the wintering grounds. If one can use Greenberg's findings, that a deciduous-breeding species is more stereotyped than a coniferous-breeding species, Bachman's Warblers should be more vulnerable to modifications to their winter habitat than are Kirtland's or Golden-cheeked warblers. But the issue becomes clouded because I have linked specialization by Kirtland's and Golden-cheeked warblers to breeding requirements, which are problems of the breeding grounds only. These specializations could be part of a more general manifestation of stereotypy, but this argument, otherwise unsupported, is more tenuous for the latter two species than for the Chestnut-sided Warbler.

12 Population Limitation and Species Diversity

Birds that live in a variable environment, or that travel through several environments, are subject to limitation at one or more times and locations. By "limitation" I am referring to any factor that could constrain population size; the term includes both density-dependent and density-independent factors. "Regulation" refers to changes of population size as a consequence of interactions between organisms, typically in a density-dependent context, and thus has narrower connotations than does "limitation."

An implicit bias, that population sizes are routinely regulated by interactions on the breeding grounds, probably results from students concentrating their attention on the breeding cycle but interpreting their results in an unjustifiably broad context. Phenomena on the wintering grounds may be important in the population dynamics of migratory species. Lack (1966) and Fretwell (1972) have emphasized the importance of winter conditions in limiting bird populations, but they were concerned with resident species of the temperate zone. Lack and Lack (1972) did extend their view to North American warblers wintering in Jamaica, however. The migratory period itself provides a further possibility for population loss. Although mortality during migration is less likely to be density-dependent than that on breeding and wintering sites, catastrophic losses can damp otherwise limiting conditions on breeding or wintering grounds, or both (see Holmes et al. 1986).

The Changing Character of Breeding and Wintering Areas

Although it is now widely accepted that many Neotropical migrants winter in primary lowland forests, numerous Neotropical migrants nevertheless occupy second growth, so increases of this habitat would provide additional opportunities for some migrant species. Census data are still grossly deficient, however. Second-growth forest can regenerate after cutting, but primary forest is decreasing more rapidly than second-growth forms (Diamond 1986; World Resources Institute 1986).

These patterns of habitat change are not uncommon in geologic time. There is ample evidence of striking changes in the amounts of forest, grasslands, and arid areas over the Pleistocene and up to the present (Haffer 1969). More recently, the extent of warbler breeding grounds has fluctuated as a result of land being cleared for agriculture and other human activities. Considerable areas in the eastern United States may also have been cleared in precolonial times (Cronon 1983), although the extent to which Native Americans cleared land remains a contentious issue (Russell 1983). Prior to European settlement the area available for some migrant species to breed must have been greater, although those of others have increased since that time. Also, wintering areas have not been constant over the past. Pre-Columbian civilizations undertook major land-clearing schemes, although their projects probably did not encompass as great an area or occur at as rapid a rate as today. Recent studies have shown that humans have made major modifications to their environment for longer than was once thought (Rue 1987). Of equal or greater importance than the large-scale changes we associate with the modern era may have been the traditional slash-and-burn techniques used prior to European colonization by Native Americans, for this type of agriculture produces a continuous supply of second-growth forest. Subsequent to the Spanish Conquest, the number of Indians in Mesoamerica declined to less than 10 percent of previous levels (Sanders 1971a,b). Human populations in this area did not return to their pre-Columbian level for several hundred years. Simultaneously, though, the introduction of slave-based agriculture into the West Indies drastically decreased the extent of forested areas, possibly detrimentally affecting the wintering habitat of Kirtland's Warbler in the Bahamas and Bachman's Warbler in Cuba (Rappole et al. 1983).

The face of temperate North America also changed drastically after the first Western settlers arrived in the 1500s and 1600s. For example, although New England's forests still exceeded its cleared lands in 1800, the remaining forests had been altered by grazing, burning, and cutting, especially near settled areas (Cronon 1983). This change along the Atlantic seaboard was but the beginning of a tide of habitat change that swept across much of eastern North America, only to be reversed by the eventual abandonment of large areas of open land to second-growth woodland, albeit of altered physical size and species composition.

Even from this brief sketch one gains the impression that the size and characteristics of both breeding and wintering areas of warblers have fluctuated markedly and irregularly over the past several hundred years or more as a consequence of human land-use patterns. This conclusion suggests selection for opportunistic behavior by these birds, a trait to which they may have been preadapted by contingencies met during migration.

Evidence for Limitation on the Breeding Grounds

A large body of literature, including MacArthur's (1958) study and my own work on spruce-forest species (Morse 1968, 1971a, 1976b), is consistent with the view that breeding populations are limited in size, at least in favored habitat. Although maintenance costs may be low when birds are nesting, breeding introduces additional costs. Intense territoriality and aggressive behavior (Morse 1967a, 1976a), and rapid occupation of sites from which territory holders have been removed (Hensley and Cope 1951; Stewart and Aldrich 1951), further support the argument of limitation on the breeding grounds. I could not, however, separate effects of resource competition from those of predation pressure in censuses of small islands along the Maine coast and the adjacent mainland.

If resources suitable for migrants have increased faster in the tropics than in the temperate zone as a result of vigorous second-growth forest, migrant populations should now become denser on their breeding grounds, if they were previously limited there. Censuses of breeding birds in mature (nonsuccessional) habitats carried out over a period of rapid change in tropical wintering habitats provide an opportunity to test this hypothesis. Wilcove (1988) ran such a test by comparing 1947–48 and 1982–83 census results from large forested areas in the Great Smoky Mountains of Tennessee. In general, Neotropical migrant populations remained stable over this period. Fluctuations of Neotropical migrants at Hubbard Brook, New Hampshire, over the period 1969–1984 showed strongly individualistic patterns from species to species and no clear pattern. Poor weather or caterpillar outbreaks affected many species simultaneously, but few pairs of species responded similarly (Holmes et al. 1986).

Members of a population may not breed either because they are physiologically incapable of doing so or because they cannot obtain a territory or other critical resources. These birds live apart from breeding individuals, in different habitats, the interstices of territories, or on territories of birds that are breeding (S. M. Smith 1978, 1984). Surpluses of both sexes, if able to breed, would indicate that numbers of territories were limiting the breeding population's size, if not the total population's size. Birds incapable of breeding might include immatures, but small passerines such as warblers are probably always capable of breeding in their first year, if given an opportunity. Information of this type for migratory species is sparse at best, although S. M. Smith (1978, 1984) proposed that floaters are widespread among passerine birds. Where floaters are sought, nonbreeders often are found, as in Stewart and Aldrich's (1951) removal of spruce-woods *Dendroica* warblers followed by massive incursions into the spaces made available. In few instances do we know the numbers of both male and female nonbreeders, however.

There is less contentious indirect evidence for periodic limitation of populations on their breeding grounds; but it is so scarce as to leave open the question of whether limitation on the breeding grounds is the norm. The close relation between population density and territory size of some Ovenbird populations and their food supply suggests that food supply limits the breeding population's size, and the change in territory size that accompanies experimental manipulation of the food supply verifies the causality of this relation (Featherstone, in MacArthur and Wilson 1967; Zach and Falls 1975). Further, the coincident increase in numbers of Bay-breasted and Cape May warblers with Spruce Budworms in northern coniferous forests implies that numbers of birds are affected by budworm densities; in turn, outbreak conditions may affect nonoutbreak species of the forests, such as Black-throated Green and Blackburnian warblers, in a negative way (Kendeigh 1947; Morris et al. 1958). In other instances, close relations between habitat structure and territory size can be related to pressures of nest predation (Martin 1988b).

The Chestnut-sided Warbler provides a striking illustration that limitation probably once occurred on the breeding range. Its rapid increase in response to expanding second-growth habitat (see Chapter 11) makes it unlikely that its abundance was limited on wintering grounds prior to the European settlement of northeastern North America. On its wintering grounds in Panama it is one of the most territorial of the wintering warblers (Greenberg 1984b), which is consistent with current limitation there. Numbers of Chestnut-sided, as well as Prairie, Nashville, Golden-winged, and Blue-winged Warblers waxed and then waned in response to the changed vegetational stage of their breeding areas.

Hints about prevailing pressures of the past come from trophic (beak) morphology and foraging patterns of these warblers (Greenberg 1984b), both of which could be adaptations to efficient food exploitation during the summer. These include large beaks, characteristic of coniferous habitat (Ficken et al. 1968; Grant 1972), and flexible foraging, characteristic of unpredictable conditions. Beak shapes of migratory warblers in general may be adaptations to caterpillar feeding during the summer, when the birds are feeding these important prey to their young (Greenberg 1981).

The discovery that adult small passerines may reach a marginal body condition during the nesting cycle (Yarbrough 1970, 1971) strengthens the argument that the period is a critical time of the year. Any worsening of climatic conditions during this time may result in the death of young or abandonment of eggs, as is clear from reports on losses of eggs and nestlings during inclement weather. Adult mortality appears to be low during the breeding season (e.g., Nolan 1978), but offspring may be the resulting losers.

These observations cannot establish that warblers are always limited on their breeding grounds, but they do suggest that limiting factors have strongly affected their activities there in the recent past. If parulines evolved in a tropical environment as generally assumed (Lönnberg 1927; Mayr 1946), changes from primary broadleaf adaptations would not be random events. Therefore, if characteristics such as large size were the consequence of random factors, or not associated with the breeding season, we would expect size to be distributed randomly among warbler species, rather than the largest species being characteristic of coniferous nesting areas.

Evidence for Limitation on the Wintering Grounds

The Neotropical region is currently undergoing rapid deforestation, which may or may not limit warbler populations on their wintering grounds. Conversion of primary forest to different habitats may lower the number of wintering sites available to some migrants, but even increase them for others.

No satisfactory examples of winter population limitation currently exist for paruline warblers. Special cases of winter limitation (e.g., the White-throat of the Old World: Winstanley et al. 1974) may not be typical of Neotropical wintering populations. Yet, the Whitethroat example emphasizes the possibility that populations may be limited on wintering grounds. Several characteristics of the wintering season are also consistent with limitation, at least at a local level: the ubiquity of winter territories among many migrant species, the presence of floaters, and correlations between morphology and behavior on wintering grounds. Combined with the evidence from the breeding season, the results suggest that major changes on either wintering or breeding grounds may have a quick and major effect on population size (Morse 1980c). In that major habitats are now changing much faster on wintering grounds than on breeding grounds, winter limitation could be widespread.

Initial impressions of tropical deforestation (Monroe 1970) suggested that second-growth habitat following the removal of primary forests would provide improved habitats for migrants, but several problems arose from this assessment. First, are preferences for secondary forest widespread? No simple answer emerged; for even though several warblers frequent secondary tropical forests, others are unlikely to adapt to the new environment (Keast and Morton 1980).

The failure of new second-growth tropical forest to keep pace with the loss of primary forest suggests that space may be limiting there now. Large areas are used in such endeavors as cattle ranching (Terborgh 1980), which make them unsuitable for all forest-dwelling birds. Other kinds of intensive land use, agricultural or industrial, will have a similar effect. Turning land over to plantation monocultures, such as citrus, pine, or eucalyptus, may not produce as drastic an effect on habitat structure as cattle raising or industry, but the plantings have notoriously depauperate faunas (Terborgh 1980).

I have noted that many migrants frequent highland areas. Although debate on the importance of these areas for migrants has followed the same lines as the one for second growth, it leaves little room for complacency, for deforestation of the highlands has been especially rapid.

Two-thirds of the breeding bird populations of many parts of North America winter in the tropics, and as many as half of them winter in a limited area (West Indies, Mexico, and northern Central America) that is no more than one-seventh or one-eighth the size of the breeding grounds from which they have come. The wintering ranges of many Neotropical migrants fall entirely within these areas. Densities of migrants are often high, sometimes exceeding those of the resident fauna. Given the differences in available area on breeding and wintering grounds, the impact on birds from clearing one hectare of wintering ground may be comparable to that of clearing 5–8 hectares on their northern breeding grounds (Terborgh 1980). There is still doubt about how the migrants manage to situate themselves during winter, even though their per capita demands do not match those of the breeding season. The data on habitat change can only heighten one's impression of the impact that this habitat change could have on population size.

Although numerous migrants travel beyond northern Central America, their rapid drop in density (Figure 10.1) as one proceeds southward raises doubt about the potential of more southerly areas to provide adequate wintering sites. Whether any of the species wintering in the northern Neotropics could shift their ranges farther south is unclear (even if that tactic offered a long-term advantage, given the rapid rate of deforestation to the south as well). Species already wintering to the south may have saturated these habitats. The widespread occurrence of allopatry in large genera, such as *Dendroica* and *Vermivora* (Keast 1980b), in itself consistent with competitive origins, suggests that even if the more northerly wintering species found these areas and could adjust to them, migrant species already present might exclude them. Lack and Lack (1972) and Terborgh and Faaborg (1980) maintained that West Indian communities that they studied were saturated and that distribution patterns of wintering warblers in the West Indies and Central America were consistent with this interpretation. The question also remains whether more northerly-wintering tropical migrants could sustain additional migration in such a shift.

These arguments indicate that we can anticipate sizable decreases in populations of some species in the near future (Terborgh 1980), if they have not already commenced. Diamond (1986) noted that long-term census results are equivocal thus far, and Robbins et al. (1986) suggested that recovery from the effects of pesticide programs in the 1960s has even led to an increase in some species. Terborgh (1980) estimated that at the current rate of exploitation all the primary forest in Central America outside of reserves will have disappeared by the year 2000, so the process seems inevitable—thus the main question is not if, but when. Reports of recent declines in numbers of warblers and other Neotropical migrants on their breeding grounds (e.g., Ambuel and Temple 1982) may well be indicative of habitat loss on wintering grounds, but since they fall within the long-term range of variation of such populations, they cannot be marshaled as definitive evidence. Any such analysis is bound to face difficulties for this reason alone (Rappole et al. 1983).

The Effect of Migration on Population Levels

Up to now, perhaps the most convincing data in support of wide-scale limitation of North American migrants come from catastrophes during migration, after which population levels are severely depressed. Declines are usually only of short term, however, which emphasizes the normal stability of breeding numbers (see Griscom 1941; Batten 1971). The migratory Scarlet Tanager *(Piranga olivacea)* recently suffered such a depression, the consequence of a May period of rain and unusual cold in northern New England, and this event probably depressed their numbers for at least two years after (Zumeta and Holmes 1978). Although few warblers were apparently lost in Zumeta and Holmes's area, the cold spell may have contributed to the low breeding numbers of warblers in my study areas 150 kilometers to the east (Morse 1976b).

Perhaps the biggest permanent effect of migration is the set of constraints required for survival over a yearly period. Factors assuring migratory success may decrease efficiency during the breeding season. For example, long-distance migrants have longer and more pointed wings than permanent residents, which are bound to compromise maneuverability and perhaps their foraging repertoire (Winkler and Leisler 1985). Energy and time spent in migration could hypothetically be spent otherwise; therefore, one might predict that in lieu of migration greater fecundity might ensue. Nevertheless, the rich resources available on the northern breeding grounds may permit migrants to attain a higher fecundity than would be possible if they spent the full year somewhere else (Cox 1968, 1985). Fecundity of migratory warblers is much higher than that of tropical resident warblers, whose clutch sizes are small and which suffer high predatory pressure on their nests and fledglings (Skutch 1949, 1954).

A Dynamic Equilibrium

These observations lead to the conclusion that population limitation is not a phenomenon that can be said to take place entirely on *either* the breeding or the wintering grounds (Williams 1966; Morse 1980c). Indeed, there may be a dynamic equilibrium between the two areas (Morse 1980c), and limiting pressures could even act simultaneously much or all of the time as a consequence of the components of fitness that will be selected for in the two areas. On the breeding grounds maximizing output of young will be the most important factor (either by providing maximum energy to young or possibly by minimizing nest predation: Fretwell 1972, 1980), especially since adult mortality is usually low at this season (Mayfield 1960; Nolan 1978). On the wintering grounds, and in migration, the focus is on survival until the next breeding season (Bennett 1980; Rappole and Warner 1980). As a result of these premiums, optimal foraging patterns can differ in the two seasons.

The two areas where a bird lives can differ markedly. Rappole and Warner (1980) noted that floaters occurred in several species they studied during winter and summer, as well as in migration; the presence of birds without their own territories is highly suggestive of resources potentially limiting throughout the year. To argue this point rigorously, though, one would have to demonstrate that the same populations are subjected to limitation in more than one season. There is no evidence of this from the wintering and breeding grounds of a single population (Morse 1980c). Bennett (1980) suggested that American Redstarts are most strongly limited during winter. Not only are they more widely dispersed and more different from coexisting species than they are on their breeding grounds, but they hawk for insects more frequently on wintering than on breeding grounds. Redstarts have a broad beak and large rictal bristles similar to those of flycatchers, and atypical of warblers. The tendency to make heavy use of these morphological traits during winter raises the possibility that they are a response to winter contingencies. Bennett noted that under suboptimal conditions on the breeding grounds, individuals hawk for insects more frequently.

Conflicting pressures on wintering and breeding areas might constrain the rate and extent of adjustments made by birds in response to changes, thereby introducing a conservative element to their evolutionary pathways. Species probably differ in their ability to change, however, as Greenberg (1984b) found for the Chestnut-sided and Bay-breasted warblers' exploitation systems. More attention to these complex patterns is needed; a better picture of exploitation systems might also improve our understanding of factors that drive migration (see Cox 1968, 1985).

Survivorship

Not surprisingly, there are few life tables for warbler populations. Like passerines in general, they have usually been considered short-lived species, with high yearly turnover (Farner 1955). Chapman (1917) stated that the mortality of North American warblers is higher than that of any other family of North American birds. But this interpretation raises problems, for warblers are single-brooded, with few exceptions (e.g., Nolan 1978). The curtailed time for migrants to breed mitigates against multibroodedness; where reported, only the earliest individuals produce the occasional successful second clutch (Nolan 1978; Eliason 1986). With infrequent exception (certain species during insect outbreaks), warblers do not lay unusually large clutches. Thus, their recruitment rate is modest, which is incompatible with high mortality. These patterns make it questionable whether warbler populations could maintain themselves if survivorship is as poor as has been generally believed.

The results of Roberts (1971) are therefore most interesting, for they reveal that the survivorship of several species of parulines, once they become adults, is surprisingly high for a bird of their size, considerably surpassing that of most resident species of small passerines (Table 12.1). The high survival rates cast the dangers of long-distance migration and limited migration or residency in a different light.

To be sure, there is still great uncertainty about migratory losses— mainly the question of juvenile mortality, which is high. If Ralph's (1978) postulation that as many as 10 percent of all coastal migrants are lost in fall migration by being blown offshore, usually on their first migration, is accurate, this factor could in part account for juveniles' low rate of return to their natal grounds. In their studies of a sylviine warbler, the Eastern Great Reed Warbler, Nisbet and Medway (1972) touched on another source of disproportionate juvenile mortality. Many more immature birds on their study area in Malaysia disappeared shortly after arrival on the wintering grounds than did adults. Although one might attribute this to the young birds being pushed from the adults' favored areas, it may have had especially serious implications for the young, since they arrived at a time of low food availability. Extensive studies in adjacent habitats failed to record these marked birds. Although none of the Neotropical studies refer to differential disappearance, birds in Central America also arrive on their wintering grounds at a time of low food availability, and juveniles are especially vulnerable because of inexperience and low social status.

The fact that catastrophic climatic losses in migration are exceptional, and that populations that have suffered catastrophes in migration appear to rebound, suggests that migratory factors do not usually influence population size, although they are bound to influence the strength with which density-dependent factors act. Wiens's (1977) arguments about periodic "crunches" controlling the upper limits of population size suggest a pattern rather different from the ones seen in censuses of warblers over several years (e.g., Morse 1976b; Holmes et al. 1986). Populations of most warblers are surprisingly constant in density from year to year, although

Table 12.1 Yearly survival of juvenile and adult passerine birds.

Common name	Scientific name	Yearly survival[a] (percent)	Authority
Paruline warblers			
Yellow Warbler	*Dendroica petechia*	53	Roberts 1971
Kirtland's Warbler	*Dendroica kirtlandi*	62	Walkinshaw 1983
Prairie Warbler	*Dendroica discolor*	65	Nolan 1978
Ovenbird	*Seiurus aurocapillus*	85	Roberts 1971
Northern Waterthrush	*Seiurus noveboracensis*	72	Roberts 1971
Common Yellowthroat	*Geothlypis trichas*	54	Roberts 1971
Black-and-white Warbler	*Mniotilta varia*		
Canada Warbler	*Wilsonia canadensis*	71	Roberts 1971
American Redstart	*Setophaga ruticilla*		
Permanent residents or short-distance migrants			
Barn Swallow	*Hirundo rustica*	37	Lack 1949
Blue Jay	*Cyanocitta cristata*	55	Hickey 1952
Marsh Tit	*Parus palustris*	53	Plattner and Sutter 1947
Blue Tit	*Parus caeruleus*	46	Plattner and Sutter 1947
Great Tit[b]	*Parus major*	54	Plattner and Sutter 1947
Great Tit[b]	*Parus major*	47	Plattner and Sutter 1947
European Redstart	*Phoenicurus phoenicurus*	38	Ruiter 1941
European Robin	*Erithacus rubecula*	38	Lack 1948
American Robin	*Turdus migratorius*	48	Farner 1949
Common Starling[b]	*Sturnus vulgaris*	48	Lack and Schifferli 1948
Common Starling[b]	*Sturnus vulgaris*	37	Lack and Schifferli 1948
Common Starling[b]	*Sturnus vulgaris*	42	Lack and Schifferli 1948
Northern Cardinal	*Cardinalis cardinalis*	52	Farner 1952
Song Sparrow	*Melospiza melodia*	60[c]	Nice 1937

a. Yearly survival highly significantly greater for paruline warblers than for permanent residents and short-distance migrants ($p < 0.001$ in two-tailed Mann-Whitney U-test).

b. Each sample taken from a different population.

c. Results when habitat not disturbed; yearly survival much lower after disturbance.

they experience temporary and short-term depressions, probably in response to environmental factors. Major fluctuations in my warbler populations were related to two poor breeding seasons, combined with probable heavy mortality in spring migration resulting from sustained rain and cold immediately after the birds arrived on their breeding grounds (Morse 1976b, 1980c). Over most of the study, numbers were constant. Holmes et al. (1986) also found that periods of stability were punctuated by pulses in response to outbreaks of caterpillars.

Forest Fragmentation

Great changes in population size are brought about by gross modifications to habitat, but changes can also occur at sites that have undergone little physical change. For example, fragmentation of the forests of the eastern United States into isolated segments has led to local declines of many migratory warblers and other passerines that winter in the Neotropical region.

Whitcomb et al. (1977, 1981) carried out one of the most detailed studies of this phenomenon on the Maryland Piedmont and coastal plain and compared their results with censuses taken at the same sites 30 years earlier. No warblers characteristic of the forest interior nested in forests of less than 15 hectares, and most species were missing from forests of over 50 hectares. Smaller forests contained only Yellowthroats and Yellow-breasted Chats, species typical of edge habitats. Nearby extensive forests maintained high densities of several other warblers, including Black-and-white Warblers, Hooded Warblers, Kentucky Warblers, Ovenbirds, and Louisiana Waterthrushes *(Seiurus motacilla)*; some Worm-eating Warblers were present as well. All of these species had similar densities 30 years earlier, except for the Worm-eating Warbler, which declined regionally. In another formerly extensive site, which had suffered fragmentation over the past 30 years, Worm-eating and Black-and-white warblers had disappeared, and most other forest-interior species had declined. Studies conducted to the north, in New Jersey (Galli et al. 1976), which included only small or medium-sized woodlots, nearly lacked the forest-interior species, leading Whitcomb et al. (1981) to surmise that the birds had already been extirpated from sites of limited size.

This effect is not primarily one of habitat change. Most of these warblers occupy a moderate range of habitats, although densities vary. For example, the density and diversity of most warblers are lower in upland xeric oak-hickory sites than in adjacent mesic and lowland deciduous areas, although Black-and-white Warblers and Ovenbirds maintain their population densities throughout these forest types. The latter species are particularly sensitive to size of area, however.

Several factors could play a role in this loss of species (Table 12.2). The different possibilities have been tested using the theory of island biogeography. Habitat islands, separated by other terrestrial habitats, differ from

Table 12.2 Factors that could affect abundance and diversity of warblers on habitat islands. (From Whitcomb et al. 1981; Lynch and Whigham 1984; and Wilcove 1985.)

Habitat factors
Island size
Isolation of island
Quality of habitat (food, microhabitat, etc.)
Heterogeneity of habitat
Traits associated with migratory strategy
Other insularization-related traits
Psychological avoidance of small tracts
Brood parasitism
Nest predation
Diffuse competition with permanent residents
Mortality during nonbreeding season
Colonization potential

islands separated by intervening water bodies in being "leakier": vagrants can survive temporarily in sites between the habitat islands. For both habitats we are interested in knowing whether a population is large enough to escape random extinction and whether dispersal will be great enough for colonization. It is questionable whether many species representations on habitat islands qualify as populations; they may be part of a larger breeding group of conspecifics that would qualify as a population in traditional island-biogeography theory. Alternatively, the continuing presence of a species on a habitat island could be solely the result of continual "recolonization" by individuals from the outside. If such individuals do not contribute to local gene pools, they are not a viable part of a population.

Size of area and isolation probably play important roles in the species composition of bird communities (Lynch and Whigham 1984); however, few workers have related species abundance to characteristics of the vegetation, isolation, and area. Those ignoring vegetation factors have generally concluded that area exerts a major effect on the composition of breeding bird communities. In a study testing all of these variables, however, Robbins (1980a) found that canopy height and isolation were the most consistently important predictors, and forest area was less frequently so.

Lynch and Whigham (1984) also related floristic, structural, isolation, and area variables of forest fragments to the presence of Neotropical migrants, including eight warblers, on the coastal plain of Maryland, an area adjacent to the Piedmont-based studies of Whitcomb et al. (1981). They included areas of 5–1,000 hectares, with much shorter distances between sites than those in the Piedmont study. Their technique, which enabled them to separate area from covarying factors, led to the conclusion that forest area, independent of its covariates, had little effect on most species' presence or abundance—though it strongly affected Kentucky Warblers. Neither did isolation play a significant part for most species, the Ovenbird being an exception. Instead, plant physiognomy and plant diversity played the most important roles for most species. Since average distances among sites were an order of magnitude smaller than in the Piedmont study (0.4 vs. 4.7 kilometers), the size effect might appear under more severe constraints. Askins and Philbrick (1987) found that greater insularity decreased the numbers of Neotropical migrant species over time, but this trend reversed itself as forest regenerated on previously deforested areas.

These differences raise the question of which factors determine the decline of warblers and other long-distance migrants following the fragmentation of their habitats into isolated patches. Whitcomb et al. (1981) related this decline to several variables, first of which is the opportunity for colonization if a population cannot sustain itself without recruitment from outside. If offspring from a small area disperse widely prior to breeding, a common feature in many species (Nolan 1978; Probst 1986), these isolated birds could hypothetically contribute to the regional population while they survived, but the population, paradoxically, could eventually become extinct because of the dispersal of the young. Whitcomb et al. suggested that small areas could be behaviorally rejected by would-be colonists, and this habit is seen in Kirtland's Warblers (Probst 1986).

Having modest clutch sizes, usually only one brood, and open nests placed near to or on the ground, Neotropical migrants are especially vulnerable to fragmentation, in spite of their high survivorship. Competition and predation from adjacent habitats is enhanced in fragmented habitats (Whitcomb et al. 1981). Although concern has been expressed that deforestation and habitat change on tropical wintering grounds (Keast and Morton 1980) lead to fragmented habitats, Whitcomb et al. argued that if this were so declines should occur at both large and small sites, which is not the pattern observed.

In order to conclude that fragmentation accounts for the low frequency of many species on small habitat islands, one would have to establish that populations on large islands or mainland areas are at saturation. Otherwise, decreases in island populations may simply be the consequence of island individuals filling vacant mainland sites as they become available.

Whitcomb et al.'s study areas met this criterion, judging from their census results over a long period. Yet sometimes source populations may not meet that requirement. For instance, two poor breeding seasons decreased the number of Black-throated Green Warblers so much that populations on true islands and other suboptimal habitats declined, probably because birds immigrated from them to more preferable areas (Morse 1976b, 1977b). Had the study not commenced well before declines occurred and continued until a subsequent increase, one might have easily interpreted the results as Whitcomb et al. did.

Isolation may not have to be extensive for declines to take place (Whitcomb et al. 1981). After Cabin John Island, Maryland (an island of 7.3 hectares in the Potomac River that is separated from mainland forest by only a few meters), was isolated from the nearest mainland forest by the construction of a highway, Neotropical migrant species declined precipitously and there were compensatory increases by such permanent residents as starlings. This habitat change was initially accompanied by a slight increase in density by two species, the Northern Parula Warbler and the American Redstart. The extra birds may have come from the area of forest cut down for the highway, but in subsequent years recruitment was inadequate to balance off losses. The populations of the two warblers leveled off several years later at a new low density. Simultaneously, Kentucky Warblers and certain other nonparuline Neotropical migrants disappeared altogether.

Nest predation by species of open and edge habitats (primarily Blue Jays, Common Crows, and Common Grackles), often assumed to decrease population size, should be greater as access to edge and open areas increases. Wilcove (1985) investigated the role of forest size on nest predation by placing out artificial aviary nests, containing quail eggs, in forests of varying size in the eastern United States. The effect, as shown in Figure 12.1, was strong. The only extensive area of primeval forest in Wilcove's study, the Great Smoky Mountains of Tennessee (2,100 square kilometers), had by far the lowest rate of nest predation (2 percent). Predation was significantly higher in a 900-hectare forest in Maryland (18 percent), but lower than that of several smaller tracts (4–13 hectares) and a tract of 280 hectares with extensive frontage onto other habitats. Among the small forests, rural sites had lower predation (48 percent) than did suburban ones (71 percent). Also, nests placed on the ground sometimes had higher predation rates than ones placed 1–2 meters off the ground.

This experiment suggests that nest predation is far more pervasive in isolated woodlots than in large or virtually tractless areas. Large areas have

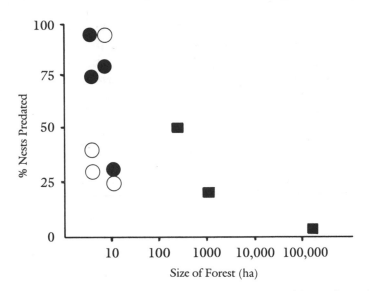

Figure 12.1 Experiments on nest predation as a function of forest size, using artificial nests with quail eggs. Squares = large forest tracts; white circles = rural fragments; black circles = suburban fragments. (Modified from Wilcove 1985.)

a lower edge effect, and since the most important avian nest predators (Blue Jay, Common Crow, Common Grackle) are edge species, they may have a disproportionately heavy impact. Also, large predators, such as Bobcats *(Lynx rufus),* Black Bears *(Ursus americanus),* and Cooper's Hawks, absent at smaller sites and completely represented only in the Smokies in this study, may regulate numbers of smaller predators (Matthiae and Stearns 1981). Wilcove sampled tracks about his experimental nests and found that several mammals took the eggs, including Opossums *(Didelphis virginiana),* Striped Skunks *(Mephitis mephitis),* and Raccoons *(Procyon lotor).* In addition, dogs and cats, which are typically associated with small lots because of their dependence on humans, preyed on nests. Blue Jays were also observed taking eggs at two sites.

Although these nests were probably more conspicuous to the predators than natural nests are, the results could be proportional to natural rates. Further, these rates of predation do fall within the range of those reported in other studies (e.g., Nolan 1963; Willis 1973). Among the birds most vulnerable to mammalian predators are some Neotropical species, including Black-and-white Warblers, Worm-eating Warblers, Ovenbirds, and Hooded Warblers, all ground or low-elevation nesters. All of them have declined when their habitats became fragmented. In contrast, there was no predation on eggs placed in artificial cavities with openings similar to that of a Downy Woodpecker (*Picoides pubescens*) hole, which is consistent with the abundance of hole nesters, mostly short-distance migrants or resident species, in these forest fragments.

Another possible basis for declines in Neotropical migrants is parasitism by Brown-headed Cowbirds. Cowbirds require open areas for feeding and social displays, thus confining their distribution during the breeding season. Given the importance of open areas, adjacent sites should be most favorable hunting grounds for them, especially since forest edges support the highest densities of host nests. Past 15 meters from the edge, numbers of host nests remain quite constant as one moves inward (Gates and Gysel 1978). Therefore, edge habitats should be most satisfactory for cowbirds, and areas would become less satisfactory as one moves into the forest because of the increasing distance from feeding and display sites.

To test this hypothesis, Brittingham and Temple (1983) divided study nests into areas that were less than 100 meters, 100–199 meters, 200–299 meters, and over 300 meters from an opening. Their results fitted the prediction of greater parasitism near the openings (Figure 12.2), and thus could also account for why small forest fragments support few high-risk species; they are edge habitats as far as cowbirds are concerned.

Brittingham and Temple's study does not determine whether cowbirds are a major factor driving the decline in numbers of Neotropical migrants in forest fragments, but it appears plausible. A combination of brood parasites, predators, and competitors probably all play a role in regulating the composition of these communities (Ambuel and Temple 1983). Under some situations, as with the Kirtland's Warbler (Walkinshaw 1983), cowbirds definitely have a major impact on population size.

Diversity and Partitioning

MacArthur (1972) noted that several explanations could account for the diversity of a species group: opportunities for speciation, hazards causing loss of species, levels of packing by competing species, the benignity or stability of climates, the complexity and divisibility of environments, the productivity of environments, the effect of predation on competition and consequent diversity, and the removal of species by predators. At least four of MacArthur's explanations for diversity directly include some aspect of resource partitioning among species; others include indirect as-

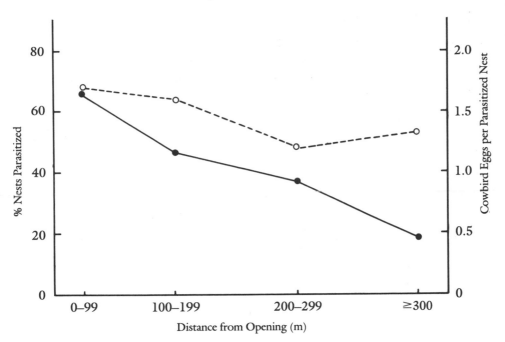

Figure 12.2 Percentage of nests of small passerine birds parasitized by Brown-headed Cowbirds (black circles), and number of cowbird eggs per parasitized nest (white circles), as a function of distance from nearest opening of 0.2 hectare or larger. (Modified from Brittingham and Temple 1983.)

pects. All but speciation (which was treated extensively in Chapter 1) operate on an ecological time scale. Certain of these factors are more important than others, and some have not been addressed as they relate to warblers. In discussing questions of diversity and partitioning I will focus on the genus *Dendroica,* making appropriate comparisons with other warblers and other insectivorous gleaners.

The rich assemblage of wood warblers in the northern spruce forests led MacArthur to ask why so many species occupied a single area. Fits between the number of bird species and the structure of the habitat (foliage height diversity) were generally good in several eastern forests (MacArthur and MacArthur 1961) but not elsewhere. Subsequently he attacked the diversity problem by developing species-packing theory (MacArthur 1969, 1970), in which the question of limiting similarity became central; that is, how similar can species be and still coexist? Using matrix techniques, he (MacArthur 1968) could predict the conditions in which varying numbers of species should coexist and the consequences of a species disappearing from the mix. The similarity of coexisting species' demands is closely linked to the variance in resource availability (May and MacArthur 1972). Since the assemblages of spruce-woods warblers are temporary summer groupings, which do not reoccur in the same form on their wintering grounds (MacArthur 1959), the period of variance is circumscribed.

The Spruce-Woods Warblers

Although the diversity of spruce-woods warblers is unusually high for a temperate-zone bird assemblage, differences in their species composition on small islands suggest that there are constraints on the number of species that may coexist in a given area. Fewer species occupy the smaller islands, implying that biologically significant overlaps between species are greater on the small islands than on the mainland. One can predict which species will drop out—it is not a random assortment; and overlap increases as island size decreases (Morse 1971a). Small islands may not provide enough discrete spaces to accommodate the missing species, even though these species occur within a short flight of the islands.

Other evidence suggesting limitations on assemblages of spruce-woods warblers comes from three additional coniferous-dwelling *Dendroica* species that do not occupy many of these forests in the northeastern United States and southeastern Canada. Two of them, the Bay-breasted and Cape May warblers, are characteristic of Spruce Budworm outbreaks, and their numbers apparently wax and wane with the insects. Both nested on Hog Island, along the central coast of Maine, in the 1930s (Cadbury and Cruickshank 1937–1958), which means there may have been an earlier lepidopteran outbreak on the island. At least the Bay-breasted Warbler still occurred 90 kilometers to the east of Hog Island at MacArthur's study site when he worked there in the mid-1950s (MacArthur 1958). However, they do not now nest at that site: the Cape May Warbler was not present in 1965 (Morse, unpubl.), and the Bay-breasted Warbler was gone by 1967 (R. H. MacArthur, pers. comm., 1968). Since Blackburnian Warblers and others may decline when Bay-breasted and Cape May warblers increase (Morris et al. 1958), equilibrium conditions may have been violated. Blackburnian Warblers increased noticeably on MacArthur's sites after the Cape May and Bay-breasted warblers disappeared.

We do not know whether declines in Blackburnian Warbler abundance result from shifts in vulnerability to nest predation, aggressive interactions, or whether budworms either do not constitute an adequate diet for the warbler species present every year or deplete the food supply for other insect prey. The normal foraging areas of Bay-breasted Warblers overlap the Blackburnian Warbler's (MacArthur 1958; Morse 1978a), enhancing the plausibility of partial exclusion. If the changes involve the quality of the food supply, this would be of great interest, for quality, not quantity, would be of prime importance. Alternatively, if interspecific aggression comes into play, we would want to know if it occurs when food is abundant. Aggressive behavior requires energy and time and thus is likely to be selected against unless the costs opposing it are lower than its overall benefits. If interspecific aggression provides major benefits at all other times, that alone could favor its routine incorporation in the bird's behavioral repertoire. The advantages could occur mainly under conditions in which they face a food shortage—the poor seasons between the good, when their numbers decline.

Some evidence suggests that certain warblers fare better on monotonous diets of lepidopteran larvae than do others. In pilot studies (Morse 1971b) captive Bay-breasted Warblers remained in good condition when fed a diet of Wax Moth *(Galleria mellonella)* larvae over a few months, but Yellow-rumped Warblers did not maintain their weight on this monotonous diet and would probably have eventually succumbed if not removed from it.

The distribution of a closely related spruce-woods species, the Blackpoll Warbler, further indicates limitations on diversity, since it seldom coexists with either Bay-breasted or Cape May warblers. The Blackpoll Warbler is morphologically similar to the Bay-breasted Warbler, with which it is sometimes considered a superspecies (Hubbard 1971), and it feeds in similar parts of the vegetation, namely in the upper parts of spruce and fir trees, especially near the trunks (Morse 1979). Toward the southern edge of its range, it is limited to mountainous areas. Its relative abundance in the avifauna is greatest on the tops of the mountains, where stunted spruce-fir forest grows. There, the only co-occurring *Dendroica* species is the Yellow-rumped Warbler, a species that forages in lower parts of the vegetation, especially on dead limbs. Blackpoll and Bay-breasted warblers do not coexist as breeding species in many areas (Morse 1979), although they both occur in the Green River watershed of northern New Brunswick and adjacent Quebec (Erskine 1977, 1980). They share a range that is tortuously contiguous, with only modest overlap, in many mountainous

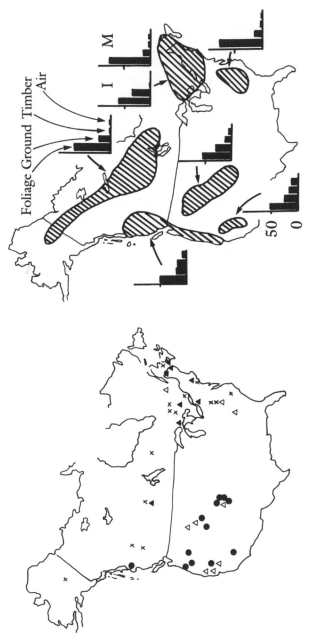

Figure 12.3 *Left*: Proportion of paruline warblers in breeding avifaunas of coniferous forests in different parts of North America. Circle = 0–10%; white triangle = 11–25%; cross = 26–50%; black triangle = 50–75%. *Right*: Proportions of different ecological guilds of breeding birds in coniferous forests of different regions. Divided into immature (I) and mature (M) stages of forests in the northeastern region. (Modified from Wiens, 1975.)

areas of the northeastern United States and adjacent Canada. Maps and general regional descriptions give the impression of their being more widely sympatric than they are. I do not know what factors permit their coexistence, for the evidence consists of censuses of breeding birds, rather than studies of their relations. However, Erskine's (1977) data show that most areas of coexistence are at sites with many budworms. Another southern population of Blackpoll Warblers nests in the coastal spruce forest, from southwestern New Brunswick eastward, and at times as far west as the coast of central Maine. There it coexists with Bay-breasted Warblers only occasionally, if ever. But most Blackpoll warblers nest to the north of the Bay-breasted Warbler in the northern part of the coniferous forests across Canada and Alaska.

Contrary to the popular theme that stability promotes diversity, the highest numbers of spruce-woods *Dendroica* species are linked to fluctuations of Spruce Budworms or other defoliating insects. To maintain an equilibrium, species apparently require a stable pattern (minimum dependable conditions) provided by the predictable insects in the spruce forests (Morse 1977b). Outbreak insects can be so unpredictable that it is impossible for birds to track them in time. If that is true, we must conclude that maximum diversities are linked to temporary (nonequilibrium) conditions. This pattern is consistent with a recent explanation of high tropical diversity, that areas with maximum diversity are not in equilibrium: local extirpation has not run its course (Connell 1978). Spruce-woods warblers are interesting, though, for being unusually diverse in constant (nonoutbreak) conditions where food supply and warbler density remain rather constant from summer to summer (Morse 1976b, 1977b). It is important to keep in mind, though, that influential events could occur at other times of the year.

The diversity of *Dendroica* in western coniferous forests is far lower than that of the northeastern forests (Figure 12.3). The basis for this difference is not clear, but it has often been attributed to the time available for colonization. Nor is it clear how quickly these diversity patterns can change, although I argue in Chapter 1 that the diverse community of the eastern forests may not be new; it could therefore have been in place well before western races and sibling species split off during recent Pleistocene glaciations. Wiens's (1975) analysis of coniferous forest communities indicates that proportions of foliage-frequenting birds in the east are higher than those in other parts of the country; therefore, rather than other groups of birds taking warblers' places in the west, it looks as though fewer opportunities are available for foliage-gleaning species there (Wiens 1975). Wiens found that this difference could result from there being less prey in western than eastern coniferous foliage. Ficken et al. (1968) noted that the northeastern forests have much larger standing crops of insects than do the western ones.

Pine-Forest Dendroica *Warblers*

Few *Dendroica* species nest in the hard-pine forests of the southeastern United States. The Pine Warbler is usually the only *Dendroica* species present, even though it often attains a density that constitutes a high percentage of bird abundance. The addition of the Yellow-throated Warbler into the forests of the Delmarva Peninsula is an unusual phenomenon. In a similar sense, southwestern and tropical hard-pine forests support only a single *Dendroica* species, the Grace's Warbler (Howell 1971; Cody 1974). This pattern correlates with the low abundance and diversity of insect prey on these pine trees (Wahlenberg 1946); the forests are in fact a major source for turpentine natural products which protect the trees from insect attack.

That this pattern is not simply the result of the dominance of a single *Dendroica* species can be seen by the co-occurrence of Pine Warblers with Black-throated Green Warblers in northern White Pine forests (Erskine 1977; pers. observ.). White Pines harbor more insects than do the southern hard pines (see Wahlenberg 1946). Yellow-throated Warblers occupy a far more specialized niche where they co-occur with the Pine Warbler on the Delmarva Peninsula than do the Black-throated Green and Pine warblers in their areas of co-occurrence.

Warbler diversity thus also appears to be due to differences in food abundance in nonoutbreak conditions. As predicted by Pianka (1966) and others, a more productive habitat should support more species than a less productive one. But some species have exploited special features of the environment, such as the use of pine cones by the Delmarva Yellow-throated Warblers, and diversity has increased as a result. This adaptation parallels that of tropical forest birds that specialize on resources inadequate for populations in temperate zones—for example, dead-leaf foraging (Greenberg 1987a)—a factor attributed to the constancy of conditions in tropical situations.

Since many of the great southeastern pine forests are composed of Longleaf Pines, it seems unlikely that Yellow-throated Warblers would find a comparable refugium there, for the cones of the Longleaf Pine are much larger and easier to probe than those of the Loblolly Pine. It stretches the imagination to suggest that the dimensions of open Longleaf Pine cones would exclude Pine Warblers. Although I have observed Pine Warblers hunting for arthropods in Longleaf Pine cones, during years of heavy mast production (the seeds from these trees are used as animal feed) they regularly take pine seeds, which entails probing deeply within the cones (Morse 1967c, 1970a, unpubl.). I do not know of Yellow-throated Warbler populations in Longleaf Pine forests. Yet another factor might reduce the value of Longleaf Pine cones as refugia, even if they proved to be more readily exploited by Yellow-throated Warblers than by Pine Warblers. In the Longleaf Pine, several years often pass between seasons of substantial cone production. Even though cones may remain on the trees long after production, overall numbers are likely to be greatly reduced after several years have followed a major crop. Loblolly Pine exhibits a less highly cyclic pattern of cone production.

The Twig Gleaners, Vermivora *and* Parula

In comparison to the variability of coexisting *Dendroica* species, only one species of *Parula* or *Vermivora* usually occupies a habitat. These birds are considerably smaller than most *Dendroica*, and they concentrate their foraging on tips of branches. They do not partition this area among themselves, even though after assessing *Dendroica* foraging one might imagine that they could separate out by height. In coastal spruce forests of Maine the Northern Parula Warbler is the local representative. Although placed in its own genus, the Parula Warbler resembles the *Vermivora* behaviorally and ecologically and is placed between *Vermivora* and *Dendroica* in standard classifications of the Parulinae.

Nashville Warblers are the common *Vermivora* in young deciduous second-growth forests of the same region and hence are in minimal contact with the spruce-forest species. Tennessee Warblers inhabit mixed coniferous-deciduous forests across much of Canada, and they recruit to budworm outbreaks in great numbers (Kendeigh 1947; Erskine 1977). Erskine's censuses suggest that their role is similar to that of the Northern Parula Warbler. I have found no information on whether they interact with kinglets, as does the Northern Parula Warbler (Morse 1967a).

Elsewhere, Northern Parula Warblers are common and widespread in the great swamp forests of the southeastern United States, but they are

not in contact with any *Vermivora* species, although the Bachman's Warbler occupied a twig-gleaning niche in this region.

Diversity, Partitioning, and Population Limitation of Warblers

No simple explanation accounts for all of the relationships among diversity, partitioning, and population limitation of warblers, which is not surprising in light of the large number of possible explanations for even the
diversity of a species group (see MacArthur 1972). The combination of
two basic strategies—species that live in relatively constant combinations
and densities, and species that are largely confined to insect outbreaks—
further guarantees a wide variety of response to these variables.

The evidence for competition playing a role in population limitation
and partitioning of warblers is strong. Aggressive interactions take place
regularly, both within and among species, and maximum population levels tend to be quite constant in most species, suggesting that the birds are
at or near carrying capacity. Also, since competitive interactions may take
place at an individual level, even if a population is not at carrying capacity,
some members may experience competition, probably those occupying
the most favorable sites (Martin 1986). This conclusion follows from
Fretwell and Lucas's (1970) model of ideal free space, in which the most
favorable sites are filled first and others are filled only when the costs
involved exceed the benefits gained from contesting for the best sites.
Obviously competition will have the strongest effect on community structure when populations are at carrying capacity. Although typically associated with food abundance, competition at this time could be associated
with spacing out to minimize nest predation (see Martin 1988b). The
relative importance of these two alternative forces is not yet understood.

The warblers are clearly very good at partitioning their habitats. Tall
vegetation with high prey abundance appears necessary for a population
to attain a high level of diversity, as suggested by the warblers found in
the northeastern, southern, and western coniferous forests. Of particular
interest is the genus *Dendroica*, which dominates exploitation of foliage
and branches and even ventures out to feed in the air column. But it does
not attain primary use of the tips of the vegetation, which is exploited
primarily by *Vermivora*. *Vermivora*, on the other hand, attains a diversity
of only one species per habitat. This difference in diversity between the
two genera is not simply the consequence of deficient exploitation abilities
on the part of *Vermivora*, because where *Parula* becomes the tip-foraging

warbler in the coastal spruce forests it plays a similar role as the only species specializing on this segment of the habitat. The strong size constraints associated with tip foraging may prevent *Vermivora* from exploiting more stable substrates in the presence of *Dendroica*, which almost completely overlaps it geographically. In rich habitats *Dendroica* species numbers are three or more times that of *Vermivora*, which also matches the ratio of species in the two genera (27 to 9), implying that the pattern is a widespread one.

For the most part warblers partition the habitat spatially, rather than by exploiting different food resources, as seen among species that partition resources by size (Hutchinson 1959; Schoener 1965). Schoener found that the *Dendroica*-like pattern predominated among groups of species that fed on common foods, as these high-density species do. Further, the range of insect sizes within temperate-zone warbler substrates may be too small to favor a range of predator sizes, since these prey can be readily picked from the small limbs and foliage. The *Vermivora-Dendroica* partitioning provides the major exception. No *Vermivora* penetrate the southern pine forests, which could be a test of this proposal, since the robust foliage of these trees could favor tip foraging by *Dendroica*. The nature of this relation is blurred, however, by the low food abundance of these areas, which argues for low diversity, also. Excavators do exhibit size-related patterns of separation in the same habitats as warblers, and show correlated differences in beak size providing differing abilities of excavation; Downy and Hairy *(Picoides villosus)* woodpeckers illustrate this pattern. But no warblers show analogous traits.

Maurer (1984) reasoned that interference competition should lead to spatial partitioning and that exploitation competition should lead to resource partitioning among species. The way that spruce-woods *Dendroica* species partition their habitats is consistent with this hypothesis, since they are spatially separated and show aggressive interference (Morse 1976a). In contrast, exploitation systems could, according to Maurer, lead to size-graded species groups. If opportunities for size-graded partitioning are limited because of an inadequate range of resource variation—if, for example, the size range of insect prey on coniferous foliage is limited—the opportunity for spatial partitioning may consequently be enhanced. Spatial partitioning could be based either directly on resource availability or on spacing out to avoid predators. MacArthur and Wilson (1967) predicted that foragers confronted with more competitors should reduce the variety of spaces they fill, rather than the types of resources they use. In mobile animals like birds, changes in spatial separation occur efficiently and are subject only to behavioral constraints, but efficient resource shifts, which can depend on morphological characters as well, seem more likely to develop over evolutionary time. MacArthur and Wilson's arguments were generated from the Lotka-Volterra competition equations (Maurer 1984), which would automatically lead to this prediction. Thus, although Maurer's hypothesis of niche-partitioning may predict the same results for active, aggressive animals like warblers as does MacArthur and Wilson's compression hypothesis, it raises different predictions for other animals and therefore deserves attention.

The usual diversity of warblers is enhanced during insect outbreaks by the insinuation of species like Cape May and Bay-breasted warblers into the community, thereby overriding normal limitations. These are species that appear to depend heavily on the outbreaks for their existence over much of their ranges. This resource-fueled species increase suggests that the normally observed spatial partitioning limits diversity only when it operates within a restricted set of conditions (for example, certain food abundances or predator pressures at nest sites), but that major changes—here the augmentation of food resources—may trigger a new level of species diversity and density of individual species. At such times, the relationship could even shift from one in which numbers of safe nesting sites are of major importance (Martin 1988b) to one in which food abundance or residual aggression directly affect the numbers. If these outbreaks are not predictable and frequently repeated, nest predators should not make a large enough inroad into warbler numbers to suppress the warblers' overall response to the greatly enhanced resources. However, the high overall numbers of warblers that follow such a change in resource conditions could dictate new population densities for species that are sensitive to nest predation. The decline of certain species, such as Blackburnian Warblers, during outbreaks could be a direct response either to the invaders themselves or to nest predation pressures in the face of high prey densities. Since the insects are usually contained rather quickly, either by resource depletion or by natural enemies, they are unlikely to continue long enough for these birds to reach a stable new equilibrium.

13   Hybrids

Closely related species of parulines frequently share the same habitat or geographic area, yet they seldom hybridize, suggesting that strong isolating mechanisms are at work. Even so, hybridization sometimes occurs when species' ranges barely overlap and the population density of one or the other is low. Under these conditions individuals encounter unfamiliar neighbors and have little chance of finding a conspecific mate. They may then form mixed pairs—in some instances the two members may even be of different genera.

Following Mayr (1963), I define hybridization as the crossing of individuals belonging to two unlike natural populations. This definition thus includes hybrids produced when two populations completely merge into intermediate forms along lines of contact as well as hybrids between two parental populations that retain their morphology and genetic integrity at sites of contact. The complete merging of characters of the western and eastern representatives of the Yellow-rumped Warbler, the Audubon's and Myrtle warblers, where they come into contact in the northern Rockies, indicates that the parental populations do not qualify as different species, for they do not remain discrete when they come together; instead, a cline of characters may be seen where the two races meet (Hubbard 1969; Barrowclough 1980b). This example lies at one end of a continuum of biological discreteness that spans from races that massively hybridize and completely merge characters to species that do not even hybridize. In contrast to the merging of traits in Myrtle and Audubon's warblers, Blue-winged and Golden-winged warblers retain their discrete parental types, despite some interbreeding (e.g., Gill 1980); hence their parental forms constitute discrete species. The term *hybridization* as I use it here does not indicate the position of populations in the speciation process.

Here I analyze several examples of warbler hybridization and concentrate on ecological factors that maintain or break down barriers to interbreeding. Where possible, I consider the traits of hybrids, a study that can help to unravel the nature of species differences over a broad spectrum—from resource exploitation to species discrimination.

Golden-winged and Blue-winged Warblers

For over a century, the closely related (Gill 1987) Golden-winged and Blue-winged warblers have attracted great ornithological interest because of their hybrids. Indeed, they are probably the most intensively studied pair of hybridizing bird species in North America (Gill 1980). At one time, two of the intergrades, Brewster's and Lawrence's warblers, were treated as distinct species, receiving their own scientific names. The parental species introgress (genes from one population infiltrate the gene pool of another) considerably more than originally believed and create a wide range of intermediate colorations, some detectable only when in the hand (Short 1963, 1969). Although the parental species regularly hybridize if they come in contact, mixed pairs usually do not exceed 5–10 percent (Gill and Murray 1972a). Interactions between the two species result from recent contact, probably caused by human-induced habitat change, which has triggered rapid local and regional shifts in abundance over the past 150 years and, with them, greatly increased opportunities for contact.

Gill (1980) documented the pattern of change in Connecticut, which parallels changes elsewhere. First the Golden-winged and then the Blue-winged warbler increased in the late 1800s and early 1900s. Within 50 years local distribution changed from 100 percent Golden-winged to nearly 100 percent Blue-winged warbler phenotypes throughout the area, a pattern similar to that in other areas where Blue-winged Warblers have invaded.

The basis for this predictable shift is not clear. Competition could play a part, since their habitats overlap broadly (Ficken and Ficken 1968d). The two species do not have noticeable foraging differences, and neither species excludes the other from its territories (Ficken and Ficken 1968d; Murray and Gill 1976). Confer and Knapp (1979) noted, though, that in central New York, Golden-winged Warblers require younger vegetation than do Blue-winged Warblers; thus the loss of satisfactory habitat could diminish Golden-winged Warbler abundance, even in the absence of Blue-winged Warblers. But Blue-winged Warblers nest in most places that Golden-winged Warblers do, in a pattern that resembles the included-niche model, which predicts that a completely overlapped species will be eliminated unless it is the superior to others in exploiting an area (Miller 1967; Morse 1974b). Gill concluded that waves of colonization were influenced by the formation of young second-growth forest in fields abandoned in the mid-1800s, when access to rich farmland to the west became available. As Confer and Knapp (1981) pointed out, habitat modification caused by human settlement produced an abnormally large amount of satisfactory

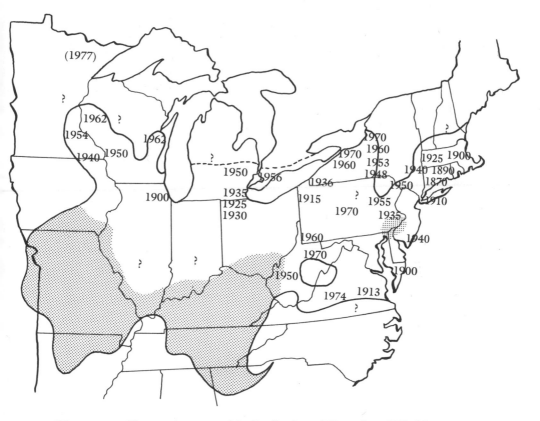

Figure 13.1 Changes in geographic distribution of Blue-winged Warblers. Dates indicate when the species first became established at a locality. Darkened area is approximate range in the mid-1800s. (Modified from Gill 1980.)

habitat, but current land-use practice has resulted in a decline of satisfactory habitat. Similar, later waves of colonization to the west, in such areas as central New York, northern Ohio, and Michigan (Figure 13.1), indicate that the Connecticut story is probably typical of the relations between these two species.

Originally, Blue-winged Warblers occupied areas largely west of the Appalachians (Gill 1980, 1987). In the late 1800s only two local populations were known in the east, in the Hudson and Delaware valleys; when they arrived is not known. They spread rapidly, pushing the Golden-winged Warbler in front of them, both northward, as in Connecticut, and to higher elevations, into which they did not spread as rapidly as along river valleys and lake shores.

As Blue-winged Warblers occupied its former territories, the Golden-winged Warbler moved northward. It now occurs as far north as northern Wisconsin and Minnesota. Nevertheless, Gill predicted that if the pattern continues, the Golden-winged Warbler may be rare or even extinct within a century. Confer and Knapp (1981) suggested that habitat modification may be necessary to maintain them, since large amounts of old-field succession are unlikely to be available in the future, and standard forestry practices do not suffice because of dense stands of sprouts that occur after an area has been logged.

If this rapid displacement is related to the ecological similarity of the two species, they may be too similar to coexist. This pattern is seen in closely related species-pairs that are reproductively isolated, such as Reed (*Acrocephalus scirpaceus*) and Sedge (*A. schoenobaenus*) warblers in England (Catchpole 1973) or the Willow and Alder (*Empidonax alnorum*) fly-catchers in the northern United States (Stein 1958). However, these species differ in being interspecifically territorial where they come together, and they have extensive ranges, ecological or geographical, in which they are allopatric.

Hybridization and introgression could also play a part in this rapid shift to Blue-winged Warbler phenotypes, although data currently available do not enable one to evaluate their role. As Gill pointed out, hybridization and introgression can siphon off a fraction of each generation of "pure" Golden-winged Warbler genotypes. If Blue-winged Warblers increased more rapidly than Golden-winged Warblers for other reasons, this difference could contribute to a decline in Golden-winged Warblers if resources are limiting. Stabilizing selection could eliminate conspicuously intro-gressed forms, thereby restoring the integrity of the type after the Blue-winged Warbler engulfs an area. Even so, there is little conclusive evidence that hybrids and backcrosses are less successful than parental species, except for Ficken and Ficken's report (1968a) that hybrids obtain mates less readily. Replacements of Golden-winged Warblers by Blue-winged Warblers occur so fast, though, that time is probably inadequate to reinforce premating isolating mechanisms. In general, selection against hybridization is not strong.

Ficken and Ficken based their arguments for the existence of isolating mechanisms on several lines of evidence. First, hybrid birds required twice as long to obtain mates as did either of the parental species (Ficken and Ficken 1968a). They also reported that 54 percent of hybrid males went unmated, versus 9 percent of the parental forms in the same areas (the combined results from several studies) (Ficken and Ficken 1968a). Hybrid females chose the commoner of the two parental species as mates, and more parental forms mated with hybrids than with the opposite parental species (Ficken and Ficken 1968c). A major reason for the disproportionate frequency of female backcrossing is that male hybrids are not as aggressive as either parental type in their interactions with females during early courtship, and this high level of aggression is a fundamental part of the pairing repertoire.

If Ficken and Ficken's results can be applied to all interactions between Blue-winged and Golden-winged warblers, they would go a long way toward establishing the existence of isolating mechanisms; however, more work is needed. It is critical to know whether individuals breeding in an area differ from ones that cannot find sites at that area and move on, or ones that move on for other reasons. The identity of the individuals is also crucial. Lastly, a sense of the birds' genetic makeup is needed; examinations in the hand would be useful, as a first step, in this regard (Short 1969).

It is not clear whether this system works the same in all areas. In other instances in which "semispecies" come into secondary contact in more than one area, they are apt to behave differently from one place to another (Mayr 1963). Information on differences among these populations is inadequate to explore this crucial question in depth. Gill and Murray's comparisons show that Maryland and Michigan populations differ in key characteristics, such as singing behavior, territory size, and movements of males outside the territory. Some differences might be environmentally induced; others could result from normal between-population genetic differences; still others could arise from varying introgression and time of contact.

Short (1969) stated that Ficken and Ficken's arguments for selection against hybrids are inadequate and subject to other interpretations because the researchers did not use marked birds, nor assess levels of introgression in birds they studied, nor study the makeup of birds that failed to establish themselves in a population. But none of these critiques explicitly refutes the Fickens' arguments, and the discrimination they reported may indeed take place. Gill's conclusions are based more fully on his and Murray's collaborating work.

Perplexing factors govern the level of hybridization, which varies from place to place (Gill and Murray 1972a) and probably reflects the stage of interaction. In central New York, Ficken and Ficken (1968a) found that

Blue-winged Warblers arrive several days before Golden-winged Warblers; this means that opportunities for mixed-pair formation are lower because most Blue-winged Warblers have mated before Golden-winged Warblers arrive. Yet differences in arrival dates in Michigan are not as great as in New York (Berger 1958; Murray and Gill 1976).

These warblers have an extremely short courtship period, 1–2 days, compared with several days for members of other warbler genera (Ficken and Ficken 1968a). This factor might enhance the formation of mixed pairs, since discrimination would have to occur over a short period of time. One might expect more "mistakes" than if there were a longer period, but if isolating mechanisms do not inhibit the formation of mixed pairs (Short 1969), the length of the pair-bonding period may be irrelevant.

Mixed pairings are most common when one or both species are at low densities and thus unlikely to provide conspecific mates. As Blue-winged Warblers move northward into Golden-winged Warbler populations, mixed pairs occur more frequently on first contact than when numerous Blue-winged Warblers are present. Gill and Murray (1972a) found in playback experiments that male Blue-winged Warblers discriminated between their songs and Golden-winged Warbler songs more accurately in areas where they had substantial contact with Golden-winged Warblers (Michigan) than where they did not encounter Golden-winged Warblers (Maryland). Sympatric birds also varied less in their songs than did Blue-winged Warblers in allopatry (Lanyon and Gill 1964; Gill and Murray 1972b). Although one might attribute this result to selection for improved discrimination, Gill and Murray attributed the high discrimination of the individuals in sympatry to experience, that is, learning that a certain song represents a certain plumage type. As an example, they cited the responses of a Blue-winged Warbler to a Golden-winged Warbler phenotype that sang Blue-winged Warbler songs. Although initially responding strongly at a distance, eventually this Blue-winged Warbler ceased to respond to the other bird's Blue-winged Warbler song but did respond to other Blue-winged Warbler songs in a normal way.

Discrimination might be auditory, visual, or both. Blue-winged and Golden-winged warblers do not respond aggressively to their opposites' songs or plumage. Ficken and Ficken (1968d) asserted the primary importance of visual cues, particularly the head-on colors (head, neck, breast), which differ strikingly in these birds. They based their conclusions on individuals, mainly hybrids, with varying songs and plumages. If plumage and song differed, they gave little response to each other, and territorial overlap was extensive. If plumage was similar and songs differed, encounters occurred. But if plumage was different and song similar, no interactions took place. The results in the latter two categories are based on only 1 or 2 individuals, however.

Murray and Gill (1976) concluded, from another small sample, that auditory cues have an important role, as judged from birds with song patterns characteristic of the other species. A bird with Golden-winged Warbler plumage and Blue-winged Warbler song responded to Golden-winged Warbler vocalizations but not its own playback. In contrast, Blue-winged Warblers responded aggressively to this bird's songs, continually approaching it but not attacking. Apparently these cues function as one would anticipate—song at long range, plumage at short range.

Vocalizations show much less sign of mixing than do the plumage characters, although aberrant songs were inevitably associated with hybrids (Gill and Murray 1972b). This similarity might make vocalizations more trustworthy cues than plumage. In common with their parental species, male hybrids have one Accented Ending Song and one Unaccented Ending Song. Their Accented Ending Song is related to courtship, and thus is important in such situations.

These studies have been done on the discrimination of males. It is assumed that females, who exert the discrimination associated with pairing, respond in similar ways. Experiments are needed to test that assumption.

Sutton's Warbler

A second instance of recurrent hybridization is between Yellow-throated and Northern Parula warblers, two species normally placed in different, though adjacent, genera. Regardless of generic allocation, the two species are not as closely related as Blue-winged and Golden-winged warblers.

Hybridization between these two species happens rarely—less frequently than between Blue-winged and Golden-winged warblers. Hybrids, called Sutton's Warblers, were first reported in 1939 in West Virginia by Karl Haller, who, during one spring weekend, collected both a male and a female, the two some 30 kilometers apart. Carlson (1981) reported 15 observations of Sutton's Warblers through the summer of 1980, 11 of them in West Virginia, Virginia, and Washington, D.C., and another 3 on migration in the southeastern United States. The most comprehensive set of observations, however, and the only one including detailed behavioral and ecological information, was made in Indiana, west of other summer records. Most authorities doubt the migratory records, given the similarity of Sutton's Warblers to Yellow-throated Warblers when not in breeding plumage, and also given that their most conspicuous difference lies in their songs, which are much like Northern Parula Warbler songs.

As with the Blue-winged and Golden-winged warblers, hybrids occur where one or both parental species are rare. The timing of these reports during the past 45 years may not be coincidental, because Northern Parula Warblers are decreasing over the range in question, and Yellow-throated Warblers are spreading into this area. During this period Northern Parula Warblers have been less common in this area than to the north or south.

The Yellow-throated–Northern Parula warbler hybridization story remains a hypothetical case because no pair, plus offspring, has yet been observed. Even so, the morphological characteristics of Sutton's Warblers leave little serious question about their identity, although Haller originally described these birds as a new species. Most specimens or other reputable records probably result from interspecific pairings, and until recently there was no evidence for a backcross or other second-generation birds.

The plumage of the hybrids resembles the Yellow-throated Warbler closely, though lacking some of the breast striping of that species; in contrast, they sing songs that resemble Northern Parula Accented Ending Songs, the song given repeatedly during courtship. Most hybrids differ from typical Northern Parula Warblers in repeating the song twice in rapid repetition. Carlson (1981) suggested that the easiest way to find Sutton's Warblers is to listen for this characteristic vocalization. In a single instance, the range extension in Indiana (Ulrich and Ulrich 1981), experiments and observations were made on a hybrid male over most of a season and revealed a nonrepeated Accented Ending Song. Later in the season this bird began to sing another Northern Parula song, under circumstances comparable to those of mated Northern Parula Warblers. It responded strongly to a playback of its own song and to the song of a Northern Parula Warbler. In a preliminary test this bird did not respond to playbacks of Yellow-throated Warbler songs, although these playbacks may have been inadequate, thus making this conclusion tentative. Hence, this situation probably resembles that of most Blue-winged–Golden-winged warbler hybrids, which respond only to their own song type, independently of their own plumage type (e.g., Murray and Gill 1976). This Sutton's Warbler vigorously attacked a male Yellow-throated Warbler that established a territory before it arrived, eventually displacing the Yellow-throated Warbler. The Ulrichs made no observations of the two males' song responses to each other during the encounters.

This Sutton's Warbler was seen with a female Northern Parula Warbler and close to it several times over a summer, but neither nest nor young were ever found. Nevertheless, the Ulrichs assumed that the female had built a nest, for several times she took yarn placed out for nesting birds, and she remained in the area for weeks afterward.

The Indiana observation also provides information on foraging patterns of the male Sutton's Warbler. It resembled Yellow-throated Warblers in this regard, concentrating on small branches and reaching for insects in the leaves above and below, hovering at branch tips and hanging upside-down like a chickadee. Early in the season this bird foraged on the bark like a Yellow-throated Warbler (see Ficken et al. 1968).

There is now at least one report of a second-generation bird, which, if valid, means that some hybrids are viable. This bird's song was like a Northern Parula Warbler's, except in being repeated, and it had a greenish-yellow wash on the back characteristic of Northern Parula Warblers but not Yellow-throated Warblers. This bird provides the strongest morphological evidence that the two species hybridize, that the crosses are fertile, and, if it is a second-generation bird, that some hybrids are fertile. Birds previously reported as hybrids, resembling Yellow-throated Warblers but with Northern Parula Warbler songs, are still open to the alternative, if unlikely, explanation of being aberrant Yellow-throated Warblers that learned Northern Parula Warbler songs in an area of sympatry.

Other Hybrids

Most hybrid combinations are of two types: members of closely related species-pairs (e.g., Blue-winged–Golden-winged warblers) and members of different genera (e.g., Parula–Yellow-throated warblers). Few hybrids among other pairs of congeners have been recorded—a most impressive dearth, considering the close congeneric contact among many warblers in diverse paruline faunas, as in the coniferous forests of northeastern North America. At least 10 intergeneric hybrid combinations are known (Table 13.1), some having been reported more than once. In contrast, only four intrageneric hybrids are known, other than for closely related species-pairs. Hybridization is known within five closely related species-pairs, including Myrtle and Audubon's warblers, which are now treated as con-

Table 13.1   Hybridization between members of species-pairs, other intrageneric pairs, and intergeneric pairs (after Gray 1958; Parkes 1978; Bledsoe 1988; and other sources).

Species-pairs

  Bay-breasted (*Dendroica castanea*) × Blackpoll (*D. striata*)
  Hermit (*Dendroica occidentalis*) × Townsend's (*D. townsendi*)
  Blue-winged (*Vermivora pinus*) × Golden-winged (*V. chrysoptera*): known as
    Brewster's Warblers and Lawrence's Warblers
  Mourning (*Oporornis philadelphia*) × MacGillivray's (*O. tolmei*) [but see Hall
    1979]

Other intrageneric pairs

  Nashville (*Vermivora ruficapilla*) × Tennessee (*V. peregrina*)
  Yellow-rumped (*D. coronata*) × Bay-breasted
  Yellow-rumped × Grace's (*D. graciae*)
  Yellow-rumped × Pine (*D. pinus*)

Intergeneric pairs

  Blue-winged × Kentucky (*Oporornis formosus*): known as Cincinnati Warblers
  Blue-winged × Mourning
  Northern Parula (*Parula americana*) × American Redstart (*Setophaga ruticilla*)
  Northern Parula × Yellow-throated (*Dendroica dominica*): known as Sutton's
    Warblers
  Black-and-white (*Mniotilta varia*) × Cerulean (*Dendroica cerulea*)
  Black-and-white × Blackburnian (*Dendroica fusca*)
  Northern Waterthrush (*Seiurus noveboracensis*) × Blackpoll
  Mourning Warbler × Common Yellowthroat (*Geothlypis trichas*)
  Mourning Warbler × Canada Warbler (*Wilsonia canadensis*)

---

specifics. These range in frequency from the complete intergradation of Myrtle and Audubon's warblers in hybrid zones, to well-marked introgression among Golden-winged and Blue-winged warblers, to single or infrequent hybrids, as between Blackpoll and Bay-breasted warblers.

Parkes maintained that this pattern is unlikely to be a random one, given the absence of hybrids in such other genera as *Seiurus* or *Vermivora*, apart from the commonly acknowledged species-pairs in Table 13.1. He pointed out that no intrageneric hybrids have been reported in the large tropical genera of *Basileuterus* and *Myioborus*, although they have not been as thoroughly studied as the North American species. He concluded (1961, 1978) that selective pressures favoring the evolution of reproductive isolating mechanisms would be stronger among sympatric species than between noncongeners. (The closely related species-pairs do not fit this criterion, probably because of limited contact, also.) Less closely related species would normally have fewer chances of reproductive "accidents" resulting in hybridization. For this reason little selective pressure may exist against more remotely related crosses, simply because the opportunity would arise so infrequently. Banks and Johnson (1961) reached a similar conclusion for hummingbirds.

Short and Phillips (1966) criticized this conclusion, specifically as it related to hummingbirds, claiming that it was the consequence of taxonomic categorization: hummingbirds had overzealously been split into too many genera. But Parkes (1978) observed that this criticism is invalid for warblers, in that the variety of intergeneric pairs far exceeds any generic boundaries proposed (e.g., *Seiurus* × *Dendroica*, *Mniotilta* × *Dendroica*). Even if all of these birds were placed in a single genus, the distance between them should be considered in excess of putative crosses within members of the genera as now constituted. Electrophoretic studies have corroborated these conclusions (Barrowclough and Corbin 1978; Avise et al. 1980). Further, since Parkes's conclusion was formulated to explain patterns of intergeneric hybridization among manakins, it is not strictly a post hoc hypothesis in its efforts to explain patterns among the parulines.

Parkes's hypothesis is attractive and could answer this initially unlikely distribution of hybridization, although it does not completely eliminate an alternative, that hybrid pairings simply occur where one of the species is rare (the theory of unlikely contacts). The studies on Blue-winged and Golden-winged warbler interactions have indicated that these interactions usually happen if one of the species is rare, or new in an area; the two conditions generally happen together. This explanation could also hold for certain, though not all, of the other interspecific pairings. In a way the latter explanation is attractive; for if it is sustained it would account for the observations without assuming that the warblers responded in more than one way to nonconspecifics. This alternative might fit the Northern Parula × Yellow-throated cross (Sutton's Warbler) (Carlson 1981). Moreover, the Blue-winged Warbler, a species that has undergone a great range expansion (Gill 1980), has formed intergeneric hybrids with two *Oporornis* warblers: the Kentucky, at the southern edge of the Blue-winged Warbler's range, and the Mourning, at the extreme northern edge of its range (Parkes 1978). In both instances, one or both of the species may have been rare, although if evidence for hybridization consists only of specimens taken during migration, that information is not readily forthcoming, especially if ranges overlap greatly. Information on the breeding pairs that produce these hybrids is most valuable for determining whether one or both of the species producing the hybrid were rare in that area.

Knowledge of the range and habitat of parental species can help one to make educated guesses about the pairing site. Thus, although Short and Robbins (1967) emphasized that the Blackpoll Warbler and Northern Waterthrush overlap in a wide area of North America, it is also true that they are allopatric over a vast area, with a large zone of occasional contact at the southern edge of the Blackpoll Warbler's range and the northern edge of the waterthrush's range. These two species frequent very different habitats, such that, over most zones of overlap, Blackpolls are confined to coniferous forests (at considerable elevations at the southern end of the range) and waterthrushes to lowland watercourses. As noted earlier, Sutton's Warblers are in a boundary area where both parent species are uncommon, an area where the Northern Parula Warbler is becoming progressively rarer and the Yellow-throated Warbler more common.

Another possible hybridization, involving an intrageneric pair, Bay-breasted and Blackburnian warblers, took place in West Virginia (Hurley and Jones 1983), where the Bay-breasted Warbler had not even been known as a breeding species. Unfortunately, even though considerable behavior was observed from the partners of this putative pairing, including a nest with a single nestling thought to be that of the mixed pair, the pairing was not unequivocally established, and characteristics of the single offspring were not determined. The fact that only one nestling (one day old) was present suggests limited viability. Low viability could account for the failure to produce hybrids, but would not support Parkes's argument about the intensity of selection for isolating mechanisms.

If patterns of sympatry among intergeneric combinations of warblers more frequently result in one of the species being rare than do intrageneric combinations, some of the criteria for explaining the pattern of warbler hybridization may exist. Since members of different genera use resources more differently than members of the same genus, this condition could increase their effective rarity relative to congeners. Elements of both factors (usual distance within a habitat, rarity of one or both species within area of hybridization) could combine to produce the observed pattern of hybrid formation. Regardless, the striking intrageneric-intergeneric pattern requires explanation, and Parkes's hypothesis could help to explain the surprising relation. One can most readily assess these alternatives in light of behavioral and ecological information from the parental individuals as well, but this is hard to obtain for occasional interspecific pairs.

Bledsoe (1988) issued a caveat relevant to this discussion, noting that it depends on the assumption that these warbler genera are natural ones. Members of several of these genera are very similar to each other, a conclusion corroborated by electrophoretic studies. Their similarity raises the probability that they are not completely monophyletic, which could complicate the interpretation of these hybridizations.

Hybridization within a Species Complex:
The Yellow-rumped Warbler

Hubbard's (1969, 1970) and Barrowclough's (1980b) analyses of the Yellow-rumped Warbler *(Dendroica coronata)* complex provide a textbook picture of hybridization and introgression between two closely related populations when they come into secondary contact. The two parental populations, a western one (Audubon's Warbler) and an eastern one (Myrtle Warbler), meet in the northern Rockies of Alberta, British Columbia, and southeastern Alaska. Wing pattern, tail pattern, throat color, auricular color, presence or absence of a supraloral spot, and presence or absence of a postocular line all differ in the two parental populations (Figure 13.2). Evidence for massive hybridization and introgression in the plumage characters appears in latitudinal transects across the Rockies, with roughly intermediate birds occurring over a distance often not exceeding 60 kilometers, although clear evidence of introgression occurs across the entire transect distance of 480 kilometers (Hubbard 1969).

Audubon's Warblers are characteristic of the western cordilleran forest, which is dominated by Douglas Fir *(Pseudotsuga menziesii),* Engelmann Spruce *(Picea engelmannii),* and Western White Spruce *(P. glauca albertiana);* Myrtle Warblers of the eastern boreal forest, which is dominated by White *(P. glauca glauca)* and Black *(P. mariana)* spruces in the range of introgression. Intermediate warbler hybrids are characteristic of forests with both cordilleran and boreal vegetational elements. Whether the differences of vegetation types account for the abundance of the two *coronata* types is unclear. But the change in the birds' plumages is abrupt, considering that this habitat has existed for 7,500 years (Barrowclough 1980b)—at least the lower elevation sites have been unglaciated for that long.

Figure 13.2 Plumage patterns of western (*above:* Audubon's) and eastern (*below:* Myrtle) races of the Yellow-rumped Warbler. (Illustrations by Brian Regal.)

The area of intermediate hybrids is primarily in mountain valleys and passes. The steep cline could be accentuated by the limited opportunity for contact between the two populations in this area: only about 15 percent of the habitat along the line of potential contact between Audubon's and Myrtle warblers consists of habitat satisfactory for warblers; the rest consists of montane rock, glaciers, or alpine tundra. The zone of hybridization and introgression is rather symmetrical, implying that one form is not an inherently superior colonizer, in contrast to the colonization patterns of Golden-winged and Blue-winged warblers. Low levels of variation within the zone of hybridization led Hubbard to conclude that this zone is stable; indeed, specimens collected by Barrowclough 10 years later in these areas showed no morphological change from Hubbard's specimens. Hubbard proposed that this high stability implied the action of strong selection. If so, this cline could be maintained if populations are differently adapted to habitats they occupy. Although differences among coniferous trees might appear trivial to the birds, especially since several of these trees are spruces (all of the genus *Picea*), Hubbard observed that no *coronata*-like birds range farther to the west, in an area dominated by yet another spruce, the Sitka Spruce *(P. sitchensis)*, and the Douglas Fir. Morse (1976b) also noted that Red Spruce forests supported higher densities of warblers than did White Spruce forests, a consequence of the ease of exploiting different needle structures. Thus, it is plausible that Audubon's and Myrtle warblers, and their hybrids, respond to these conifer species differently as substrates. However, behavioral data such as those gathered by Morse are needed to determine whether the parent types and hybrids vary in their tree choice and whether their choice affects their success.

Hubbard regularly found "hybrid" traits far from areas of current intergradation. These traits might be the result of long-distance introgression, but they may also be common ancestral traits that occur at strikingly different frequencies in different populations. For instance, in eastern North America, thousands of kilometers from the site of interaction, small proportions of Myrtle Warblers have yellow throat feathers or large single wing patches, or lack a supraloral spot, all traits characteristic of Audubon's Warblers.

In populations closer to the zone of current interaction, hybrid traits may represent the recent swamping of a relict population by the dominant local type. For instance, the Cypress Hills population of Audubon's Warblers, now well within the Audubon's Warbler's geographic range, on the southern end of the Alberta-Saskatchewan boundary, has a slight intergradation of Myrtle Warbler characters, especially a high frequency of black (vs. gray) auricular color and more supraloral spots than usual.

Hubbard treated Audubon's and Myrtle warblers as semispecies—that is, populations that have achieved some attributes of species rank (Mayr 1963)—and the AOU *Check-list* (1983) accordingly treats them as a single species (Yellow-rumped Warbler). Hubbard hypothesized that they originated from the separation of a common ancestral population into eastern and western parts by Wisconsin glaciation, which drove a wedge southward along the Rockies, producing glaciers or alpine tundra at least as far south as New Mexico.

More recent palynological information has raised problems with Hubbard's supposition that the Rockies provided the critical barrier (Wright 1971; Ritchie 1976), although these data are not extensive. Refugia might have existed within this range, but we need more information (Barrowclough 1980b). Since the region is also the contact zone for eastern and western representatives of other birds, butterflies, and trees with distributional patterns similar to the *Dendroica coronata* complex (Remington 1968), an allopatric model like the one suggested by Hubbard seems best. Sympatric differentiation models are weakly supported, at best, on the basis of information available (Barrowclough 1980b).

Barrowclough combined electrophoretic studies with plumage surveys and estimated gene flow. Genetic distance, a measure of the differences in allele frequencies, is very small between populations of pure Myrtle and pure Audubon's warblers, falling near the values for commonly recognized subspecies of other birds and far below the level for distinct species. Genetic distances between parental forms and hybrids, and among hybrids, are correspondingly lower, echoing Hubbard's conclusions based on plumages.

Barrowclough measured the hybrid zone, the overlap zone ranging from 80 percent of one parental population at one end to 20 percent of the population at the other, and found it to be 147.3 kilometers in width. Given that the zone has been forested for about 7,500 years, and using an estimate of gene flow from other passerine populations of 1 kilometer per year (Barrowclough 1980a), and assuming the alleles are neutral, he estimated a current width of 145.5 kilometers, using the method of Endler (1977). This result is remarkably close to the measured width and is consistent with the selective differences of Myrtle and Audubon's traits being small. In fact, Barrowclough thus proposed that random and selective effects are of similar magnitude. Although habitat differences could still play a role in selection, Barrowclough's results indicate that Hubbard's argument for strong selective pressure is unlikely and unnecessary. If one can relax the argument for strong selective pressures, the resulting scenario is easier to accommodate within our current understanding of gene flow and selection.

The underlying mechanisms associated with this "east-west confrontation" remain unknown; their resolution awaits information of other types. Barrowclough stressed the need for additional paleoecological data, which would address the question of refugia during Pleistocene glaciation. In my study of these papers, however, what is most evident is the dearth of work on the living birds themselves, which has contributed immensely to the understanding of Blue-winged–Golden-winged warbler interactions. Such work could answer important questions. Is there assortative mating? What is the success of hybrids and parental forms? How do the different forms use their habitat? What preferences do birds in the hybrid zone have for boreal and cordilleran elements in the transitional forests that they inhabit? What ecological and behavioral significance do differences in plumage characters and wing length have? Combined with the existing morphological and biochemical data on these populations, this would be a rewarding study indeed.

Birds of Unknown Affinity

Audubon and Wilson described another three purported warblers, but in no instance does information supplement their plates. These entities appeared on the hypothetical list of the AOU *Check-list* through the fifth edition (1957); in the current (sixth) edition (1983), they have been removed to a list of forms of doubtful status or hybrid origin that have received a formal scientific name. The Carbonated Warbler *(Sylvia carbonata),* listed as *Dendroica carbonata* in the fifth edition, is based solely on Audubon's description and a plate painted from two specimens obtained in Kentucky in May 1811. The published plate may have been partly based on memory, and the specimens have since been lost. They were most likely first-spring male Cape May Warblers (Parkes 1985a). The Blue Mountain Warbler *(Sylvia montana), Dendroica montana* of the fifth edition, is known only from the plates of Wilson and Audubon. Reported from Virginia, no specimens exist. The Small-headed Flycatcher, the Small-headed Warbler *(Muscicapa minuta), Wilsonia* (?) *microcephala* of the fifth edition, has generally been considered a paruline as well. It is also known only from plates of Wilson and Audubon, which were based on specimens, subsequently lost, from New Jersey and Kentucky.

It is difficult to evaluate these entities. Most workers have attributed them to faulty memory on the part of Audubon and Wilson, especially since Audubon worked from memory on paintings made at about the time of the report of the Carbonated Warbler. Amadon (1968) referred to two of them, the Carbonated and Small-headed warblers, in a hypothetical list, and indicated that they were confused with other species. This explanation seems most probable, but two other species first collected at about this time, the Swainson's and Bachman's warblers, were known from only one or two specimens and were not subsequently taken for another 50 years! Of further interest, Rappole et al. (1983) noted that wintering habitat was being degraded in the West Indies as early as the 1550s, a possible factor accounting for the traditional rarity of the Bachman's and Kirtland's warblers. They speculate whether other species became extinct before discovery for the very same reason. Another possibility is that one or more of these forms could be rare hybrids (Parkes 1985a), such as the Cincinnati Warbler, a once-reported entity from the 1800s, for which there is an extant specimen, and which is an intergeneric hybrid of the Blue-winged and Kentucky warblers.

The Significance of Isolating Mechanisms

The most suggestive fact about the pattern of hybridization among warblers is that it is so clear-cut, involving primarily members of allopatric conspecifics, sibling species that until recently were allopatric, and members of different genera. Although the basis for species discrimination has not been well established among these species, or for passerine birds in general, and some cues thought to be of importance (color patterns) apparently do not play a central role, the dearth of hybrids from other congeneric pairs indicates clearly that discrimination is strong. Given the diversity of congeneric, sympatric warblers in some avifaunas, one might expect this group to have a high frequency of hybridization. The pattern of improbable combinations that could result among species not in regular contact with each other has implications for the role of isolating mechanisms in more usual situations. The fact that unlikely participants from different genera can contribute many hybrid combinations suggests that lack of relatedness in itself does not suffice to prevent hybrids in this group. It also implies a wide range of interfertility, something not to be predicted among members of different genera. Since these genera are not distantly related to each other, it is possible that they do not warrant that level of taxonomic distinction. Electrophoretic studies indicate that members of warbler genera are more closely related than is usual among bird groups (Avise et al. 1980), yet the electrophoretic studies suggest that relations inferred by the classification are realistic and that the hybridization results are not a mere artifact of classification. If the taxonomic distribution of hybrids cannot be attributed to a comparable pattern of fertility, the best way to explain the distribution is that strong patterns of reproductive isolation routinely operate in most situations, including those involving species different enough that we do not normally consider reproductive isolating mechanisms to be important.

Most hybridizations between distinct species probably occur when one (or both) of the species is rare. In these cases the two species have not had much past contact, as in the example of Blue-winged Warblers invading an area where Golden-winged Warblers are common.

This example raises the question of how patterns of discrimination operate. If the critical part is female discrimination, it may be significant that several species of warblers practice limited polygyny, which could enable almost all females to acquire a conspecific mate, even if only as the second mate of a conspecific. Nevertheless, polygyny might be a preferable choice for the female, thereby minimizing interspecific pairings, except where no conspecific males are present. (In such highly interfertile species, the production of hybrid offspring, with the potential for back-crossing, might be a realistic alternative to polygyny. But the distribution of hybrids is inconsistent with this possibility, given the dearth of conge-neric hybrids—other than intraspecific or between-sibling species—which should predominate if hybridization were an adaptive option under some circumstances.) That may be why species uncommon in an area provide disproportionately more hybrids: their density is perhaps low enough that there is a good possibility that males of another species will encounter unmated females.

Male discrimination might also play a role in limiting hybridization in areas where a diverse array of warblers is likely to be found. Although tests of male discrimination of females have not been made, playback experiments indicate analogous patterns of discrimination, in which males respond quite differently to familiar and unfamiliar vocalizations.

14   Tropical Warblers

Because parulines are common in northern forests, we tend to think of them as northern species, even though most retreat far southward after the breeding season. Parulines being of tropical North American origin, more species occur in Mexico than in the United States and Canada (at least 66: Blake 1957), although most are migrants. The contribution of resident warblers in the breeding pool of species declines as one moves southward. Mexico has 29 species, of which 14 range into the United States, 7 are endemic to Mexico, and 8 range to the south. Most have affinities to the western montane temperate-zone species, doubtless a consequence of their concentration in the highlands and mountains of Mexico, which are continuous with ranges to the north.

Central America has 24 breeding species: 10 range north into Mexico, 5 are endemic, 4 range into South America, and 5 range from Mexico to South America. South America has 32 species, 25 endemic and 7 shared with Central America. A few species range as far south as northern Argentina, but the Tropical Parula Warbler *(Parula pitiayumi)* extends the farthest south, reaching the vicinity of Buenos Aires (central Argentina). Warblers are more important in the avifauna of the West Indies, which has 16 breeding species, 13 of them endemic. The prominence of warblers in the West Indies may be partly due to the depauperate nature of its fauna and the characteristically North American affinities of groups that occupy the region. Nine species are *Dendroica*s, representatives of the dominant North American genus, including Pine and Yellow-throated warblers. Although the West Indian *Dendroica* are most similar to the eastern North American fauna, two species, Olive-capped *(D. pityophila)* and Adelaide's *(D. adelaidae)* warblers, are most closely related to a western species, the Grace's Warbler. Six additional species are in 4 endemic genera, 2 on the Greater Antilles and 2 on the Lesser Antilles (Bond 1957).

South America has no endemic genera, in contrast with areas to the north and the West Indies. This may reflect both the group's tropical North American origin and the rich avifauna that South America has evolved in isolation, including tryannid flycatchers that have invaded warblerlike niches (Keast 1972). Most South American species are members of the genera *Basileuterus* (19 species) and *Myioborus* (8 species), which, though not confined to South America, have the most species there (*Basileuterus* has 22 species, with 15 species confined to South America; *Myioborus* has 10 species, with 7 species confined to South America). They represent an underbrush-ground form (*Basileuterus*) and an insect-hawking one (*Myioborus*) that converges with the tyrannid radiation. Because many species of these genera are allopatric, local warbler diversity is low. Few species range over much of the vast area that makes up their collective range; the Tropical Parula, Golden-crowned (*B. culicivorus*), and River (*B. rivularis*) warblers come the closest. Members of *Myioborus* are extremely similar in color patterns and habits, differing mainly in mixtures of yellows and oranges, traits probably controlled by a few genes. The number of species is open to question because of their highly allopatric distribution; many workers lump together adjacent populations with distinct color patterns (see Griscom and Sprunt 1957) simply because they probably have only small genetic differences.

Basileuterus has broader generic limits, including adaptive types that differ more widely in coloration, habits, and morphology than do members of *Myioborus*, although they are all ground- or brush-frequenting species. The sixth AOU *Check-list* (1983), which for the first time treats species of Mexico proper and Central America, splits this genus, placing the Buff-rumped Warbler into *Phaeothlypis* (*P. fulvicauda*), and with it, by inference, the closely related River Warbler, which falls outside the checklist's new boundaries. Slud (1964) noted that in the field the Buff-rumped Warbler in no way suggests that it belongs to *Basileuterus*, whose species also have highly allopatric distributions. A site seldom has more than two species (e.g., Buskirk 1972).

Plumage and Nesting

Most tropical species differ from their temperate-zone counterparts in being monomorphic; furthermore, both sexes of many arboreal species are bright, the equivalent of the northern parulines' male plumage. This pattern of bright monomorphism is not unique to parulines—tropical orioles and tanagers have it as well (see Hamilton and Barth 1962). Tropical warblers are pair-bonded throughout the year, and members of *Basileuterus, Myioborus,* and *Ergaticus* molt into their adult plumage in late summer, immediately after their juvenile plumage, rather than the following spring, as most temperate-zone species do. Therefore, they spend their first winter in adult plumage. Molt coincides with the formation of pair bonds, although they do not breed until spring, more or less in synchrony with northern species. Several species are probably territorial over their first winter; at least, they occur only as singles or pairs and are aggressive to other conspecifics (Skutch 1954). This implies that adult coloration is related to territorial defense.

Since many workers associate dull female coloration with nest duties, with only the females incubating and building the nest, the tropical pattern is of particular interest, as nest predation is often intense in the tropics (Foster 1974a). One compensating factor is that, in contrast to most temperate species, many tropical species build covered nests; all of the *Basileuterus, Myioborus,* and *Ergaticus* species have covered nests, usually domed (Skutch 1957). It would be interesting to compare the nests and nesting behavior of bright and dull *Basileuterus;* however, in species studies thus far, only females incubate, and associated behavior is consistent.

Other warbler genera provide opportunities for comparison, because this trait for bright-colored monochromatic forms reaches its apex in the two species of the genus *Ergaticus:* the Red *(E. ruber)* and Pink-headed *(E. versicolor)* warblers. Pink-headed Warblers build a domed nest (Skutch 1954), and Red Warblers build both open and domed nests, a trait that differs even within a single population. Red Warblers, high-mountain nesters, live in a temperate habitat but experience high predation rates in their home sites (Elliott 1969). In contrast, Crescent-chested Warblers *(Parula superciliosa)* and Flame-throated Warblers, members of a genus with northern representatives, do not build domed nests, though their nests are well concealed (Skutch 1954, 1967b). In Guatemala, Crescent-chested Warblers often cover their nests with large, fallen leaves. The sexes of both *Parula* species are similar, and only females build the nests or incubate. At least the Central American race of the tropical Masked Yellowthroat *(Geothlypis aequinoctalis)* also builds an open nest and has a dimorphic color pattern comparable to that of its relative of the temperate zone, the Common Yellowthroat.

The genus *Parula* presents a picture that reverses the trend for tropical warblers to have covered nests. Temperate-zone representatives of the genus, Northern Parula Warblers, build nests—uncharacteristically, for a high-latitude species—that are extremely well hidden and worked into festoons of *Usnea* lichen (in the northern part of their range) or Spanish

Moss (in the southern part). Their southern counterparts, Tropical Parulas, also hide their nests. Skutch (1967b) reported a nest from Costa Rica with an entrance through a hole in moss growing epiphytically on a branch, and hence completely invisible from the outside. Nests from Trinidad (Belcher and Smooker 1937) and southern Texas (Bent 1953) were also covered.

Bright plumage coloration in birds is often related to aggression, which becomes especially high in migratory species during territory formation and declines later in the season as territorial boundaries become stabilized (see Morse 1976a for northern warblers). Considerable aggression also takes place during the early phases of courtship in migratory species (Ficken and Ficken 1962a). The drawbacks of bright monomorphy might be minimized for tropical species because of their stable situations (permanent pair-bonding and territories are characteristic), although courtship clearly has strong aggressive elements in even the permanently paired Red Warbler (Elliott 1969). These species experience year-round pressure on their territories; in this instance, bright monomorphy might be an added benefit. If bright monomorphy plays an important part in territorial defense, one would expect females to take a greater part in this activity than do females of temperate-zone species, whose males play the primary role (Morse 1976a). Whether constant territoriality is equally critical for eventual reproductive success in the bright monomorphic permanent resident warblers is not known.

Basileuterus, Myioborus, and *Ergaticus* are tropical genera without representatives north of Mexico. The molt patterns of at least some West Indian parulines differ fundamentally from them: three species of *Dendroica*—the Elfin Woods Warbler *(D. angelae),* Arrow-headed Warbler *(D. pharetra),* and Plumbeous Warbler *(D. plumbea)*—and the closely related Whistling Warbler *(Catharopeza bishopi)* all retain their immature plumage for nearly a year, fully as long as do the migratory North American species (Kepler and Parkes 1972). Anderle and Anderle (1976) suggested that the Whistling Warbler retains this plumage through its first breeding season, as the American Redstart does. The difference between tropical mainland species and West Indian species probably lies in the temperate North American affinities of the West Indian birds. The four West Indian species are essentially monomorphic (Parkes 1985b), although all are black-and-white or gray-and-white species. It would be of interest to know whether they hold permanent territories and form pair bonds at the same early age as birds of the tropical mainland genera do; these data would provide insight to patterns of monomorphism and space use. In contrast, West Indian species conspecific with North American forms are dimorphic.

It thus looks as though the suite of characters typical of certain genera of tropical warblers and other tropical nine-primaried groups—bright monomorphism, permanent territoriality, and covered nests—are conservative ones reflecting the origin of their bearers. Tropical representatives of primarily temperate-zone paruline genera show varying trends toward some characters, but not the level of development achieved by tropical genera. By contrast, both representatives of *Parula* that range into the temperate zone have covered nests, although the Tropical Parula Warbler's dimorphism is comparable to that of most temperate-zone parulines. Tropical representatives of *Dendroica* and *Geothlypis* show modest changes of this sort, if any at all. None of the West Indian species builds a covered nest (Bond 1957).

Clutch Size and Tempo of Reproduction

Tropical parulines have smaller clutches than any temperate-zone representative recorded in this family (see Figure 2.3). Clutch sizes of two are the norm for some of these species; Skutch (1957) mentioned that a Pink-headed Warbler with a clutch of four in the Guatemalan highlands was the largest tropical warbler clutch that he had ever seen. Skutch also observed that tropical warblers normally produce only a single clutch, although probably not without several efforts. West Indian species, including those conspecific with North American species, also have small clutches; again, two is the norm (Bond 1957).

Tropical species differ markedly from their temperate-zone counterparts in how rapidly they nest. For instance, the Buff-rumped Warbler in Costa Rica builds nests in a week or more, has an incubation period of 16–17 days, and young that remain in the nest for 12–15 days, even though they are ground nesters (Skutch 1954). The breeding season of West Indian species is also prolonged (Bond 1957) and involves extended periods devoted to nesting activities. These might be contrasted with warbler timetables in the north, where nest building takes 3–4 days, incubation 11–14 days, and time between hatching and fledging 8–11 days (see Bent 1953). Temperate-zone species thus bring a successful brood to fledging in roughly two-thirds the time required by their tropical equivalents. Warblers in the temperate zone experience severe time constraints, either relative or absolute, which may have driven them to shorter periods.

The highly pulsed spring flush of insect life in the north is not matched in the tropics, a factor usually advanced as the reason migration developed (Cox 1968). In a tropical territory whose size is probably honed to a succession of factors that impinge on the birds throughout the year, space is probably barely adequate for gleaning insects to feed even a small clutch. Warblers have to renest regularly, often several times.

Differences in clutch size and fledging time among warblers are unlikely to result entirely from enlarged clutch size and faster pace in the north, however; certain pressures have probably led tropical species to reduce their clutch size or increase their nesting period. Nest predation is generally thought to be especially high in the tropics (e.g., Ricklefs 1969; Skutch 1985). Although telescoping the season might minimize predation by reducing the period of vulnerability, low visibility of nests, enhanced by low rates of nest visitation, reduces the impact of predators that do not search randomly. Some tropical species may have extended their nesting period in this way to minimize nest predation. Small clutches do not require as many feeding visits to a nest, and subsequent selection for slow growth would cut the frequency of feeding visits.

The roles of resource availability, predation, and other factors in determining clutch size have been a point of major debate over the past 40 years. Lack (1947, 1954) argued that birds rear as many young as possible and that the small clutch sizes of low latitudes are due to the shorter amount of daylight available to feed them. Skutch (1949, 1976) countered that many tropical species do not spend all of their time feeding young, and hence clutch size is not limited by extrinsic factors. These two arguments take the diametrically opposed positions that reproductive abilities regulate population size (Lack's argument), or that mortality rates, high in the tropics, regulate reproductive rate (Skutch's argument). Skutch's argument is attractive in that it accounts for the probably saturated populations that prevail in many low-latitude habitats, but it has the problem that individuals reducing their reproductive effort may not contribute as many offspring to the next generation as those with larger clutches. This objection would be overridden if individuals with low reproductive effort managed to bring more young into the next generation, as might be the case if the offspring were of higher quality or if predation were reduced. I know of nothing, however, that indicates whether members of tropical passerine populations with small clutches have different rates of nest predation or nestling survival than do those with larger clutches. This pattern holds if one samples an entire avifauna (Foster 1974a), but at the critical level for individual selection, the within-population level, this result is unknown. This information would resolve the question of whether low clutch size is an adaptation to lowering predatory rate; it would also resolve the argument between Lack's and Skutch's proposals.

Willis and Oniki (1978) and Oniki (1979) have argued that, in contrast to conventional wisdom, predatory rates are not necessarily high in tropical avifaunas. They base these conclusions on data from the Amazon, where rates are strikingly lower than those reported from such areas as Barro Colorado Island (Willis 1974) and Skutch's (1966) isolated forest in Costa Rica. Willis and Oniki suggested that the sites with high rates of nest predation may suffer from the loss of large predators; this results in the release of small predators in a pattern similar to that of temperature-zone habitat islands. Willis and Oniki's thesis complicates the issue of clutch size and points out the consequences of human encroachment.

If Willis and Oniki are correct that predation rates do not differ latitudinally, producing small clutches may not be a way of minimizing predation rates, as Skutch (1976a, 1985) has proposed. Such a discounting of predation would be consistent with alternative resource limitation arguments for clutch reduction, although support is needed for that argument. More work is needed to confirm the observations of Skutch and others that tropical species do not spend all their time feeding young.

Despite the impression of stability, most tropical areas undergo marked seasonal changes and periods of abundant or scarce food supplies, both animal and vegetable. Many tropical birds rear young during a time of plenty. Insectivorous species concentrate their efforts about the rainy season, or at the end of the dry season if it is not severe (periods that provide a flush of food). They do not breed at the height of the rainy season in extremely wet areas, since the weather itself decreases foraging time (Foster 1974b) or insect productivity (Janzen 1973; Buskirk and Buskirk 1976). The breeding season in the Buskirk's Costa Rican study area had a three-fold flush in insect abundance. Thus, if the breeding season is appropriately timed, parents may have time to spare from foraging, although this in itself does not explain why they do not rear a larger clutch. Small clutches could be related to the carrying capacity at other seasons, when food supplies are demonstrably lower; nevertheless, feedback is required to select for a small clutch size. If more young survived from small broods, reduction of clutch size, or retention of small clutch size, would be selected. Clutch size may not be variable enough to test the efficacy of small broods with simple marking studies, but one might test it by artificially increasing clutch sizes. It would be profitable to determine whether young birds forage constantly during times of minimum resource availability, and whether this variable differs on territories that can be independently characterized as high-quality or low-quality ones (young of long-established pairs versus young of pairs that have held a territory for the first time; young of dominant versus subordinate pairs, as characterized by outcomes of boundary disputes, etc.).

Social Systems

Social systems of resident tropical warblers differ from those of their migratory counterparts, probably as a consequence of their permanent resident status and also of space limitation at certain times. Species maintaining year-round territories often do so as pairs. Thus, their social systems are simple, but since the young of some species remain with their parents through much or all of the year, as in the Three-striped *(Basileuterus tristriatus)* and Golden-crowned warblers studied by Buskirk (1972) in Costa Rica, more complex systems are possible. Delayed independence of the young does not ensure that a large group will form, however, because of the small clutch sizes of these birds, even when a large proportion of nestlings survive. Buskirk reported newly formed pairs, pairs that were unsuccessful in rearing any young the previous breeding season, pairs that had reared a single offspring, and single birds all holding territories; that is, one to three birds per territory. Although these warblers have clutches of two, none of the breeding pairs in Buskirk's study area succeeded in fledging both offspring. It is not clear how successful lone individuals are in holding territories, and what the usual fate is of individuals that do not secure territories.

These birds can play an important role in the formation of mixed-species flocks. Three-striped and Golden-crowned warblers were the core or nuclear species (flock leaders) in flocks of the forest interior at middle elevations in Costa Rica. Most flocks had both species in them; under these circumstances, Golden-crowned Warblers foraged higher than Three-striped Warblers, although both species foraged low in the vegetation. Usually only one *Basileuterus* species participated in a flock in the region of Buskirk's study area, but his site (at an elevation of 1,550 meters) lay in a narrow area of altitudinal overlap (1,525–1,575 meters) of the two species. In a similar study at the same locality, but conducted 30 meters higher on the mountainside (1,580 meters), Powell (1979) found Golden-crowned Warblers in only about one-third as many flocks as Three-striped Warblers. The two species did not respond in a hostile way to each other, despite being intraspecifically territorial. The consequence of this limited sympatry was that the *Basileuterus* warblers had a more important quantitative effect on the flocks than would be likely in other places. Flocks with both species present attracted more attendants than those with only one, a consequence of the different species composition that followed the two species, in turn resulting from one foraging higher in the canopy than the other. Buskirk did not study flock behavior in areas of allopatry up or down the elevational gradient, so a more detailed comparison is not possible. The density of *Basileuterus* warblers in Buskirk's study area differed between years because of differences in nesting success.

Mixed-species flocks form about these two *Basileuterus* species over much of the year, although during the breeding season the central-place foraging and other nesting activities of parents disrupts the directionality of these groupings. Both warblers may be nuclear species because they maintain larger territories than do most others that participate with them. Consequently, they may exhibit more directionality in patrolling their large boundaries (territories were 3–3.5 hectares in Buskirk's study area) than species with territories only a fraction of this size, which continually drop out and rejoin flocks depending on whether the flock remains within their territorial area. The *Basileuterus* species are active and noisy, as are nuclear species in other geographic areas (Morse 1977a). Such traits might be associated with the formation and retention of family groups.

Slate-colored Redstarts *(Myioborus miniatus)* also participate in the flocks, but they do not play a strong nuclear role; that is, they are seldom joined or followed by other species. This difference might be a consequence of territory size and foraging behavior. At 1–1.5 hectares, their exclusive areas are less than half the size of the *Basileuterus* species. The redstarts forage primarily by sallying into the open for flying insects from a perch, and consequently they make more nondirectional movements (relative to the flock's direction) than do the *Basileuterus* species. Also, much of the redstart's activity is higher in the forest than that of the *Basileuterus* species, even the Golden-crowned Warbler in the presence of the Three-striped Warbler, which at such times forages extensively in the subcanopy. Although Slate-colored Redstarts also flick and fan their wings and tail, and call, they do not perform these motions as frequently as the *Basileuterus* warblers. Buskirk proposed that wing and tail movements of Slate-colored Redstarts flush insects from hiding places.

The Golden-crowned Warbler, which ranges from Mexico to Argentina, was also the nuclear species in mixed flocks studied by Davis (1946) in Brazil. There, too, the birds usually were in pairs and led the flocks, consisting of woodcreepers, furnariid ovenbirds, antbirds, and tyrannids.

Moynihan (1962) found the Black-cheeked Warbler to be a prominent member of flocks that formed about the Sooty-capped Bush-tanager (*Chlorospingus pileatus*) in edge and second-growth habits at middle elevations in western Panama. The Black-cheeked Warbler resembles bush-tanagers in color patterns and behavior, which Moynihan (1962, 1968) attributed to social mimicry, that is, adaptations facilitating interspecific gregariousness. Moynihan provided no strong basis for this argument, however. Regardless of the basis for similarity, Black-cheeked Warblers are also noisy and active, which attracts members of mixed-species flocks to them, although other species joined and followed bush-tanagers more frequently than they joined warblers. Slate-covered Redstarts also participated in bush-tanager flocks, but they were joined and followed by other species much less than they joined or followed other species, especially bush-tanagers and Black-cheeked Warblers.

A second species of *Myioborus,* Collared Redstarts *(M. torquatus)*, also participated in Moynihan's bush-tanager flocks and played a more important role than did Slate-colored Redstarts. Although behaving much like the latter, Collared Redstarts were joined and followed more often by other species than were Slate-colored Redstarts, and if the flocks were high in the trees, Collared Redstarts were even likely to be the leaders of the flocks. Collared Redstarts were more common at high elevations than were Slate-colored Redstarts, even though the altitudinal ranges of the two species overlap.

Thus, the behavior of tropical redstarts, especially Slate-colored Redstarts, in flocks is similar to the flocking behavior of migrant species that form territories, although migrants form single-individual territories rather than pair or family territories. But there is little sign that permanent residents have the high level of gregariousness that typifies some migratory species, such as Bay-breasted and Tennessee warblers, during the winter (Morton 1980; Greenberg 1984b). Large accumulations of resident species may occasionally occur immediately after nesting, as Pink-headed Warblers were reported to do by Skutch (1954); they probably involve mergings of family groups. Reports of more than two individuals of tropical species usually involve small family groups or pairs at the boundaries of their territories.

It is not clear what benefits warblers obtain by remaining in family groups over much of the year. Young staying with their parents for an extended time may act as nest helpers during the breeding season (Brown 1978). Neither Buskirk nor Moynihan dismissed this possibility, but since these birds reared only one successful brood a year, it is improbable that members of old broods assist later ones. Still, if the young remained with their parents as long as do members of complex social groups, such as Florida Scrub Jays (*Aphelocoma c. coerulescens:* Woolfenden and Fitzpatrick 1984), helping would be possible. More likely, remaining with the parents enhances their chances of obtaining their parents' territory. Buskirk proposed, but provided no evidence, that members of Three-striped Warbler family groups divide sentinel services. If predatory dangers are important, which they could be because of the presence of small bird-hunting hawks in these forests (Buskirk 1972; Powell 1985), this tactic would benefit both adults and young—the young in learning these techniques, and the adults in having an additional pair of eyes to watch. Time spent surveying for predators declines with increasing numbers of individuals in other flocking species, and feeding rates correspondingly increase (Murton et al. 1971; Kenward 1978). Interspecific advantages probably lie primarily in avoidance (Morse 1977a).

Competition with Other Animals

Wright (1979, 1981) has suggested that insectivorous birds and lizards compete for insect resources in the West Indies and elsewhere. By implication, this speculation includes warblers, since they are among the few groups of small insectivorous birds in the West Indies. Wright (1981) noted the extraordinary abundance of anoles on some West Indian islands (sometimes nearly one per square meter), and reported an inverse relation between the biomass of insectivorous birds and the biomass of anoles. This pattern need not be causal, but simply a consequence of differences in extinction rates. Island extinctions are regular among the birds (Ricklefs and Cox 1972; Wright 1981) but rare in the anoles (Williams 1969), a consequence of the large population sizes they attain. If anoles reach high population densities in the absence of birds, and overlap with birds in resource exploitation patterns, they may make invasions considerably more difficult for insectivorous birds.

To the contrary, Adolph and Roughgarden (1983) have argued that anoles and birds do not oppose each other for resource exploitation on St.

Eustatius, Netherlands Antilles. There, competition, or the potential for it, was far higher among anole species than between anoles and the low-foraging insectivore, the Yellow (Golden) Warbler, which exploited proximal parts of branches and paid little heed to the ground. Thus, anoles occupied areas that Golden Warblers did not exploit. St. Eustatius has a depauperate land-bird fauna, as do the Lesser Antillean islands on which Wright based his arguments. Adolph and Roughgarden acknowledged that anoles might prevent the establishment of ground-foraging insectivorous birds, but they contended that Wright had not made a strong argument for the interactions of birds and anoles.

An Aberrant Life-style: The Wrenthrush

Although numerous aberrant species have been assigned to the parulines (four genera currently assigned to the parulines are considered to be of questionable affinities by the 1983 AOU *Check-list*), the Yellow-breasted Chat is without doubt the one that has attracted the most attention by virtue of its unusual size, behavior, and morphological characteristics. Yet this species lies far nearer the norm of paruline life-style than does the Wrenthrush, which has been assigned a paruline or near-paruline affinity by Sibley (1968, 1970), a stance since adopted by the recent *Check-list*.

The Wrenthrush, a resident of the mountains of Costa Rica and western Panama, has adopted a nearly mouselike existence in the heavy understory (Morse 1966a; Hunt 1971). In keeping with this life-style, the rounding of its wings and reduction of its tail have progressed to a degree not approached in any other species among the parulines. Indeed, the species has nearly lost its ability to fly. I never saw Wrenthushes fly, even when I repeatedly tried to flush them, and in more extensive observations Hunt noted that these birds seldom flew and that any flight involved down-slope movements, either by weak flapping or gliding.

Although long understood to have affinities with the Neotropical nine-primaried assemblage, the Wrenthrush was previously assigned either to thrushes or to a family of its own, the Zeledoniidae. This placement acknowledged its similarities, in traits now considered to be convergent, to the latter groups. Sibley (1970) suggested that the different groups of New World, nine-primaried oscines represent clusters of species with similar feeding adaptations; thus, members of his tribe Parulini (basically the same birds I have included here under the Parulinae) are gleaning insectivores: Coroebini, nectar feeders; Thraupini, fruit eaters; and so on.

Sibley recognized the Wrenthrush as a monotypic tribe, Zeledoniini, in recognition of its specialized characters for a skulking existence.

Other characteristics of these birds are consistent with those of typical tropical paruline genera. They are monomorphic (although dull), they are ground nesters (each nest found by Hunt was built into a banking), they build covered nests, and their young have the yellow mouth linings (Hunt 1971) of the tropical paruline genera, as opposed to other parulines (Ficken 1965).

Wrenthrushes also resemble other tropical passerines in having small clutches (each of Hunt's three nests contained two eggs or young), a long nest-building time (8 or 9 days for the one nest that Hunt followed over most of the period), a hiatus of 8 days from completion of the nest to laying of the first egg, a day between the laying of the first and second eggs, and a fledging time of at least 17 days. Hunt did not obtain any data on the length of the incubation period.

The Long-Term Effects of Permanence

Key aspects of the tropical warblers' life-style, including their nesting habits and social systems, are probably related to permanent residence. Since they remain on location permanently, their ability to occupy a site at the earliest possible time is not a consideration; therefore, the first parts of the nesting sequence can be commenced well in advance of the most demanding aspects of rearing young. Their small clutch size should also minimize demands on accumulating additional resources, thereby avoiding the delays that would ensue if resources had to be found for a large clutch. Thus, long development time is a realistic option, and the birds are free to respond to selective pressures such as strong predatory pressure.

Permanent residence also facilitates the existence of prolonged family groups, which may enhance survival and provide offspring with their best opportunity to obtain a territory, largely from being on site and superseding a parent that may die. Although advantages of such groups have received considerable discussion (e.g., Powell 1985), no evidence indicates that individuals support other adult members, although that does not devalue the casual benefits resulting from association. Further, since members of these groups are closely related, advantages of association would accrue to all through kin selection.

The breeding systems of tropical warblers are not known, but females are the dispersing sex in bird social systems (e.g., Brown 1978). Whether dispersal prevents inbreeding depression is not known; indeed, such effects have seldom been investigated in territorial systems such as these. Bulmer (1973) demonstrated an inbreeding depression in resident, territorial Great Tits in England, however, and tropical warblers would be even stronger candidates for inbreeding than Great Tits, given their long life span and site tenacity.

The common characteristic of bright monomorphy may be associated with permanent pairing and permanent territoriality. Male-female aggression may be a lesser concern in permanent pairing than in the recurrent formation that characterizes temperate-zone species, although it occurs. Bright monomorphy may be important if both male and female take an active role in defending a territory. Under these circumstances one would predict a higher proportion of intersexual aggressive encounters than in temperate-zone breeders. The fact that neither dimorphic migratory species nor monomorphic tropical species hybridize regularly decreases the value of color as an isolating mechanism, thereby implicitly strengthening the argument that distinctive coloration plays its major role in nonsexual interactions. This observation supports arguments in earlier chapters which questioned the utility of species-specific coloration as an isolating mechanism. Bright female coloration should nevertheless enhance visual species-specificity, although at the expense of intersexual distinctiveness. Unfortunately, we have no basis for comparing the response of temperate and tropical species to color, display, and vocal cues. Although one might initially not think them useful, given the differences in age at pairing, time of pairing, and rates of reproductive activities, comparisons between tropical and temperate species may be useful for evaluating sexual differences in avian breeding systems.

Profitable comparisons may be made among mainland tropical residents, temperate-zone migrants, and West Indian residents. Many West Indian species differ from others in being tropical residents that are much more closely related to temperate-zone species than to mainland tropical residents, a majority being *Dendroicas*, the dominant temperate-zone genus. Some West Indian *Dendroica* species may date to the Pleistocene periods, when the West Indies were separated from the North American mainland by water gaps narrower than the current ones. In addition to distinct endemic forms within the genus, several West Indian residents are conspecific or closely related to mainland species. The endemic species share a mixture of characters found in mainland tropical and temperate species, which appear to reflect both local ecological pressures and common ancestry. This comparison may give a sense of which characteristics are conservative and which respond to environmental differences. The strong tendency toward monomorphy in the endemics, the mode among tropical mainland species, does not occur among forms that are conspecific with North American species. But all of the species have small clutches with open nests. As island inhabitants, the birds presumably experience less intense predatory pressure than do those on the mainland, yet they have a clutch size comparable to that of tropical mainland species and also a protracted season in common with them. The combination of these traits, plus the open nests, highlights the importance of taking into account the energy demands in the nesting cycle, and indicates that concealment may be a simple response to predator avoidance, not that the length of the nesting season is part of an antipredatory strategy. The pattern of monomorphy and dimorphy suggests that monomorphy is associated with permanent residency but is not quickly attained over evolutionary time.

15 Ecological Equivalents

Any examination of parulines in their natural communities must take into account their relation to other insectivorous species that co-occur with them. Since parulines are of tropical origin (Mayr 1946), their migratory patterns are probably a consequence of their colonization of the temperate zone. Even so, is their failure to become year-round residents in the north affected by their breeding area being preempted by other species during the winter? Do any taxa adopt distinctly paruline foraging adaptations in other geographic areas, and how is this "warbler adaptive zone" filled? The two related questions will be explored simultaneously to determine the extent to which the warbler life-style exists in other biogeographic provinces, and whether it parallels the one that I have presented.

Lack (1968), Cody (1974), and Morse (1975) have considered the question of ecological equivalents in a more general way. Eighteenth- and nineteenth-century taxonomists routinely described parulines and certain South American warblerlike tyrannids as sylviines: even the Galápagos Warbler Finch *(Certhidea olivacea)* was described as a sylviine! Although these determinations were based on the morphology available on museum skins, the taxonomic designations suggest that the similarities among the groups were great enough to indicate considerable ecological similarity.

Ecological equivalents to the parulines are found in all other major arboreal communities of the world. Keast (1972) considered small-bodied foliage gleaners to be the occupants of the warbler adaptive zone—an area dominated numerically by paruline warblers in North America, sylviine warblers in the Palaearctic region and Africa, sylviine and acanthizine warblers in Australia, and tyrant flycatchers in South America. I will briefly survey some of these equivalents and relate them to the parulines. Comparably detailed data are not available for most of the groups or geographic areas; therefore, I will focus on comparisons between parulines and Old World sylviines, of which several members have been studied in detail. The dominant forms by no means preempt the gleaning life-style within any geographic area; also, some parulines have diverged from this life-style. Thus, North American kinglets (Sylviinae) play a much greater role in northern forests than would be predicted from their two species, and the paruline "redstarts" *(Setophaga, Myioborus)* hawk insects much as Old World flycatchers (Muscicapinae) and New World flycatchers (Tyrannidae) do.

Old World Warblers

Sylviine warblers of the Old World provide an interesting comparison to the parulines. The European component alone consists of about 37 breeding species and therefore invites comparison with North American parulines. It is part of a much larger group that numbers well over 300 species and occupies all of the major landmasses of the Eastern Hemisphere. Since the sylviines and parulines arise from different major passerine assemblages (the sylviines are part of the 10-primaried group, belonging to the family Muscicapidae), many of their similarities are probably adaptations to comparable conditions. In common with the parulines, European sylviine warblers often are numerically dominant breeding members of their communities, although in mature forests their numbers are equalled or exceeded (especially in pure coniferous forests) by the resident titmouse (Paridae) populations (Simms 1985), some of whose members glean heavily from leaves or needles (Perrins 1979).

Most sylviines differ from parulines in being relatively dull-colored. Perhaps in compensation, sylviines have more complex vocalizations than do parulines, which depend more on their coloration for advertisement, at least during the breeding season. Simms (1985) noted that the verb *warble,* which means "to sing or utter in a trilling, vibratory or quavering manner, to modulate, to carol, or to produce any melodious succession of pleasing sounds," applies well to the songs of several sylviine species but is much less appropriate for describing paruline song.

The sexes are also usually similar in sylviines, although there are exceptions (e.g., *Cisticola*). Associated with this trait—in contrast with the parulines—males in several sylviine genera participate in building the nest and incubating the eggs; indeed, males of several species form brood patches. Male sylviines are also more likely to care for young nestlings than are male parulines.

Sylviines are largely insectivorous and have a variety of feeding adaptations that facilitate hunting for and capture of prey. In common with parulines, some species take a lot of fruit in the summer and fall (e.g.,

Debussche and Isenmann 1983). Given their conservative coloration, it is not surprising that many of the species are extremely similar and can be distinguished most readily by vocalization, geography, or habitat.

Most sylviines in Britain and elsewhere in Europe have, over the past few thousand years, experienced striking shifts in the availability of breeding habitat as a result of changes wrought by humans; however, several species have been resilient in using habitats that become available to them. For example, certain sylviines, in common with other insectivorous species, colonized the royal parks of mid-London after the Clear Air Acts of the 1950s, and some are well-known for their ability to use vacant lots in urban areas (Simms 1985). One therefore infers that they are not as sensitive to island-biogeographic effects as their North American equivalents, perhaps a consequence of the sylviines having been under human habitat pressure much longer than the birds in North America. Vulnerable species may have already disappeared from marginal areas. Simms's (1985) effort to reconstruct the waxing and waning of sylviine distribution in concert with recent postglacial episodes indicates that the birds experienced considerable flux. Some must have had limited geographic ranges during phases of Pleistocene glaciation, and the species that survived the interludes may have emerged well able to cope with the diminution of range observed today. Blondel (1985) suggested that several species evolved in small isolated patches of scrub vegetation.

The frequency with which sylviine vagrants from eastern Europe and Asia reach western Europe is impressive (Simms 1985), and these birds might fill niches vacated by species not well adapted to island-biogeographic effects. Populations of easternmost Europe might also augment or recolonize the ranges of their more westerly conspecifics in the same way. Other species have rapidly expanded their breeding ranges in recent times, which would provide new entities for restocking. For instance, the European Barred Warbler *(Sylvia nisoria)* has extended its range into southern Sweden (Cody 1985), and Cetti's Warbler *(Cettia cetti)* has expanded to Britain from the Mediterranean in 50 years (Bonham and Robertson 1975). Cody (1985) has summarized these and other striking changes.

Given the similar number of "warbler" species in Europe and eastern North America, and the presence of genera with several species each, one might predict that warblers of the New and Old World would have similar resource-use patterns. The common pattern, however, as outlined by Lack (1971), is fundamentally different in the two assemblages. A substantial number of congeners are geographic replacements of each other. Congeners that do occupy the same geographic range often frequent different habitats, thus minimizing the potential for ecological overlap. The exceptions, those species with more than one congener in the same habitat, usually use different parts of the habitat, as is common with the *Dendroica* assemblage of northeastern North America. But Lack did not list any instances in which more than two such species coexisted in this way. Since male and female *Dendroica* further partition the habitat at an intraspecific level, it seems quite clear that the level of niche partitioning in the paruline assemblages is far more complex than the one for the sylviine genera (Lack 1971). Most strikingly, where congeneric sylviines come together, their habitats are often not clearly separated, and instead they are interspecifically territorial (e.g., Catchpole 1973). This pattern is found in several North American nonparuline species-pairs that are extremely similar morphologically and ecologically (see Willson and Orians 1964; Murray 1971), but not in the parulines. Thus, high-diversity congeneric assemblages of sylviines appear to be the exception.

Cody's (1978) work on English and Swedish *Phylloscopus* and *Sylvia* warblers demonstrates the situation usually reported for sylviines, in which members of a genus overlap in habitats only to a limited degree, especially if the species are very similar ecologically to each other. Interspecific territoriality is common in these situations, and as vegetational structure becomes more complex—with progressively taller vegetation—increases in sylviine species diversity are primarily due to new genera being added to the fauna. Genera of European warblers group themselves in habitats according to a vegetational gradient, often successional in nature (Figure 15.1): *Cisticola* in grassy areas; *Acrocephalus* and *Locustella* in marshes, edge, and open areas; *Sylvia* in low scrub and woodland; *Hippolais* in open woodland and edge; and *Phylloscopus* in taller scrub and woodland (Cody 1985). Although paruline genera show differences in habitat choice, their preferences are not as clearly delineated among most genera, and large genera such as *Dendroica*, and probably *Vermivora* to a lesser extent, will collectively course over much of the total vegetational gradient occupied by European sylviines.

Cody attributed this difference in partitioning to a variety of factors. Many of the sylviines occupy highly transitory successional habitats, agricultural regions, and areas disturbed by logging, glacial retreat, or fire (primarily in the mediterranean habitats of southern Europe). Furthermore, these birds are migrants that must reclaim a site each year. Cody

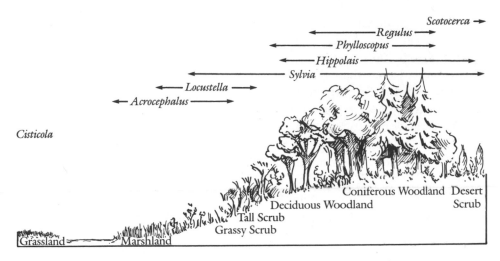

Figure 15.1 Habitat ranges of sylviine warbler genera in western Europe. (Modified from Cody 1985.)

suggested that these factors make a habitat so unpredictable that specialization to a narrow part of the habitat is unlikely; instead, a response to a broader range of habitats that will coincidentally cast species into interaction with one another, especially among congeners, will be favored. These are the same problems that many parulines encounter, however, and it is possible that the North American studies were conducted in relatively undisturbed areas and therefore may not include species or situations in which parulines would be so constrained. It might be more profitable to compare European sylviines with paruline species that have prospered in second-growth situations to determine if a similar pattern has emerged in the New World. Consistent with the argument that disturbance dictates the habitat distribution of European sylviines in minimally disturbed areas, coexistence is higher even in short-growth vegetation on Sardinia (Cody and Walter 1976) than in England and Sweden (Cody 1978). The Sardinian example is not entirely comparable to the more northerly ones, though, because residents there are likely to be permanent. Cody (1985) concluded that, among the basically Palaearctic assemblage of sylviines, only in North Africa and the Canary Islands are habitat preferences distinct and the sites occupied not strongly affected by interspecific interactions. A larger assemblage occurs in North Africa than in any of the European studies, although several of the same species are noted in both areas.

Cody attributed the differences between the two areas to the stability of habitats in North Africa and the Canaries, whose populations were spared the ebb and flow of glaciation that acted on the European populations. Again, however, these African and Canary Island birds are largely resident, and the areas around the edge of the Sahara and the Sahara itself have been subject to a great deal of environmental flux over the past few thousand years (Pachur and Kröpelin 1987; Ritchie and Haynes 1987). Thus, it is difficult to separate the factors responsible for the interspecific relations observed on northern breeding grounds, but the temporary nature of their residence appears to be important.

By contrast, European populations may be able to adjust rapidly to new relationships. An example is the European Barred Warbler's colonization of Öland, an island off Sweden's Baltic coast, which required its insinuation into an established species pool of warblers. Cody (1978, 1985) reported a shift in their interactions with Garden Warblers *(Sylvia borin)* over an 8-year period. He concluded that the two species avoided each other's prime habitat over this short time, but Cody's (1985) brief treatment of his follow-up study makes it difficult to evaluate this conclusion. Cody argued that the flexibility of these species, associated with wide ranges of habitat acceptability, reflects the ability to respond to new situations.

The importance of strict habitat segregation in the European species is brought into question by a study of the relations among *Sylvia* species on Sardinia (Cody and Walter 1976). As many as six species occupy a habitat, five of them likely to be common. Although they have different habitat preferences along a gradient, overlap is substantial. In certain instances marked interspecific aggression and evidence of displacement arises between members of species-pairs, suggesting more complex patterns of segregation than those on the European mainland (Lack 1971). These patterns raise the question of whether mechanisms of ecological isolation are as slow to develop as we had thought. The argument for the distributional patterns among *Sylvia* and other sylviine genera on mainland Europe has been that sites of sympatry represent evolutionarily new zones of contact; the pattern seen on Sardinia is not entirely consistent with it. Few other warblers occur in the habitat types that Cody and Walter (1976) worked, raising the possibility that Sardinia is a true island habitat for sylviines, notwithstanding the diversity of *Sylvia*. Comparing resource exploitation of *Sylvia* warblers in southern France and Corsica, however, Blondel (1985) argued that island and mainland relations among *Sylvia* are similar. This genus warrants careful appraisal over the relation between coexistence and length of contact.

One's first impression is thus that European sylviines are in an earlier state of ecological differentiation than are North American parulines. Yet, both groups have experienced Pleistocene glaciation, and unless the glaciations differed markedly in the New and the Old World, with respect to population sizes during the glacials and locations and configurations of refugia, it is difficult to attribute differences merely to ages of populations. The presence of largely parapatric species exhibiting interspecific territoriality, regardless of the basis for that territoriality (Willson and Orians 1964; Murray 1971), suggests an ecologically new situation, which is unlikely to be stable over long periods of time, judging from North American studies.

One is thus tempted to look for extrinsic factors in accounting for ecological differences between European sylviines and North American parulines. Although sylviines make up a substantial proportion of the total individuals in forest communities (Simms 1985), that level does not approach that of many New World forests, where parulines dominate. Calculations of species per warbler genus reveal an even more striking pattern. One major difference between European and North American plots of diversity and biomass is the prominence of titmice and chickadees (Paridae) in the two areas. Although some parids feed heavily from the bark, many glean foliage and small limbs (Morse 1970a; Perrins 1979). In Europe as many as five or six members of the genus *Parus* co-occur in a single area, exhibiting the complex interspecific partitioning in resource exploitation (Hartley 1953; Gibb 1954) of the spruce-forest parulines. In North America, local parid diversity resembles that of the European sylviines; no more than two species co-occur, and they always are members of different subgenera whose partitioning is size-related (Lack 1969). Members of any North American parid subgenus exclude each other if they come in contact (Dixon 1954; Minock 1972).

Although the reciprocity between parid and paruline diversity is striking, its basis is less clear. If one confines the comparison to western Europe and eastern North America, the differences may be a consequence of variable opportunities for the groups. The parids are largely nonmigratory; parulines and sylviines are migratory at high latitudes. Influenced by the Gulf Stream, winter conditions in much of western Europe are milder than those of eastern North America, which may enhance opportunities for the resident parids in Europe (Morse 1978b). Some insects remain active during most of the English winter, a few even reproducing (Varley et al. 1973). With larger numbers of insects to supplement their seed resources, European parids may be able to exploit a broader part of the resource base than is possible in eastern North America, where, arguably, resources are more scarce. The result may be fewer opportunities for migratory species to invade the local fauna in Europe than in North America.

This argument has at least two problems, although neither may be critical. Much of the vast Eurasian range over which the tits occur has a far more continental climatic regime than the one characteristic of western Europe. Numbers of tits eventually decline with increasing latitude (e.g., Meinertzhagen 1938; Snow 1949); however, more than two co-occur in large areas of the north (e.g., Alatalo et al. 1986). Eventually, in the subarctic (Lapland and the northern Soviet Union), only two tits are widely distributed—the Lapp *(Parus cinctus)* and Willow *(P. montanus)* tits. It would be of interest to compare the resource-exploitation patterns of summering sylviine warblers of these areas with those seen to the south. My brief experience in Swedish Lapland suggests that decline in parid diversity is accompanied by a decline in the vegetation's structural complexity, relative to a temperate broadleaf forest. The prevailing vegetation over large areas in Lapland grades between stunted pure birch forests and tundra, which diminishes bird diversity (see MacArthur and MacArthur 1961). If vegetational structure plays an important role, one would not expect differences in sylviine distribution to be linked to parid diversity.

A second difficulty of comparing eastern North America with western Europe is that the low diversity of parulines in western North America, comparable to that of European sylviines, is not accompanied by a greater parid diversity. Thus, we may have to consider the distributional problem in light of both climatic and historical aspects. The family Paridae probably evolved in the Old World and is relatively new in the New World: certainly its overall diversity and range are far higher in the Old World, with about 32 species of several subgenera and representatives throughout Eurasia ranging south into tropical Asia and sub-Saharan Africa. In contrast, North America has 10 species in only two subgenera, which do not penetrate into tropical America. The local diversity of parids is similar in eastern and western North America; thus the parids and parulines combined have a lower within-habitat diversity in western North America than in eastern North America. The niche-partitioning pattern of western parulines is consistent with their being an evolutionarily young assemblage; Mengel's model, which proposes that they arose from different episodes of Pleistocene glaciation, supports this hypothesis. The present pattern of sibling paruline species, which have no clear ecological isolation, matches the early stages of ecological isolation postulated by Lack (1969) to account for differences in parid species diversity in the Palaearctic and Nearctic. Thus, a combination of historical and ecological factors could account for this difference in diversity of gleaners, to which Wiens (1975) also alluded.

Neotropical Tyrannids

In the Neotropics parulines are largely replaced by tyrant flycatchers (Tyrannidae) that have radiated into a similar life-style (Keast 1972). Only 2–3 percent of the resident passerine Neotropical genera and species are parulines (Figure 15.2). They are primarily of two genera, *Myioborus* and *Basileuterus*, many of which are allopatric replacements of each other. A prominent group of several genera, which Keast (1972) referred to as "tiny

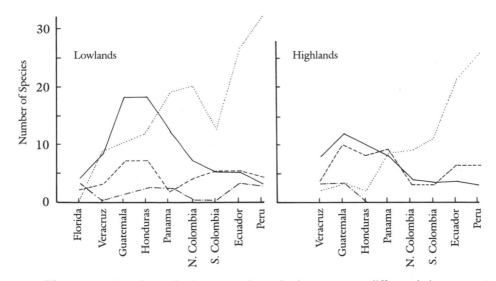

Figure 15.2 Numbers of migrant species and other groups at different latitudes. Solid line = migratory warblers; dashed line = resident warblers; dotted line = small resident tyrannids; dashed/dotted line = sylviids and parids. (Modified from Keast 1980b; used by permission of the Smithsonian Institution Press, © Smithsonian Institution, Washington, D. C.)

olivaceous tyrannids" and "small-bodied warbler equivalents," are probably the most important participants in this warbler adaptive zone. These species are mostly members of the elaenia group (Elaeniinae) (Fitzpatrick 1980, 1985). Other members of this tyrannid group are prominent residents of the dry shrubby hillsides of Argentina and Chile and act as ecological equivalents of tits, gnatcatchers, and kinglets. The tody flycatchers (*Todirostrum* and related genera), an elaeniine group characterized by their long, parallel-sided bills, although lacking close equivalents elsewhere, also employ many warblerlike tactics in feeding, including fluttering up to foliage to remove prey from the leaves and catching flying insects in the outer branches. Other tyrannids, characterized by their long legs, have even become reed-bed specialists, paralleling sylviine warblers such as *Acrocephalus* (reed warblers) of the Palaearctic and Africa and the paruline yellowthroats *(Geothlypis)*, a widespread group of marsh-dwelling species.

Keast (1972) and Fitzpatrick (1985) maintained that the tyrannids are the dominant occupants of the warbler adaptive zone because of their long (60 million years) period of isolation. The South American landmass was separated from lands to the north from the Paleocene until the end of the Miocene, a few million years ago. Although the period subsequent to reattachment may have been long enough for other groups to invade and radiate at the expense of the tyrannids, any such replacement has been minimal. Although a handful of sylviines is in the Neotropical fauna, their impact appears to be even smaller than that of the parulines. The failure of sylviines could partially result from the few species that penetrate through the North American paruline filter. Since these elements have reached South America but have not had a major impact, it is more likely that they have been suppressed by the remarkable success of the tyrannid group—a family that has been so successful that it has usurped a wide variety of adaptive zones, including the range that a full eight families or subfamilies fill in the African bird fauna. Given the isolation and the low diversity of founders in the South American avifauna (largely tyrannids, furnariids, and formacarids) (Keast 1972), one might expect them to be vulnerable to invasion, but this has not happened.

In addition to noting this relation between small tyrannids and resident parulines, Keast (1980b) pointed out an inverse pattern between the diversity of these birds and the distribution of wintering migratory parulines (Figure 15.2). Migratory birds peak in Mexico and northern Central America and then rapidly decline in numbers and species as one moves to the south. Over this same transect numbers and abundance of small tyrannids increase precipitously.

Island Faunas

Island forms are noteworthy because sometimes the usual entries to the warbler adaptive zone are absent, and other groups converge in this direction. Keast's works on the avifauna of Tasmania and the nearby Australian mainland (1968) illustrates this pattern well. Less than half as many small passerine bird species are on Tasmania as on the adjacent Australian mainland, and the sylviine, acanthizine (Australian warbler), and malurine (Australian wren-warbler) entries, which dominate the warbler life-style on the Australian mainland, are similarly deficient on Tasmania, consisting of only two thornbills (Acanthizinae). But an endemic honeyeater, *Melithreptus affinis* (Meliphagidae), on Tasmania has the shortest bill of any member of its genus, and also an extremely short hallux, which should enhance its foliage-gleaning ability and thus convergence into this zone (Keast 1968).

The radiation of the white-eyes (Zosteropodidae), small birds that are generally greenish above and grayish to yellow below (Moreau and Kikkawa 1985), is perhaps the most notable example of convergence into foliage-gleaning niches. In addition to occupying most of the Old World except the western Palaearctic, white-eyes have colonized most of Oceania, including islands so small that they are the only species of breeding land bird, often attaining prodigious densities under these circumstances (Kikkawa 1980). White-eyes have a wide range of foraging patterns, but most are insect gleaners and play roles similar to those of warblers. The genus *Zosterops* is conservative in its plumage patterns, and so uncertainty exists as to the exact relation and limits of allopatric species. The island distribution of these birds is reminiscent of Golden Warblers *(D. petechia)* on Neotropical islands. Despite their conservative plumage patterns, though, differentiation among island populations of white-eyes has proceeded further than has that of Golden Warblers.

Remarkable radiations from one, or at most a few, original colonizers on isolated island groups also warrant consideration. Only two need be mentioned, the more differentiated being that of the Hawaiian honeycreepers (Drepanidinae). These birds have radiated from a finchlike ancestor into a remarkable group of some 28 recent species (AOU 1983) that occupy most nonraptorial life-styles, including species that have adopted a paruline life-style *(Hemignathus parvus, Paroreomyza maculata)*. At least eight of these species are now extinct and others are rare or possibly extinct as a result of habitat modification, introductions, and, perhaps of greatest significance, the introduction of avian malaria (van Riper et al. 1986). At least another 15 species, known from subfossil material found in lava tubes, may have become extinct after Polynesians occupied the islands (Olson and James 1982).

The second radiation, on the Galápagos Islands, also by finch ancestors (Emberizinae) (Lack 1947; Grant 1986), has been less extreme, both in breadth of adaptations and in numbers of species (13), but it probably represents an earlier phase of adaptive radiation. A modest range of body size exists among the finches, with the Warbler Finch having a much smaller and more pointed beak than most others. The Galápagos Islands have also been colonized by Golden Warblers. Grant (1986) reported that Warbler Finches and Golden Warblers sometimes exploit similar foods, but he did not study their relations. Given the demonstration of dry-season limitation of some Galápagos finches by Grant and his colleagues (Grant 1986), it would not be surprising if the warblers limited the finches' move into a warblerlike life-style. These Golden Warblers are found in virtually all habitats of the islands, although at low densities in some areas (Snow 1966). Steadman (1985) presented evidence from lava tubes suggesting that the Golden Warbler is a recent arrival, which could account for the existence of the warbler finch in the first place.

Limitations on the Range of a Species: What Is to Be Learned from Differences among Equivalents?

Close equivalents to parulines occur in other geographic areas, but the ways they exploit and divide resources differ, at least between paruline and sylviine warblers. Recent geological histories, environmental factors, and characteristics of other ecological equivalents (e.g., the tits) could all help to account for these differences. Prevailing environmental conditions could determine social systems, which in turn dictate interaction patterns among these species.

It is unlikely that temperate-zone paruline warblers are prevented from retaining their breeding areas year-round by the presence of other groups of birds. Their winter diversity in the north does not differ greatly from that of European sylviines, which also vacate their breeding areas, probably because of food scarcity and lack of time for harvesting. Parulines have the same general pattern of migratory movement as sylviines, even though the resident contingent of titmice in western Europe contains up to three times as many species and four times as many individuals as in North America (Lack 1969). Although climatic conditions in eastern North America are more severe than those in western Europe, parid diversity in eastern North America is similar to that in North American areas with milder winter climates. Thus, parts of the eastern winter community should be vulnerable to invasion by hardy species. That parulines are not more successful in this regard suggests that they are limited by factors other than competition by parids, or, more generally, by any birds.

If we assume a tropical North American origin for parulines, their low breeding diversity in the Neotropics—in light of the high diversity of avifauna in these areas—raises questions. Of interest would be comparisons of the roles of parulines and tyrannids, especially elaeniine tyrannids, the warblerlike group that Keast (1972) proposed was stanching the diversity of wintering and resident parulines in low-latitude tropics.

Given the recent connections of Asia and North America across the Bering land bridge, and the current proximity of the two areas, it may seem strange that parulines have been unable to establish a single beachhead in northern Asia (or that none currently exists). Although conditions may never have been propitious for colonizing this bridge (see Yurtsev 1985), the presence of Blackpoll and Yellow-rumped warblers far into Alaska, and Yellow Warblers in riverine vegetation even to the north of them, suggests no dearth of vagrants. Yet the movement of these species into the extreme northwest of the North American continent may be a geologically new phenomenon. The presence of a well-developed sylviine fauna on the other side of the Bering Strait could also constrain the expansion of paruline ranges. Other than for the quite distinct kinglet, gnatcatcher, and gnatwren line, only one sylviine, the Arctic Warbler *(Phylloscopus borealis)*, has colonized Alaska, however, and it is confined to the western part. The Arctic Warbler is the only temperate-zone or northern North American sylviine that exhibits the dominant paruline leaf-gleaning habit. This limited amount of colonization suggests that either the crossing conditions were unfavorable (and remain so) or that the regions are already saturated. The latter explanation is problematic in light of the existence of a much less diverse assemblage of parulines in northwestern North America than in eastern North America. Further, parids probably are incidental to the story: their low diversity in North America should further enhance the possibility of sylviine colonization.

The kinglets may have prospered by being able to winter in the north, in that sense differing from most temperate-zone sylviines and parulines. Their hardiness could have provided them with the ability to invade North America, whereas other sylviine lines have been unsuccessful. They are twig foragers, and this trait separates them from most parulines, although they overlap in foraging patterns with *Vermivora* and *Parula* and interactions may result from these similarities (Morse 1967a).

The success of kinglets in the face of failure by typical sylviines to colonize the Western Hemisphere suggests that sylviines are limited not in the far north but in their inability to establish wintering ranges to the south. Arctic Warblers breeding in Alaska actually retrace their ancestral invasion route, returning to the Old World to winter. Retracing ancestral routes may be a standard procedure for new colonists; for example, American Redstarts that have extended their range westward in western North America do it. Yet it adds to the length of the Arctic Warbler's migration, as it would for any other Siberian colonist, and it is unlikely to provide special advantages to these birds, other than allowing them to move to less dense breeding areas. Even if the move provides a benefit, the problem arises of whether the Old World wintering grounds could support the increase in population size that would follow the spread of the Arctic Warblers to a large new breeding area.

One might thus speculate that substantial benefits for putative Siberian sylviine invaders would accrue only if they ceased migratory movement, which is improbable because of the harsh Alaskan winters, or if they established new wintering grounds in the Western Hemisphere. No sign of such activity has been reported for the Arctic Warbler; the AOU (1983) lists no records of this species in the Western Hemisphere outside of Alaska. Whether the species could insinuate itself into New World winter communities to the south is a serious question; what is more significant is that inertia for such a change in wintering grounds probably will prevent the species from ever performing the wintering experiment.

16 Opportunities for Further Research

An analysis such as I have presented in this book can establish which subjects require attention and what work needs to be done. It can provide the focus for advances on both the biology of the warblers and larger issues of ecology, behavior, and evolutionary biology. Advances in warbler biology should be encouraged as ends in their own right and also because a more complete and integrated understanding of these birds will increase their usefulness as a model group for addressing general questions. I hope I have shown that warblers offer substantial advantages as research subjects for exploring broader questions, and there is no reason why their usefulness cannot be further enhanced.

I wish in this chapter to emphasize the value of experimental studies, although sometimes they may not be appropriate—notably for answering evolutionary questions. Properly designed and interpreted experimental studies, however, may provide useful information even in this area.

In most situations inappropriate use of experiments is not the problem. The problem is that the experimental approach has as yet received inadequate attention (see Wheelwright 1986). Bird biologists have concentrated on descriptive and comparative studies, an appropriate place to start. Experiments are likely to be tedious, and the questions a biologist might ask may be answered more easily with other kinds of animals. Nevertheless, more descriptive or comparative work is unlikely to provide answers to numerous questions currently of interest and thus is not a good use of time for exploring several subjects. Researchers need to give the possibility of experimentation the attention that it deserves.

I list here a few areas in which our knowledge is inadequate. Virtually any contribution to these topics, even randomly selected, might be of importance, merely because the relevant data are so scarce. Critical areas include the following: behavior and movement patterns upon and immediately after arrival in the spring; activities while newly fledged young are still dependent on their parents; activities during the period leading up to migration; resource-exploitation patterns and other variables over the entire migration period, both spring and fall; and the entire wintering season, with few exceptions, although information is beginning to accumulate. I conclude the chapter with brief discussions of other topics that would advance understanding of parulines or make fundamental contributions to questions of broad ecological, behavioral, or evolutionary interest.

Breeding Systems

Warbler breeding systems (and those of other passerine groups) are currently believed to be more complex than the simple monogamous relations traditionally envisioned by researchers who concentrated on the behavior seen about nests. Floater individuals have been incorporated into this picture, but even though they are considered to be nonbreeding their identity has seldom been established. Further, the possibility of promiscuity or polygyny has received little attention. These wandering individuals have usually been identified as unmated males, but recently it has been established that they might be adjacent territory holders, possibly even bigamists, thus placing a different light on their position in population structure. Bigamous males could account for an absence of surplus females. Consequently, establishing the existence of "surplus" males depends on verifying the identity of wandering individuals, a difficult proposition in the absence of a marking program. Still, determining the importance of these additional birds should contribute to an evaluation of the evolutionary forces acting on warbler breeding systems. For instance, if mates are not a limiting factor, one can search elsewhere for limiting factors. The limiting factors in force could dictate the displays and behavior patterns that will evolve to address other males and females, the patterns of spatial use that develop, and the intensity of sexual selection.

Foraging Patterns and Inclement Weather

The drastic difference between foraging patterns during favorable weather and the inclement periods of the breeding season, a time of high offspring mortality and adult stress, emphasizes the value of obtaining data sets from unfavorable weather conditions. Once this information is in hand, it is then important to determine the levels of reproductive success associated with the proportions of time that individuals must indulge in foraging under unfavorable conditions or cannot forage at all. What is the importance of characteristic and uncharacteristic foraging patterns in the production of offspring? Does foraging under unfavorable conditions play a role in the survival of offspring, or adults? Or, do adults depend on these crisis periods occurring infrequently enough that no individual is likely to experience one? Adequate data sets of foraging patterns and activity patterns during inclement periods are needed for comparison with those when conditions are favorable and warblers readily gather large amounts of food. Is there a relation between the proportion of inclement weather and nesting success, and do individuals vary in their response to activity in suboptimal foraging areas? Do species differ from one another in this regard? In that the "safe" large-limb locations are favored during inclement conditions, the effect of bad weather should be less drastic for Yellow-rumped Warblers than for other spruce-woods warblers, and one might thus predict they will fare well under inclement circumstances. Yet this shift in foraging patterns by other species toward the Yellow-rumped Warbler's normal sites may push different species together, thereby crowding the Yellow-rumped Warblers as well. The scenario suggests possibilities for selective removals complementing the quantitative foraging data needed for comparison.

Spacing, Community Structure, and Habitat Selection

Given the high diversity of paruline warblers in eastern communities, and Sherry and Holmes's (1985) discovery that species space their breeding territories at different scales, work is required to establish the basis for these differences. This information could provide insight into an important aspect of community structure—habitat exploitation by the different species. These factors could be associated with intraspecific interactions, interspecific interactions, or habitat considerations, probably a combination of them. The proportions are likely to differ with the species in question and may therefore play a major role in determining the success of the species in question. The loose colonial relation noted for several warbler species suggests that social dynamics enhance the probability of finding a mate; social factors are likely to be balanced against the problems of acquiring adequate resources if packing becomes too dense. Selective removals might allow one to sort out the roles of conspecifics and other community members in establishing spatial scale.

One could address the role of vegetation structure in habitat selection by selectively modifying the habitat. Unfortunately, that problem is not easily addressed, given the large areas occupied by territorial warblers. But the problem is not intractable, if efforts are made to work cooperatively with experimental forestry programs, which modify large areas of habitat anyway (Morse 1985). In fact, such programs will permit manipulations at a magnitude likely to be impossible with most other groups; for that reason, warbler studies may be useful as model systems.

Predators and Parasites

I have repeatedly argued that resource exploitation is an important limiting factor on the size of warbler populations, with climatic factors only intermittently playing a minor role. Little evidence exists about predation on fledged individuals, which may imply that it is not a significant factor, although stronger evidence would be desirable. Since certain predators feed heavily on songbird prey, however, a more precise evaluation of prey-predator interactions is needed. It would also be of use to establish whether nest predation or parasitism significantly depresses population size, as was shown to be the case among Kirtland's Warblers before the initiation of cowbird control. Further, predatory pressure may select for antipredatory behavior even in the absence of heavy predation. The fact that some individuals with successful first nestings have reared second broods, and that Nolan (1978) recorded a depression in reproductive output from cowbird nest parasitism on Prairie Warblers, indicates that differential pressures may seriously affect numbers. If numbers of young produced still exceeded those that could be supported at some other time of the year, however, the effect might be inconsequential.

If the question of predation receives inadequate attention, the issue of ectoparasites, endoparasites, and diseases of nongame bird populations, including warblers, receives near or total neglect. Quantitative biological information about these organisms would be a major contribution. Even if they are not a direct factor in warbler population dynamics, these organisms may have shaped some of the behavior patterns, especially the breeding biology and sociality, of their hosts.

Migrants on Their Wintering Grounds

The perception of migrants on their wintering grounds has changed greatly over the past few years. Although much of the information recently gathered does not match the former view that they are primarily exploiters of edge or temporary situations, information on alternative strategies is for the most part modest. Thus, although workers have frequently reported space-related aggression, they have as yet documented very few examples of sustained territorial use by known individuals. More detailed work is needed to establish the occurrence of this and other types of winter social systems. Although the social system at least in part is related to resource opportunities, it is also clear that under similar situations species perform differently, as with Chestnut-sided and Bay-breasted warblers. Greenberg (1984b) has suggested that this difference is related to the vegetation types that the two species experience at other times (deciduous in the summer vs. coniferous in the summer), but it is necessary to test this prediction on other species. The results have implications for the extent of stereotypy and resource range and for species' vulnerability to extinction due to changes in conditions.

Specialization

The relationship between behavioral specialization and rarity may be of value in interpreting extinction and exploitation patterns. The special nesting requirements of the Kirtland's and Golden-cheeked warblers raise the question of whether they represent one extreme in the interaction between these factors. Or, is the difference merely a consequence of whether the special items or resources in question are themselves common or rare, so that given equal availability of resources, the abundance of species will be the same? What other traits do rare species have in common? For instance, will recurring patterns be related to differences that Greenberg discovered between Chestnut-sided and Bay-breasted warblers; that is, for deciduous-adapted forms to exhibit specialization and for coniferous-adapted ones not to do so? That would not hold for Kirtland's and Golden-cheeked warblers, as both are conifer dwellers in the summertime. On the other hand, the special requirements identified for them relate to nesting, not foraging.

These comments give a sense of how little we know about what determines the success (survival) or failure (extinction) of a species. Factors to consider include habitat features, both foraging-related and not foraging-related, as well as interactions among species. All of them could affect community structure.

The "Paruline" Niche

Spruce-woods warblers, and the eastern deciduous warblers as well, exhibit high diversity, a consequence of partitioning the habitat rather than exploiting the habitat differently or partitioning resources by differences in size. Although this pattern could be an expression of an old, stable relation and a resource base that does not vary enough in size to permit the size-related partitioning seen in parids, that possibility has not been adequately explored. Comparisons are of interest, but although sylviines have received considerable study, no one has undertaken comparable analyses of other paruline equivalents in other geographic areas. Such results would permit further interpretation of the differences between paruline community structure in eastern North America and sylviine community structure in western Europe, which may be due to differences in the recent climatic stability of the areas available to them. Profitable comparisons between these two groups could also be made with low-diversity paruline assemblages of coniferous forests in western North America, in which some participants are more closely related to eastern parulines. Studies of parulines and elaeniine and other tyrannid community elements over a geographic or altitudinal gradient would shed light on the dynamics of foliage gleaning in general and on ecological relations between parulines and ecologically similar species as well. Quantitative studies designed to permit ready comparison of work done in different systems would maximize understanding of the subject. Upon completion, studies of habitat manipulations may be profitable.

Summer versus Winter Limitation

One of the most difficult questions, which nevertheless is important for management and conservation programs, is whether populations are limited on their breeding or wintering areas. This information could help to establish the sensitivity of warblers to identifiable stresses. For instance, information supporting winter limitation would suggest that environmental changes on the wintering grounds might result in a corresponding change in population size: the concerns and the opportunities for the populations lie there. The presence of territorial behavior on breeding and wintering grounds among several species suggests that both areas have been subject to limitation, which complicates analysis. Although the conclusion that limitation takes place in both seasons is an interesting evolutionary conclusion, it is of little help in dealing with conservation problems associated with rapid habitat change. Further, this tendency may vary among species.

It is virtually impossible directly to test the question of which area is more crucial, although multiyear comparisons from several areas in both regions should permit one to look for trends. A data base exists on the breeding grounds because of censuses *(American Birds)* and surveys (Robbins et al. 1986) of breeding birds, but we have no comparable data base for wintering grounds.

Of greatest interest and importance, however, are the dynamics of local populations, either those from a breeding area on their wintering grounds or vice versa. Such studies are currently intractable because we do not know where individuals of Neotropical migrant species spend both summers and winters. Extensive studies of specimens from wintering grounds should permit one to narrow the range and perhaps identify likely species, populations, and areas for study. Still, what little we do know suggests that mixing is great enough to cause serious problems with this approach. Studies of species that consist of local single breeding populations, namely Kirtland's and Golden-cheeked warblers, might allow one to skirt that problem, but those species are seldom seen on their wintering grounds and would therefore be difficult subjects, notwithstanding the importance of such data for conservation purposes. Also, one may ask whether rare endemics would provide a picture representative of the whole.

Male-Female and Parent-Offspring Conflict

Warbler breeding systems provide an excellent opportunity to test predictions of parent input to offspring, from the viewpoint of benefits to parents and offspring. Opportunities for male cheating may be high early in the season because females take sole responsibility for building nests and incubating eggs, as well as assuming a major role in the care of just-hatched young. By contrast, males maintain territorial integrity at that time, although the fact that they may spend time off their territories and sometimes even set up a subsequent mating raises the question of whether territoriality severely constrains their activity.

Of particular interest, however, is the partitioning of efforts in the postfledging period (at this time males and females tend to separate, both with part of the clutch), as well as the extent of male and female contributions, although this period should not be considered separately from the earlier part of the year. The two parents appear to make similar contributions in terms of numbers of offspring reared to independence. However, explicit studies are needed to establish the magnitudes of male and female contributions with respect to the numbers and sexes of young each parent tends, extent of care, and length of period of care. Neither is it clear how this pattern varies among members of a population or species. Also, if an effort is taken to rear a second brood, the male allegedly takes over the care of the entire first brood. How is this contribution orchestrated, and is the decision to attempt another brood in any way affected by contributions of the male? Brood manipulations might permit one to address some of these problems experimentally.

Regardless of the parents' contributions, the offspring inevitably solicit far more care than they eventually receive. What factors determine the amount of care given, and how do male and female parents respond to these demands? To what extent are they influenced by the number of young?

Repertoires, Vocal and Visual Communication, and Discrimination

Warblers provide excellent opportunities for studying communication, by virtue of their wide span of repertoire sizes and patterns. Their correspondingly wide range of species diversity will evoke predictions about redundancy and repertoire size, as well as song characteristics resulting from the physical environment. Experimental studies of habitat selection that I proposed earlier should identify critical and noncritical elements for habitat choice. Noncritical elements might nevertheless affect communicatory optima and result in predictions for singing behavior.

The high density and wide diversity of warblers allows one to test whether competition for communicatory pathways occurs and whether it could affect population size. These possibilities are open to testing by way of playback experiments and selective removals.

Given the distinctive plumages of most warbler species, it may be possible to evaluate the roles of vocal and visual cues in communication, both in terms of isolating mechanisms and male-male aggressive interactions. These cues may be evaluated through study of hybrid Golden-winged and Blue-winged warblers with varied color and song patterns; and study of song patterns, combined with experimental color modifications in certain species, could lead to an understanding of the separate visual and auditory parts of communication.

Hybrids and Isolating Mechanisms

The distribution of hybrids among warblers is highly nonrandom. The implication is that many potentially hybridizing species virtually never form mixed pairs, which suggests that isolating mechanisms operate among species distantly enough related that one would not assume hybridization to be a possibility. If so, evaluation of the consequences is a worthwhile venture. Laboratory pairings would be a critical, though not definitive, part of such a study, since the natural frequency of hybrid formation is so low that it would not be practical to study this putative phenomenon in the field. Studies of mixed pairs should proceed under a variety of conditions, when the choice of a conspecific partner is and is not available; from these studies we might learn the patterns of pair formation and how isolating mechanisms prevent the occurrence of more likely hybrid pairs and offspring. Although warbler species may be closely enough related not to serve as a perfect model of all other coexisting species groups, the genetic distances of pairs of species with and without known hybrids are of interest to other systems of closely related species.

References

Adolph, S. C., and J. Roughgarden. 1983. Foraging by passerine birds and *Anolis* lizards on St. Eustatius (Netherlands Antilles): implications for interclass competition and predation. *Oecologia* 56:313–317.

Alatalo, R. V., A. Carlson, A. Lundberg, and S. Ulfstrand. 1981. The conflict between male polygamy and female monogamy: the case of the Pied Flycatcher *Ficedula hypoleuca. American Naturalist* 117:738–753.

Alatalo, R. V., L. Gustafsson, and A. Lundberg. 1986. Interspecific competition and niche change in tits (*Parus* spp.): evaluation of nonexperimental data. *American Naturalist* 127:819–834.

Alexander, W. B., and R. S. R. Fitter. 1955. American land birds in western Europe. *British Birds* 48:1–14.

Allee, W. C., A. E. Emerson, O. Park, T. Park, and K. P. Schmidt. 1949. *Principles of animal ecology*. Philadelphia: Saunders.

Amadon, D. 1968. Foreword to the Dover edition. In F. M. Chapman, *The warblers of North America*, 3rd ed. New York: Dover Publications, pp. v–vii.

Amann, H. 1949. Starke Schwankungen im Bestande des Waldlaubsängers. *Ornithologische Beobachter* 46:148–150.

Ambuel, B., and S. A. Temple. 1982. Songbird populations in southern Wisconsin forests: 1954 and 1979. *Journal of Field Ornithology* 53:149–158.

——— 1983. Area-dependent changes in the bird communities and vegetation of southern Wisconsin forests. *Ecology* 64:1057–1068.

American Ornithologists' Union (AOU). 1957. *Check-list of North American Birds,* 5th ed. Prepared by a committee of the American Ornithologists' Union. Published by the AOU. Baltimore, Md.: Lord Baltimore Press.

——— 1983. *Check-list of North American Birds,* 6th ed. Prepared by the Committee on Classification and Nomenclature of the American Ornithologists' Union. Published by the AOU. Lawrence, Kans.: Allen Press.

Anderle, R. F., and P. R. Anderle. 1976. The Whistling Warbler of St. Vincent, West Indies. *Condor* 78:236–243.

Anderson, R. M., and R. M. May. 1979. Population biology of infectious diseases: Part I. *Nature* 280:361–367.

Anderson, S. H., and H. H. Shugart, Jr. 1974. Habitat selection of breeding birds in an east Tennessee deciduous forest. *Ecology* 55:828–837.

Andersson, M. 1983. On the function of conspicuous seasonal plumages in birds. *Animal Behaviour* 31:1262–1263.

Andrew, R. J. 1969. The effects of testosterone on avian vocalizations. In R. A.

Hinde, ed., *Bird vocalizations*. Cambridge: Cambridge University Press, pp. 97–130.

AOU. *See* American Ornithologists' Union.

Apfelbaum, S., and A. Haney. 1981. Bird populations before and after wildfire in a Great Lakes pine forest. *Condor* 83:347–354.

Askins, R. A., and M. J. Philbrick. 1987. Effects of changes in regional forest abundance on the decline and recovery of a forest bird community. *Wilson Bulletin* 99:7–21.

Avise, J. C., P. C. Patton, and C. F. Aquadro. 1980. Evolutionary genetics of birds. *Journal of Heredity* 71:302–310.

Baird, J., and I. C. T. Nisbet. 1960. Northward fall migration on the Atlantic coast and its relation to offshore drift. *Auk* 77:119–149.

Baker, R. R., and G. A. Parker. 1979. The evolution of bird coloration. *Philosophical Transactions of the Royal Society of London,* ser. B, 287:63–130.

Banks, R. C., and N. K. Johnson. 1961. A review of North American hybrid hummingbirds. *Condor* 63:3–28.

Bankwitz, K. G., and W. L. Thompson. 1979. Song characteristics of the Yellow Warbler. *Wilson Bulletin* 91:533–550.

Barrowclough, G. F. 1980a. Gene flow, effective population sizes, and genetic variance components in birds. *Evolution* 34:789–798.

———— 1980b. Genetic and phenotypic differentiation in a wood warbler (Genus *Dendroica*) hybrid zone. *Auk* 97:655–668.

Barrowclough, G. F., and K. W. Corbin. 1978. Genetic variation and differentiation in the Parulidae. *Auk* 95:691–702.

Batten, L. A. 1971. Bird population changes on farmland and in woodland for the years 1968–69. *Bird Study* 18:1–8.

Batten, L. A., and J. H. Marchant. 1977. Bird population changes for the years 1974–75. *Bird Study* 24:55–61.

Beals, E. W. 1960. Forest bird communities in the Apostle Islands of Wisconsin. *Wilson Bulletin* 72:156–181.

Belcher, C., and C. D. Smooker. 1937. Birds of the colony of Trinidad and Tobago. *Ibis,* ser. 14, 1:225–249, 504–550.

Bennett, S. E. 1980. Interspecific competition and the niche of the American Redstart *(Setophaga ruticilla)* in wintering and breeding communities. In A. Keast and E. S. Morton, eds., *Migrant birds in the Neotropics.* Washington, D.C.: Smithsonian Institution Press, pp. 319–335.

Bent, A. C. 1940. Life histories of North American cuckoos, goatsuckers, hummingbirds and their allies. *Bulletin of the United States National Museum* 176:1–506.

———— 1953. Life histories of North American wood warblers. *Bulletin of the United States National Museum* 203:1–734.

———— 1958. Life histories of North American blackbirds, orioles, tanagers, and allies. *Bulletin of the United States National Museum* 211:1–549.

Berger, A. J. 1955. Six-storied Yellow Warbler nest with 11 cowbird eggs. *Jack-Pine Warbler* 33:84.

———— 1958. The Golden-winged—Blue-winged warbler complex in Michigan and the Great Lakes area. *Jack-Pine Warbler* 36:37–71.

Berger, M., and J. S. Hart. 1974. Physiology and energetics of flight. In D. S. Farner and J. R. King, eds., *Avian biology*, vol. 4. New York: Academic Press, pp. 415–477.

Black, C. P. 1975. The ecology and bioenergetics of the northern Black-throated Blue Warbler *(Dendroica caerulescens caerulescens)*. Ph.D. diss., Dartmouth College, Hanover, N.H.

Blais, J. R. 1973. Control of Spruce Budworm: current and future strategies. *Bulletin of the Entomological Society of America* 19:208–213.

Blake, E. R. 1957. The warblers of Mexico. In L. Griscom and A. Sprunt, Jr., eds., *The warblers of America.* New York: Devin-Adair, pp. 247–255.

Bledsoe, A. H. 1988. A hybrid *Oporornis philadelphia* × *Geothlypis trichas*, with comments on the taxonomic interpretation and evolutionary significance of intergeneric hybridization. *Wilson Bulletin* 100:1–8.

Blondel, J. 1985. Habitat selection in island versus mainland birds. In M. L. Cody, ed., *Habitat selection in birds*. Orlando, Fla.: Academic Press, pp. 477–516.

Bond, J. 1937. The Cape May Warbler in Maine. *Auk* 54:306–308.

———— 1957. The resident wood warblers of the West Indies. In L. Griscom and A. Sprunt, Jr., eds., *The warblers of America.* New York: Devin-Adair, pp. 263–268.

Bonham, P. F., and J. C. M. Robertson. 1975. The spread of the Cetti's Warbler in north-west Europe. *British Birds* 68:393–408.

Boxall, P.C. 1983. Observations suggesting parental division of labor by American Redstarts. *Wilson Bulletin* 95:673–674.

Boyd, E. M. 1951. The external parasites of birds: a review. *Wilson Bulletin* 63:363–369.

Bradlee, T. S., L. L. Mowbray, and W. F. Eaton. 1931. A list of birds recorded from the Bermudas. *Proceedings of the Boston Society of Natural History* 39:279–382.

Brand, A. R. 1938. Vibration frequencies of passerine bird song. *Auk* 55:263–268.

Brewster, W. 1875. Some observations on the birds of Ritchie County, West Virginia. *Annals of the Lyceum of Natural History of New York* 11:129–146.

———— 1878. The Prothonotary Warbler *(Protonotaria citrea)*. *Bulletin of the Nuttall Ornithological Club* 3:153–162.

———— 1891. Notes on Bachman's Warbler *(Helminthophila bachmani)*. *Auk* 8:149–157.

———— 1906. The birds of the Cambridge region of Massachusetts. *Memoirs of the Nuttall Ornithological Club* 4:1–426.

———— 1938. The birds of the Lake Umbagog region of Maine, part 4. Compiled by Ludlow Griscom. *Bulletin of the Museum of Comparative Zoology* 66:525–620.

Brittingham, M. C., and S. A. Temple. 1983. Have cowbirds caused forest songbirds to decline? *BioScience* 33:31–35.

Brooks, M. 1947. Breeding habitats of certain wood warblers in the unglaciated Appalachian region. *Auk* 64:291–295.

Brooks, M., and W. C. Legg. 1942. Swainson's Warbler in Nicholas County, West Virginia, *Auk* 59:76–86.

Brown, C. R., and M. B. Brown. 1986. Ectoparasitism as a cost of coloniality in Cliff Swallows *(Hirundo pyrrhonota)*. *Ecology* 67:1206–1218.

Brown, J. L. 1964. The evolution of diversity in avian territorial systems. *Wilson Bulletin* 76:160–169.

———— 1969a. The buffer effect and productivity in tit populations. *American Naturalist* 103:347–354.

———— 1969b. Territorial behavior and population regulation in birds. *Wilson Bulletin* 81:293–329.

———— 1978. Avian communal breeding systems. *Annual Review of Ecology and Systematics* 9:123–155.

Brown, W. L., Jr., and E. O. Wilson. 1956. Character displacement. *Systematic Zoology* 5:49–64.

Bulmer, M. G. 1973. Inbreeding in the Great Tit. *Heredity* 30:313–325.

Burleigh, T. D. 1927. Further notes on the breeding birds of northeastern Georgia. *Auk* 44:229–234.

———— 1944. The bird life of the Gulf Coast region of Mississippi. *Occasional Papers of the Museum of Zoology, Louisiana State University* 20:329–492.

Burtt, E. H., Jr., 1984. Colour of the upper mandible: an adaptation to reduce reflectance. *Animal Behaviour* 32:652–658.

———— 1986. An analysis of physical, physiological, and optical aspects of avian coloration with emphasis on wood warblers. *Ornithological Monographs* 38:1–126.

Burtt, E. H., Jr. and A. J. Gatz, Jr. 1982. Color convergence: is it only mimetic? *American Naturalist* 119:738–740.

Busby, D. G., and S. G. Sealy. 1979. Feeding ecology of a population of nesting Yellow Warblers. *Canadian Journal of Zoology* 57:1670–1681.

Buskirk, R. E., and W. H. Buskirk. 1976. Changes in arthropod abundance in a highland Costa Rican forest. *American Midland Naturalist* 95:288–298.

Buskirk, W. H. 1972. Ecology of bird flocks in a tropical forest. Ph.D. diss., University of California, Davis.

———— 1980. Influence of meteorological patterns and trans-gulf migration on the calendars of latitudinal migrants. In A. Keast and E. S. Morton, eds., *Migrant birds in the Neotropics*. Washington, D.C.: Smithsonian Institution Press, pp. 485–491.

Cadbury, J. M., and A. D. Cruickshank. 1937–1958. Climax Red and White spruce forest (in reports of annual breeding bird census). *Bird-lore*, vols. 39–60.

Carlson, C. W. 1981. The Sutton's Warbler—a critical review and summation of current data. *Atlantic Naturalist* 34:1–11.

Catchpole, C. K. 1973. Conditions of co-existence in sympatric breeding populations of *Acrocephalus* warblers. *Journal of Animal Ecology* 42:623–635.

Chapman, F. M. 1917. *The warblers of North America,* 3rd ed. New York: D. Appleton.

Chappuis, C. 1971. Un example de l'influence du milieu sur les émissions vocales des oiseaux: l'évolution des chants en forêt équitoriale. *La Terre et la Vie* 1971:183–202.

Charnov, E. L. 1982. The theory of sex allocation. *Monographs in Population Biology* 18:1–355.

Chipley, R. M. 1977. The impact of wintering migrant wood warblers on resident insectivorous passerines in a subtropical Colombian oak woods. *Living Bird* 15:119–141.

——— 1980. Nonbreeding ecology of the Blackburnian Warbler. In A. Keast and E. S. Morton, eds., *Migrant birds in the Neotropics*. Washington, D.C.: Smithsonian Institution Press, pp. 309–317.

Clapham, C.S. 1964. The birds of the Dahlac Archipelago. *Ibis* 106:376–388.

Clark, K. L., and R. J. Robertson. 1979. Spatial and temporal multi-species nesting aggregations in birds as anti-parasite and anti-predator strategies. *Behavioral Ecology and Sociobiology* 5:359–371.

——— 1981. Cowbird parasitism and evolution of anti-parasite strategies in the Yellow Warbler. *Wilson Bulletin* 93:249–258.

Clay, T., A. M. Hutson, and A. Baker. 1985. Ectoparasite. In B. Campbell and E. Lack, eds., *A dictionary of birds*. Calton, U.K.: Poyser, pp. 170–172.

Cody, M. L. 1974. Competition and the structure of bird communities. *Monographs in Population Biology* 7:1–318.

——— 1978. Habitat selection and interspecific territoriality among the sylviid warblers of England and Sweden. *Ecological Monographs* 48:351–396.

——— 1981. Habitat selection in birds: the roles of habitat structure, competitors, and productivity. *BioScience* 31:107–113.

——— 1985. Habitat selection in the sylviine warblers of Western Europe and North Africa. In M. L. Cody, ed., *Habitat selection in birds*. New York: Academic Press, pp. 85–129.

Cody, M. L., and J. H. Brown. 1969. Song asynchrony in neighboring bird species. *Nature* 222:778–780.

Cody, M. L., and H. Walter. 1976. Habitat selection and interspecific interactions among Mediterranean sylviid warblers. *Oikos* 27:210–238.

Collins, S. L. 1981. A comparison of nest-site and perch-site vegetation structure for seven species of warblers. *Wilson Bulletin* 93:542–547.

——— 1983. Geographic variation in habitat structure of the Black-throated Green Warbler *(Dendroica virens)*. *Auk* 100:382–389.

Collins, S. L., F. C. James, and P. G. Risser. 1982. Habitat relationships of wood warblers (Parulidae) in northern central Minnesota. *Oikos* 39:50–58.

Confer, J. L., and K. Knapp. 1979. The changing proportion of Blue-winged and Golden-winged warblers in Tompkins County and their habitat selection. *Kingbird* 29:8–14.

——— 1981. Golden-winged Warblers and Blue-winged Warblers: the relative success of a habitat specialist and a generalist. *Auk* 98:108–114.

Connell, J. H. 1975. Some mechanisms producing structure in natural communi-

ties: a model and evidence from field experiments. In M. Cody and
J. Diamond, eds., *Ecology and evolution of communities*. Cambridge, Mass.:
Harvard University Press, pp. 460–490.

———— 1978. Diversity in tropical rain forests and coral reefs. *Science* 199:1302–1310.

Connor, E. F., and D. Simberloff. 1979. The assembly of species communities:
chance or competition? *Ecology* 60:1132–1140.

———— 1984. Neutral models of species' co-occurrence patterns. In D. R. Strong,
Jr., D. Simberloff, L. G. Abele, and A. B. Thistle, eds., *Ecological communities:
conceptual issues and the evidence*. Princeton, N.J.: Princeton University
Press, pp. 316–331.

Cox, G. W. 1960. A life history of the Mourning Warbler. *Wilson Bulletin* 72:5–28.

———— 1968. The role of competition in the evolution of migration. *Evolution*
22:180–192.

———— 1985. The evolution of avian migration systems between temperate and
tropical regions of the New World. *American Naturalist* 126:451–474.

Crawford, H. S., R. W. Titterington, and D. T. Jennings. 1983. Bird predation
and Spruce Budworm populations. *Journal of Forestry* 81:433–435, 478.

Cronon, W. 1983. *Changes in the land*. New York: Hill and Wang.

Crowell, K. L. 1962. Reduced interspecific competition among the birds of
Bermuda. *Ecology* 43:75–88.

Davis, D. E. 1946. A seasonal analysis of mixed flocks of birds in Brazil. *Ecology*
27:168–181.

Davis, J. W., R. C. Anderson, L. Karstad, and D. O. Trainer, eds. 1971. *Infectious
and parasitic diseases of wild birds*. Ames, Iowa: Iowa State University Press.

Debussche, M., and P. Isenmann. 1983. La consommation des fruits chez quelques
fauvettes méditerranéennes (*Sylvia melanocephala, S. cantillans, S. hortensis
et S. undata*) dans la région de Montpellier (France). *Alauda* 51:302–308.

DeSante, D. F. 1973. An analysis of the fall occurrences and nocturnal orientations
of vagrant wood warblers (Parulidae) in California. Ph.D. diss., Stanford
University, Stanford, Calif.

———— 1983. Annual variability in the abundance of migrant landbirds on Southeast Farallon Island, California. *Auk* 100:826–852.

Diamond, A. W. 1986. *An evaluation of the vulnerability of Canadian migratory birds
to changes in Neotropical forest habitats*. Ottawa: Canadian Wildlife Service.

Diamond, A. W., and R. W. Smith. 1973. Returns and survival of banded warblers
wintering in Jamaica. *Bird-banding* 44:221–224.

Diamond, J. M. 1982. Mirror-image navigational errors in migrating birds. *Nature*
295:277–278.

Dixon, K. L. 1954. Some ecological relations of chickadees and titmice in central
California. *Condor* 56:113–124.

Dobson, A. P., and P. J. Hudson. 1986. Parasites, disease and the structure of
ecological communities. *Trends in Ecology and Evolution* 1:11–15.

Drury, W. H., and J. A. Keith. 1962. Radar studies of songbird migration in coastal New England. *Ibis* 104:449–489.

Dugmore, A. R. 1902. The increase in the Chestnut-sided Warbler. *Bird-Lore*
4:77–80.

Dunham, A. E. 1980. An experimental study of interspecific competition between the iguanid lizards *Sceloporus merriami* and *Urosaurus ornatus*. *Ecological Monographs* 50:309–330.

Dunn, E. H., and E. Nol. 1980. Age-related migratory behavior of warblers. *Journal of Field Ornithology* 51:254–269.

Dwight, J., Jr. 1900. Sequence and plumage of moults of the passerine birds of New York. *Annals of the New York Academy of Science* 13:73–360.

Eaton, S. W. 1953. Wood warblers wintering in Cuba. *Wilson Bulletin* 65:169–174.

Eldredge, N., and S. J. Gould. 1972. Punctuated equilibria: an alternative to phyletic gradualism. In T. J. M. Schopf, ed., *Models in paleobiology*. San Francisco: Freeman, Cooper, pp. 82–115.

Eliason, B. C. 1986. Female site fidelity and polygyny in the Blackpoll Warbler (*Dendroica striata*). *Auk* 103:782–790.

Elkins, N. 1979. Nearctic landbirds in Britain and Ireland: a meteorological analysis. *British Birds* 72:417–433.

Elliott, B. G. 1969. Life history of the Red Warbler. *Wilson Bulletin* 81:184–195.

Emlen, J. T. 1973. Territorial aggression in wintering warblers at Bahama agave blossoms. *Wilson Bulletin* 85:71–74.

—— 1977. Land bird communities of Grand Bahama Island: the structure and dynamics of an avifauna. *Ornithological Monographs* 24:1–129.

—— 1980. Interactions of migrant and resident land birds in Florida and Bahama pinelands. In A. Keast and E. S. Morton, eds., *Migrant birds in the Neotropics*. Washington, D.C.: Smithsonian Institution Press, pp. 133–143.

—— 1981. Divergence in the foraging responses of birds on two Bahama islands. *Ecology* 62:289–295.

Emlen, J. T., and M. J. DeJong. 1981. Intrinsic factors in the selection of foraging substrates by Pine Warblers: a test of an hypothesis. *Auk* 98:294–298.

Emlen, S. T. 1975. Migration: orientation and navigation. In D. S. Farner and J. R. King, eds., *Avian biology*, vol. 5. New York: Academic Press, pp. 129–219.

Endler, J. A. 1977. Geographic variation, speciation, and clines. *Monographs in Population Biology* 10:1–246.

—— 1987. Review of "An analysis of physical, physiological, and optical aspects of avian coloration with emphasis on wood-warblers." *Condor* 89:680–681.

Errington, P. L. 1946. Predation and vertebrate populations. *Quarterly Review of Biology* 21:144–177, 221–245.

Erskine, A. J. 1977. Birds in boreal Canada. *Canadian Wildlife Service Report Series* 41:1–73.

—— 1980. A preliminary catalogue of bird census plot studies in Canada, part 4. *Progress Notes, Canadian Wildlife Service* 112:1–26.

—— 1984. A preliminary catalogue of bird census plot studies in Canada, part 5. *Progress Notes, Canadian Wildlife Service* 144:1–34.

Faaborg, J., W. J. Arendt, and M. S. Kaiser. 1984. Rainfall correlates of bird population fluctuations in a Puerto Rican dry forest: a nine year study. *Wilson Bulletin* 96:575–593.

Farner, D. S. 1949. Age groups and longevity in the American Robin: comments, further discussion, and certain revisions. *Wilson Bulletin* 6:68–81.

——— 1952. The use of banding data in the study of certain aspects of the dynamics and structures of avian populations. *Northwest Science* 26:119–144.

——— 1955. Birdbanding in the study of population dynamics. In A. Wolfson, ed., *Recent studies in avian biology*. Urbana, Ill.: University of Illinois Press, pp. 397–449.

Ficken, M. S. 1961. Redstarts and cowbirds. *Kingbird* 11:83–85.

——— 1962. Agonistic behavior and territory in the American Redstart. *Auk* 79:607–632.

——— 1963. Courtship of the American Redstart. *Auk* 80:307–317.

——— 1965. Mouth color of nestling passerines and its use in taxonomy. *Wilson Bulletin* 77:71–75.

Ficken, M. S., and R. W. Ficken. 1962a. The comparative ethology of the wood warblers: a review. *Living Bird* 1:103–122.

——— 1962b. Some aberrant characters of the Yellow-breasted Chat. *Auk* 79:718–719.

——— 1965. Comparative ethology of the Chestnut-sided Warbler, Yellow Warbler, and American Redstart. *Wilson Bulletin* 77:363–375.

——— 1966. Notes on mate and habitat selection in the Yellow Warbler. *Wilson Bulletin* 78:232–233.

——— 1967. Age-specific differences in the breeding behavior and ecology of the American Redstart. *Wilson Bulletin* 79:188–199.

——— 1968a. Courtship of Blue-winged Warblers, Golden-winged Warblers, and their hybrids. *Wilson Bulletin* 80:161–172.

——— 1968b. Ecology of Blue-winged Warblers, Golden-winged Warblers, and some other *Vermivora*. *American Midland Naturalist* 79:311–319.

——— 1968c. Reproductive isolating mechanisms in the Blue-winged and Golden-winged warbler complex. *Evolution* 22:166–179.

——— 1968d. Territorial relationships of Blue-winged Warblers, Golden-winged Warblers and their hybrids. *Wilson Bulletin* 80:442–451.

——— 1969. Responses of Blue-winged Warblers and Golden-winged Warblers to their own and the other species' song. *Wilson Bulletin* 81:69–74.

Ficken, R. W., M. S. Ficken, and E. S. Hadaway. 1967. Mobbing of a Chuck-will's-widow by small passerines. *Auk* 84:266–267.

Ficken, R. W., M. S. Ficken, and J. P. Hailman. 1974. Temporal pattern shifts to avoid acoustic interference in singing birds. *Science* 183:762–763.

Ficken, R. W., M. S. Ficken, and D. H. Morse. 1968. Competition and character displacement in two sympatric pine-dwelling warblers (*Dendroica,* Parulidae). *Evolution* 22:307–314.

Ficken, R. W., J. W. Popp, and P. E. Matthiae. 1985. Avoidance of acoustic interference by Ovenbirds. *Wilson Bulletin* 97:569–571.

Finch, D. W. 1975. The spring migration, April 1–May 31, 1974, Northeastern Maritime Region. *American Birds* 29:125–129.

Fitzpatrick, J. W. 1980. Foraging behavior of Neotropical tyrant flycatchers. *Condor* 82:43–57.

———— 1985. Form, foraging behavior, and adaptive radiation in the Tyrannidae. *Ornithological Monographs* 36:447–470.

Folkers, K. L., and P. E. Lowther. 1985. Responses of nesting Red-winged Blackbirds and Yellow Warblers to Brown-headed Cowbirds. *Journal of Field Ornithology* 56:175–177.

Forbush, E. H. 1929. *Birds of Massachusetts and other New England states*, vol. 3. Boston: Massachusetts Department of Agriculture.

Ford, N. L. 1983. Variation in mate fidelity in monogamous birds. *Current Ornithology* 1:329–356.

Foster, M. S. 1969. Synchronized life cycles in the Orange-crowned Warbler and its mallophagan parasites. *Ecology* 50:315–323.

———— 1974a. A model to explain molt breeding overlap and clutch size in some tropical birds. *Evolution* 28:182–190.

———— 1974b. Rain, feeding behavior and clutch size in tropical birds. *Auk* 91:722–726.

Francis, C. M., and F. Cooke. 1986. Differential timing of spring migration in wood warblers (Parulinae). *Auk* 103:548–556.

Franzeb, K. E. 1983. Intersexual habitat partitioning in Yellow-rumped Warblers during the breeding season. *Wilson Bulletin* 95:581–590.

Freeman, F. J. 1950. Display of Oven-birds, *Seiurus aurocapillus*. *Auk* 67:521.

Fretwell, S. D. 1972. Populations in a seasonal environment. *Monographs in Population Biology* 5:1–218.

———— 1980. Evolution of migration in relation to factors regulating bird numbers. In A. Keast and E. S. Morton, eds., *Migrant birds in the Neotropics*. Washington, D.C.: Smithsonian Institution Press, pp. 517–527.

Fretwell, S. D., and J. L. Lucas, Jr. 1970. On territorial behavior and other factors influencing habitat distribution in birds. I. Theoretical development. *Acta Biotheoretica* 19:16–36.

Friedmann, H. 1929. *The cowbirds, a study in the biology of social parasitism*. Springfield, Ill.: Thomas.

Futuyma, D. J. 1979. *Evolutionary biology*. Sunderland, Mass.: Sinauer.

Gaddis, P. K. 1983. Composition and behavior of mixed-species flocks of forest birds in north-central Florida. *Florida Field Naturalist* 11:25–34.

Galli, A. E., C. F. Leck, and R. T. T. Forman. 1976. Avian distribution patterns in forest islands of different sizes in central New Jersey. *Auk* 93:356–364.

Gans, C. 1974. *Biomechanics*. Philadelphia: Lippincott.

Gates, J. E., and L. W. Gysel. 1978. Avian nest dispersion and fledging success in field-forest ecotones. *Ecology* 59:871–883.

Gause, G. F. 1934. *The struggle for existence*. Baltimore: Williams and Wilkins.

Gauthreaux, S. A., Jr. 1971. The radar and direct visual study of passerine spring migration in southern Louisiana. *Auk* 88:343–365.

———— 1979. Priorities in bird migration studies. *Auk* 96:813–815.

————, ed. 1980. *Animal migration, orientation, and navigation*. New York: Academic Press.

———— 1982. The ecology and evolution of avian migration systems. *Avian Biology* 6:93–168.

Geer, T. A. 1978. Effects of nesting Sparrowhawks on nesting tits. *Condor* 80:419–422.

George, W. G. 1962. The classification of the Olive Warbler, *Peucedramus taeniatus*. *American Museum Novitates* 2103:1–41.

Gibb, J. A. 1954. Feeding ecology of tits, with notes on Treecreeper and Goldcrest. *Ibis* 96:513–543.

——— 1960. Populations of tits and Goldcrests and their food supply in pine plantations. *Ibis* 102:163–208.

Gill, F. B. 1980. Historical aspects of hybridization between Blue-winged and Golden-winged warblers. *Auk* 97:1–18.

——— 1987. Allozymes and genetic similarity of Blue-winged and Golden-winged warblers. *Auk* 104:444–449.

Gill, F. B., and B. G. Murray, Jr. 1972a. Discrimination behavior and hybridization of the Blue-winged and Golden-winged warblers. *Evolution* 26:282–293.

——— 1972b. Song variation in sympatric Blue-winged and Golden-winged warblers. *Auk* 89:625–643.

Gilpin, M. E., and J. M. Diamond. 1984a. Are species co-occurrences on islands non-random, and are null hypotheses useful in community ecology? In D. R. Strong, Jr., D. Simberloff, L. G. Abele, and A. B. Thistle, eds., *Ecological communities: conceptual issues and the evidence.* Princeton, N.J.: Princeton University Press, pp. 297–315.

——— 1984b. Rejoinder. In D. R. Strong, Jr., D. Simberloff, L. G. Abele, and A. B. Thistle, eds., *Ecological communities: conceptual issues and the evidence.* Princeton, N.J.: Princeton University Press, pp. 332–341.

Gish, S. L., and E. S. Morton. 1981. Structural adaptations to local habitat acoustics in Carolina Wren songs. *Zeitschrift für Tierpsychologie* 56:74–84.

Göransson, G., G. Högstedt, J. Karlsson, H. Källander, and S. Ulfstrand. 1974. Sångens foll för revirhållandet hos naktergal *Lusinia lusinia*: några experiment med play-back-teknik. *Vår Fågelvärld* 33:201–209.

Gowaty, P. A. 1985. Multiple parentage and apparent monogamy in birds. *Ornithological Monographs* 37:11–21.

Graber, J. W., and R. R. Graber. 1983. Feeding rates of warblers in spring. *Condor* 85:139–150.

Granit, R. 1955. *Receptors and sensory perception.* New Haven, Conn.: Yale University Press.

Grant, P. R. 1968. Polyhedral territories of animals. *American Naturalist* 102:75–80.

——— 1972. Convergent and divergent character displacement. *Biological Journal of the Linnaean Society* 4:39–68.

——— 1986. *Ecology and evolution of Darwin's finches.* Princeton, N.J.: Princeton University Press.

Gray, A. P. 1958. *Bird hybrids.* Bucks., England: Commonwealth Agricultural Bureaux.

Greenberg, R. 1979. Body size, breeding habitat and winter exploitation systems in *Dendroica. Auk* 96:756–766.

——— 1980. Demographic aspects of long-distance migration. In A. Keast and

E. S. Morton, eds., *Migrant birds in the Neotropics*. Washington, D.C.: Smithsonian Institution Press, pp. 493–504.

——— 1981. Dissimilar bill shapes in New World tropical versus temperate forest foliage-gleaning birds. *Oecologia* 49:143–147.

——— 1983. The role of neophobia in determining the degree of foraging specialization in some migrant warblers. *American Naturalist* 122:444–453.

——— 1984a. The role of neophobia in the foraging site selection of a tropical migrant bird: an experimental study. *Proceedings of the National Academy of Sciences, USA* 81:3778–3780.

——— 1984b. The winter exploitation systems of Bay-breasted and Chestnut-sided warblers in Panama. *University of California Publications in Zoology* 116:1–107.

——— 1986. Competition in migrant birds in the nonbreeding season. *Current Ornithology* 3:281–307.

——— 1987a. Development of dead leaf foraging in a tropical migrant warbler. *Ecology* 68:130–141.

——— 1987b. Seasonal foraging specialization in the Worm-eating Warbler. *Condor* 89:158–168.

Griscom, L. 1941. The recovery of birds from disaster. *Audubon Magazine* 43:191–196.

——— 1957. The classification of warblers. In L. Griscom and A. Sprunt, Jr., eds., *The warblers of America*. New York: Devin-Adair, pp. 8–13.

Griscom, L., and A. Sprunt, Jr., eds. 1957. *The warblers of America*. New York: Devin-Adair.

Grubb, T. C., Jr. 1975. Weather-dependent foraging behavior of some birds wintering in a deciduous woodland. *Condor* 77:175–182.

——— 1985. *Weather and birds*. In B. Campbell and E. Lack, eds., *A dictionary of birds*. Calton, England: Poyser, pp. 646–648.

Gundlach, J. 1855. Beitrage sur ornithologie Cuba's. *Journal für Ornithologie* 3:465–480.

Gwinner, E., P. Berthold, and M. Klein. 1972. Untersuchungen zur Jahresperiodik von Laubsängern. III. *Journal für Ornithologie* 113:1–8.

Haffer, J. 1969. Speciation in Amazon birds. *Science* 165:131–137.

Hailman, J. P. 1967. Spectral discrimination: an important correction. *Journal of the Optical Society of America* 57:281–282.

——— 1977. *Optical signals: animal communication and light*. Bloomington, Ind.: Indiana University Press.

——— 1979. Environmental light and conspicuous colors. In E. H. Burtt, Jr., ed., *The behavioral significance of color*. New York: Garland STPM Press, pp. 289–354.

Hall, G. A. 1979. Hybridization between Mourning and MacGillivray's warblers. *Bird-banding* 50:101–107.

——— 1981. Fall migration patterns of wood warblers in the southern Appalachians. *Journal of Field Ornithology* 52:43–49.

——— 1984. A long-term population study in an Appalachian spruce forest. *Wilson Bulletin* 96:228–240.

Hamel, P. B. 1986. *Bachman's Warbler: a species in peril.* Washington, D.C.: Smithsonian Institution Press.

Hamilton, T. H., and R. H. Barth, Jr. 1962. The biological significance of season change in male plumage appearance in some New World migratory bird species. *American Naturalist* 96:129–144.

Hamilton, W. J., III, and G. H. Orians. 1965. Evolution of brood parasitism in altricial birds. *Condor* 67:361–382.

Hann, H. W. 1937. Life history of the Oven-bird in southern Michigan. *Wilson Bulletin* 49:145–237.

Hardin, G. 1960. The competitive exclusion principle. *Science* 113:1292–1297.

Hart, J. S., and M. Berger. 1972. Energetics, water economy and temperature regulation during flight. *Proceedings of the International Ornithological Congress* 15:189–199.

Hartley, P. H. T. 1953. An ecological study of the feeding habits of the English titmice. *Journal of Animal Ecology* 22:261–288.

Hausman, L. A. 1927. On the winter food of the Tree Swallow *(Iridoprocne bicolor)* and the Myrtle Warbler *(Dendroica coronata).* *American Naturalist* 61:379–382.

Hebard, F. V. 1961. Yellow Warblers in conifers. *Wilson Bulletin* 73:394–395.

Hendricks, E. S. 1981. Niche relationships of spruce-woods warblers: a mulitvariate statistical analysis. Ph.D. diss., Indiana University, Bloomington.

Henshaw, H. W. 1881. On some of the causes affecting the decrease of birds. *Journal of the Nuttall Ornithological Club* 6:189–197.

Hensley, M. M., and J. B. Cope. 1951. Further data on removal and repopulation of the breeding birds in a spruce-fir forest community. *Auk* 68:483–493.

Herman, C. M. 1955. Diseases of birds. In A. Wolfson, ed., *Recent studies in avian biology.* Urbana, Ill.: University of Illinois Press, pp. 450–467.

Herrera, C. M. 1978. Ecological correlates of residence and non-residence in a Mediterranean passerine bird community. *Journal of Animal Ecology* 47:871–890.

Hespenheide, H. A. 1980. Bird community structure in two Panama forests: residents, migrants, and seasonality during the nonbreeding season. In A. Keast and E. S. Morton, eds., *Migrant birds in the Neotropics.* Washington, D.C.: Smithsonian Institution Press, pp. 227–237.

Hickey, J. J. 1952. Survival studies of banded birds. *Special Scientific Report, Wildlife, United States Department of the Interior, Fish and Wildlife Service* 15:1–177.

Hildén, O. 1965. Habitat selection in birds. *Annales Zoologici Fennici* 2:53–75.

Hilty, S. L. 1980. Relative abundance of North Temperate Zone breeding migrants in Western Colombia and their impact at fruiting trees. In A. Keast and E. S. Morton, eds., *Migrant birds in the Neotropics.* Washington, D.C.: Smithsonian Institution Press, pp. 265–271.

Hinde, R. A. 1956. The biological significance of the territories of birds. *Ibis* 98:340–369.

Hofslund, P. B. 1957. Cowbird parasitism of the Northern Yellow-throat. *Auk* 74:42–48.

Högstedt, G., and C. Persson. 1982. Do Willow Warblers *Phylloscopus trochilus* of northern origin start their autumn migration at an earlier age than their southern conspecifics? *Holarctic Ecology* 5:76–80.

Holmes, R. T. 1976. Body composition, lipid reserves and caloric densities of summer birds in a northern deciduous forest. *American Midland Naturalist* 96:281–290.

————— 1986. Foraging patterns of forest birds: male-female differences. *Wilson Bulletin* 98:196–213.

Holmes, R. T., C. P. Black, and T. W. Sherry. 1979. Comparative population bioenergetics of three insectivorous passerines in a deciduous forest. *Condor* 81:9–20.

Holmes, R. T., R. E. Bonney, Jr., and S. W. Pacala. 1979. Guild structure of the Hubbard Brook bird community: a multivariate approach. *Ecology* 60:512–520.

Holmes, R. T., and H. F. Recher. 1986. Search tactics of insectivorous birds in an Australian eucalypt forest. *Auk* 103:515–530.

Holmes, R. T., and R. H. Sawyer. 1975. Oxygen consumption in relation to ambient temperature in five species of forest-dwelling thrushes (*Hylocichla* and *Catharus*). *Comparative Biochemistry and Physiology* 50A:527–531.

Holmes, R. T., T. W. Sherry, and S. E. Bennett. 1978. Diurnal and individual variability in the foraging behavior of American Redstarts (*Setophaga ruticilla*). *Oecologia* 36:171–179.

Holmes, R. T., T. W. Sherry, and F. W. Sturges. 1986. Bird community dynamics in a temperate deciduous forest: long-term trends at Hubbard Brook. *Ecological Monographs* 56:201–220.

Hooper, R. G., and P. B. Hamel. 1977. Nesting habitat of Bachman's Warbler—a review. *Wilson Bulletin* 89:373–379.

Horn, H. S. 1968. The adaptive significance of colonial nesting in the Brewer's Blackbird (*Euphagus cyanocephalus*). *Ecology* 49:682–694.

Houston, C. S., and S. J. Shadick. 1974. The spring migration, April 1–May 31, 1976, Western Great Lakes Region. *American Birds* 28:814–817.

Howe, H. F. 1974. Age-specific differences in habitat selection by the American Redstart. *Auk* 91:161–162.

Howell, T. R. 1971. An ecological study of the birds of the lowland pine savanna and adjacent rain forest in northeastern Nicaragua. *Living Bird* 10:185–242.

Hubbard, J. P. 1969. The relationships and evolution of the *Dendroica coronata* complex. *Auk* 86:393–432.

————— 1970. Geographic variation in the *Dendroica coronata* complex. *Wilson Bulletin* 82:355–369.

————— 1971. The avifauna of the Southern Appalachians: past and present. *Virginia Polytechnic Institute and State University Research Division Monograph* 4:197–232.

Hunt, G. L. Jr., 1972. Influence of food distribution and human disturbance on the reproductive success of herring gulls. *Ecology* 53:1051–1061.

Hunt, J. H. 1971. A field study of the Wrenthrush, *Zeledonia coronata*. *Auk* 88:1–20.

Hurley, G. F., and J. W. Jones, II. 1983. A presumed mixed Bay-breasted × Black-burnian Warbler nesting in West Virginia. *Redstart* 50:108–111.

Hussell, D. J. T., and A. B. Lambert. 1980. New estimates of weight loss in birds during nocturnal migration. *Auk* 97:547–558.

Hutchinson, G. E. 1959. Homage to Santa Rosalia *or* why are there so many kinds of animals? *American Naturalist* 93:145–159.

Hutto, R. L. 1980. Winter habitat distribution of migratory land birds in Western Mexico with special reference to small, foliage-gleaning insectivores. In A. Keast and E. S. Morton, eds., *Migrant birds in the Neotropics*. Washington, D.C.: Smithsonian Institution Press, pp. 181–203.

———— 1981a. Seasonal variation in the foraging behavior of some migratory western wood warblers. *Auk* 98:765–777.

———— 1981b. Temporal patterns of foraging activity in some wood warblers in relation to the availability of insect prey. *Behavioral Ecology and Sociobiology* 9:195–198.

———— 1985a. Habitat selection by nonbreeding migratory land birds. In M. L. Cody, ed., *Habitat selection in birds*. New York: Academic Press, pp. 455–476.

———— 1985b. Seasonal changes in the habitat distribution of transient insectivorous birds in southeastern Arizona: competition mediated? *Auk* 102:120–132.

Huxley, J. S. 1934. A natural experiment on the territorial instinct. *British Birds* 27:270–277.

Ickes, R. A., and M. S. Ficken. 1970. An investigation of territorial behavior in the American Redstart utilizing recorded songs. *Wilson Bulletin* 82:167–176.

James, F. C. 1971. Ordinations of habitat relationships among breeding birds. *Wilson Bulletin* 83:215–236.

Jansson, R. B. 1976. The spring migration, April 1–May 31, 1976, Western Great Lakes Region. *American Birds* 30:844–846.

Janzen, D. H. 1973. Sweep samples of tropical foliage insects: effect of seasons, vegetation types, elevation, time of day, and insularity. *Ecology* 54:687–708.

———— 1980. Heterogeneity of potential food abundance for tropical small land birds. In A. Keast and E. S. Morton, eds., *Migrant birds in the Neotropics*. Washington, D.C.: Smithsonian Institution Press, pp. 545–552.

Janzen, D. H., and T. W. Schoener. 1968. Differences in insect abundance and diversity between wetter and drier sites during a tropical dry season. *Ecology* 49:96–110.

Johnson, A. W. 1967. *The birds of Chile and adjacent regions of Argentina, Bolivia, and Peru*, vol. 2. Buenos Aires: Platt Establecimientos Graficos.

Johnson, T. B. 1980. Resident and North American migrant bird interactions in the Santa Marta highlands, Northern Colombia. In A. Keast and E. S. Morton, eds., *Migrant birds in the Neotropics*. Washington, D.C.: Smithsonian Institution Press, pp. 239–247.

Johnston, D. W. 1975. Ecological analysis of the Cayman Island avifauna. *Bulletin of the Florida State Museum, Biological Sciences* 19:235–300.

Kale, H. W., II. 1967. Aggressive behavior by a migrating Cape May Warbler. *Auk* 84:120–121.

Karr, J. R. 1971. Structure of avian communities in selected Panama and Illinois habitats. *Ecological Monographs* 41:207–233.

——— 1976. On the relative abundance of migrants from the North Temperate Zone in tropical habitats. *Wilson Bulletin* 88:433–458.

Karstad, L. 1971. Pox. In J. W. Davis, R. C. Anderson, L. Karstad, and D. O. Trainer, eds. *Infectious and parasitic diseases of wild birds.* Ames, Iowa: Iowa State University Press, pp. 34–41.

Keast, A. 1968. Competitive interactions and the evolution of ecological niches as illustrated by the Australian honeyeater genus *Melithreptus* (Meliphagidae). *Evolution* 22:762–784.

——— 1972. Ecological opportunities and dominant families, as illustrated by the Neotropical Tyrannidae (Aves). *Evolutionary Biology* 5:229–277.

——— 1980a. Migratory Parulidae: what can species co-occurrence in the north reveal about ecological plasticity and wintering patterns? In A. Keast and E. S. Morton, eds., *Migrant birds in the Neotropics.* Washington, D.C.: Smithsonian Institution Press, pp. 457–476.

——— 1980b. Spatial relationships between migratory parulid warblers and their ecological counterparts in the Neotropics. In A. Keast and E. S. Morton, eds., *Migrant birds in the Neotropics.* Washington, D.C.: Smithsonian Institution Press, pp. 109–130.

——— 1980c. Synthesis: ecological basis and evolution of the Nearctic-Neotropical bird-migration system. In A. Keast and E. S. Morton, eds., *Migrant birds in the Neotropics.* Washington, D.C.: Smithsonian Institution Press, pp. 559–576.

Keast, A., and E. S. Morton, eds. 1980. *Migrant birds in the Neotropics.* Washington, D.C.: Smithsonian Institution Press.

Kendeigh, S. C. 1941. Birds of a prairie community. *Condor* 43:165–174.

——— 1945. Nesting behavior of wood warblers. *Wilson Bulletin* 57:145–164.

——— 1947. Bird population studies in the coniferous forest biome during a Spruce Budworm outbreak. *Ontario Department of Lands and Forests, Biological Bulletin* 1:1–100.

——— 1952. Parental care and its evolution in birds. *Illinois Biological Monograph* 22:1–356.

Kenward, R. E. 1978. Hawks and doves: factors affecting success and selection in Goshawk attacks on Woodpigeons. *Journal of Animal Ecology* 47:449–460.

Kepler, C. B., and A. K. Kepler. 1970. Preliminary comparison of bird species diversity and density in Luquillo and Guanica forests. In H. T. Odum, ed., *A tropical rain forest.* Oak Ridge: U.S. Atomic Energy Commission, Technical Information Extension, pp. E183–186.

Kepler, C. B., and K. C. Parkes. 1972. A new species of warbler (Parulidae) from Puerto Rico. *Auk* 89:1–18.

Ketterson, E. D., and V. Nolan, Jr. 1976. Geographic variation and its climatic correlates in the sex ratio of eastern-wintering Dark-eyed Juncos (*Junco hyemalis hyemalis*). *Ecology* 57:679–693.

—— 1985. Intraspecific variation in avian migration: evolutionary and regulatory aspects. *University of Texas Contributions in Marine Science, Supplement* 27:553–579.

Kikkawa, J. 1980. Winter survival in relation to dominance classes among Silvereyes *Zosterops lateralis chlorocephala* on Heron Island, Great Barrier Reef. *Ibis* 122:437–446.

Kilgore, B. M. 1971. Response of breeding bird populations to habitat changes in a giant forest. *American Midland Naturalist* 85:135–152.

King, A. P., and M. J. West. 1977. Species identification in the North American Cowbird: appropriate responses to abnormal song. *Science* 195:1002–1004.

King, J. R. 1974. Seasonal allocation of time and energy resources in birds. *Publications of the Nuttall Ornithological Club* 15:4–85.

Klaas, E. E. 1975. Cowbird parasitism and nesting success in the Eastern Phoebe. *Occasional Papers of the Museum of Natural History, University of Kansas* 41:1–18.

Klimkiewicz, M. K., R. B. Clapp, and A. G. Futcher. 1983. Longevity records of North American birds: Remizidae through Parulinae. *Journal of Field Ornithology* 54:287–294.

Klopfer, P. H. 1965. Behavioral aspects of habitat selection: a preliminary report on stereotypy in foliage preferences of birds. *Wilson Bulletin* 77:376–381.

—— 1967. Behavioral stereotypy in birds. *Wilson Bulletin* 79:290–300.

Klopfer, P. H., and R. H. MacArthur. 1960. Niche size and faunal diversity. *American Naturalist* 94:293–300.

—— 1961. On the causes of tropical species diversity: niche overlap. *American Naturalist* 95:223–226.

Krebs, J. R. 1971. Territory and breeding density in the Great Tit, *Parus major*. *Ecology* 52:2–22.

—— 1977a. The significance of song repertoire: the Beau Geste hypothesis. *Animal Behaviour* 25:475–478.

—— 1977b. Song and territory in the Great Tit *Parus major*. In B. Stonehouse and C. Perrins, eds., *Evolutionary ecology*. Baltimore: University Park Press, pp. 47–62.

Krebs, J. R., and R. Dawkins. 1984. Animal signals: mind-reading and manipulation. In J. R. Krebs and N. B. Davies, eds., *Behavioural ecology: an evolutionary approach*. Sunderland, Mass.: Sinauer, pp. 380–402.

Krebs, J. R., M. H. MacRoberts, and J. M. Cullen. 1972. Flocking and feeding in the Great Tit *Parus major*—an experimental study. *Ibis* 114:507–530.

Krebs, J. R., D. W. Stephens, and W. J. Sutherland. 1983. Perspectives in optimal foraging. In A. H. Brush and G. A. Clark, Jr., eds., *Perspectives in ornithology*. Cambridge, England: Cambridge University Press, pp. 165–221.

Kroll, J. C. 1980. Habitat requirements of the Golden-cheeked Warbler: management implications. *Journal of Range Management* 33:60–65.

Kroodsma, D. E. 1977. Correlates of song organization among North American wrens. *American Naturalist* 111:995–1008.

—— 1981. Geographical variation and functions of song types in warblers (Parulidae). *Auk* 98:743–751.

———— 1983. The ecology of avian vocal learning. *BioScience* 33:165–171.

Kroodsma, D. E., V. A. Ingalls, T. W. Sherry, and T. K. Werner. 1987. Songs of the Cocos Flycatcher: vocal behavior of a suboscine on an isolated oceanic island. *Condor* 89:75–84.

Kroodsma, D. E., W. R. Meservey, and R. Pickert. 1983. Vocal learning in the Parulinae. *Wilson Bulletin* 95:138–140.

Lack, D. 1940. Courtship feeding in birds. *Auk* 57:169–178.

———— 1945. The ecology of closely related species with special reference to Cormorant *(Phalacrocorax carbo)* and Shag *(P. aristotelis)*. *Journal of Animal Ecology* 14:12–16.

———— 1946. Competition for food by birds of prey. *Journal of Animal Ecology* 15:123–129.

———— 1947. *Darwin's finches.* Cambridge, England: Cambridge University Press.

———— 1948. Notes on the ecology of the Robin. *Ibis* 90:252–279.

———— 1949. Vital statistics from ringed Swallows. *British Birds* 42:147–150.

———— 1954. *The natural regulation of animal numbers.* Oxford: Clarendon Press.

———— 1966. *Population studies of birds.* Oxford: Clarendon Press.

———— 1968. *Ecological adaptations for breeding in birds.* London: Methuen.

———— 1969. Tit niches in two worlds; or homage to Evelyn Hutchinson. *American Naturalist* 102:43–49.

———— 1971. *Ecological isolation in birds.* Oxford: Blackwell.

———— 1976. *Island biology illustrated by the land birds of Jamaica.* Oxford: Blackwell.

Lack, D., and P. Lack. 1972. Wintering warblers in Jamaica. *Living Bird* 11:129–153.

Lack, D., and S. Schifferli. 1948. Die Lebensdauer des Stares. *Ornithologischer Beobachter* 45:107–114.

Landres, P. B., and J. A. MacMahon. 1983. Community organization of arboreal birds in some oak woodlands of western North America. *Ecological Monographs* 53:183–208.

Lanyon, W. E., and F. B. Gill. 1964. Spectrographic analysis of variation in the songs of a population of Blue-winged Warblers *(Vermivora pinus)*. *American Museum Novitates* 2176:1–18.

Laursen, K. 1978. Interspecific relationships between some insectivorous passerine species, illustrated by their diet during spring migration. *Ornis Scandinavica* 9:178–192.

Lawrence, L. deK. 1953. Notes on the nesting behavior of the Blackburnian Warbler. *Wilson Bulletin* 65:135–144.

Lein, M. R. 1972. Territorial and courtship songs of birds. *Nature* 237:48–49.

———— 1978. Song variation in a population of Chestnut-sided Warblers *(Dendroica pensylvanica)*: its nature and suggested significance. *Canadian Journal of Zoology* 56:1266–1283.

———— 1980. Display behavior of Ovenbirds *(Seiurus aurocapillus)*: Non-song vocalizations. *Wilson Bulletin* 92:312–329.

———— 1981. Display behavior of Ovenbirds *(Seiurus aurocapillus)*: II. Song variation and singing behavior. *Wilson Bulletin* 93:21–41.

Lemon, R. E., R. Cotter, R. C. MacNally, and S. Monette. 1985. Song repertoires and song sharing by American Redstarts. *Condor* 87:457–470.

Lemon, R. E., S. Monette, and D. Roff. 1987. Song repertoires of American warblers (Parulinae): honest advertising or assessment? *Ethology* 74:265–284.

Lemon, R. E., J. Struger, M. J. Lechowicz, and R. F. Norman. 1981. Song features and singing heights of American warblers: maximization or optimization of distance. *Journal of the Acoustical Society of America* 69:1169–1176.

Leopold, N. F., Jr. 1924. The Kirtland's Warbler in its summer home. *Auk* 41:44–58.

Lewis, D. M. 1972. Importance of face mask in sexual recognition and territorial behavior in the Yellowthroat. *Jack-Pine Warbler* 50:98–109.

Lewke, R. E. 1982. A comparison of foraging behavior among permanent, summer, and winter resident bird groups. *Condor* 84:84–90.

Loftin, H. 1977. Returns and recoveries of banded North American birds in Panama and the tropics. *Bird-banding* 48:253–258.

Loftin, H., D. T. Rogers, and D. C. Hicks. 1966. Repeats, returns and recoveries of North American migrant birds banded in Panama. *Bird-banding* 37:35–44.

Lönnberg, E. 1927. Some speculations on the origin of the North American ornithic fauna. *Kungliga Svenska Vetenskapsakademiens Handlingar*, ser. 3, 4:1–24.

Lowery, G. H., Jr. 1945. Trans-gulf spring migration of birds and the coastal hiatus. *Wilson Bulletin* 57:92–121.

——— 1946. Evidence of trans-gulf migration. *Auk* 63:175–211.

——— 1974. *Louisiana birds,* 3d ed. Baton Rouge, La.: Louisiana State University Press.

Lowery, G. H., Jr., and R. J. Newman. 1966. A continent-wide view of bird migration on four nights in October. *Auk* 83:547–586.

Lynch, J. F., E. S. Morton, and M. E. Van der Voort. 1985. Habitat segregation between the sexes of wintering Hooded Warblers *(Wilsonia citrina). Auk* 102:714–721.

Lynch, J. F., and D. F. Whigham. 1984. Effects of forest fragmentation on breeding bird communities in Maryland, USA. *Biological Conservation* 28:287–324.

Lyon, B. E., and R. D. Montgomerie. 1985. Incubation feeding in Snow Buntings: female manipulation or indirect male parental care? *Behavioral Ecology and Sociobiology* 17:279–294.

——— 1986. Delayed plumage maturation in passerine birds: reliable signaling by subordinate males? *Evolution* 40:605–615.

MacArthur, R. H. 1958. Population ecology of some warblers of northeastern coniferous forests. *Ecology* 39:599–619.

——— 1959. On the breeding distribution pattern of North American migrant birds. *Auk* 76:318–325.

——— 1968. The theory of the niche. In R. C. Lewontin, ed., *Population biology and evolution.* Syracuse, N.Y.: Syracuse University Press, pp. 159–176.

——— 1969. Species packing, and what interspecies competition minimizes. *Proceedings of the National Academy of Sciences, USA* 64:1369–1371.

———— 1970. Species packing and competitive equilibrium for many species. *Theoretical Population Biology* 1:1–11.

———— 1972. *Geographical ecology*. New York: Harper and Row.

MacArthur, R. H., and J. W. MacArthur, 1961. On bird species diversity. *Ecology* 42:594–598.

MacArthur, R. H., J. W. MacArthur, and J. Preer. 1962. On bird species diversity. II. Prediction of bird census from habitat measurements. *American Naturalist* 96:167–174.

MacArthur, R. H., and E. C. Pianka. 1966. On optimal use of a patchy environment. *American Naturalist* 100:603–609.

MacArthur, R. H., H. C. Recher, and M. L. Cody. 1966. On the relation between habitat selection and species diversity. *American Naturalist* 100:319–332.

MacArthur, R. H., and E. O. Wilson. 1967. The theory of island biogeography. *Monographs in Population Biology* 1:1–203.

Marten, K., and P. Marler. 1977. Sound transmission and its significance for animal vocalization. I. Temperate habitats. *Behavioral Ecology and Sociobiology* 2:271–290.

Martin, T. E., 1986. Competition in breeding birds. On the importance of considering processes at the level of the individual. *Current Ornithology* 4:181–210.

———— 1988a. Habitat and area effects on forest bird assemblages: is nest predation an influence? *Ecology* 69:74–84.

———— 1988b. On the advantage of being different. *Proceedings of the National Academy of Sciences, USA* 85:2196–2199.

Martin, T. E., and J. R. Karr. 1986. Temporal dynamics of Neotropical birds with special reference to frugivores in second-growth woods. *Wilson Bulletin* 98:38–60.

Matthiae, P. E., and F. Stearns. 1981. Mammals in forest islands in southeastern Wisconsin. In R. L. Burgess and D. M. Sharpe, eds., *Forest island dynamics in man-dominated landscapes*. New York: Springer-Verlag, pp. 55–66.

Maurer, B. A. 1984. Interference and exploitation in bird communities. *Wilson Bulletin* 96:380–395.

Maurer, B. A., and R. C. Whitmore. 1981. Foraging of five bird species in two forests with different vegetation structure. *Wilson Bulletin* 93:478–490.

May, R. M. 1973. Stability and complexity in model ecosystems. *Monographs in Population Biology* 6:1–235.

May, R. M., and R. M. Anderson. 1979. Population biology of infectious diseases: Part II. *Nature* 280:455–461.

May, R. M., and R. H. MacArthur. 1972. Niche overlap as a function of environmental variability. *Proceedings of the National Academy of Sciences, USA* 69:1109–1113.

May, R. M., and S. K. Robinson. 1985. Population dynamics of avian brood parasitism. *American Naturalist* 126:475–494.

Mayfield, H. F. 1953. A census of Kirtland's Warblers. *Auk* 70:17–20.

———— 1960. The Kirtland's Warbler. *Bulletin of the Cranbrook Institute of Science* 40:1–242.

———— 1962. 1961 decennial census of the Kirtland's Warbler. *Auk* 79:173–182.

———— 1965. The Brown-headed Cowbird, with old and new hosts. *Living Bird* 4:13–28.

———— 1983. Kirtland's Warbler, victim of its own rarity? *Auk* 100:974–976.

Maynard Smith, J. 1982. *Evolution and the theory of games.* Cambridge, England: Cambridge University Press.

Mayr, E. 1935. Bernard Altum and the territory theory. *Proceedings of the Linnaean Society of New York* 45–46:24–38.

———— 1946. History of the North American bird fauna. *Wilson Bulletin* 58:3–41.

———— 1963. *Animal species and evolution.* Cambridge, Mass.: Harvard University Press.

———— 1964. Inferences concerning the Tertiary American bird faunas. *Proceedings of the National Academy of Sciences, USA* 51:280–288.

McAtee, W. L. 1912. Methods of estimating the contents of bird stomachs. *Auk* 29:449–464.

McClintock, E. P., T. C. Williams, and J. M. Teal. 1978. Autumnal bird migration observed from ships in the western North Atlantic Ocean. *Bird-banding* 49:262–277.

McDonald, M. A. 1987. Distribution of *Microligea palustris* in Haiti. *Wilson Bulletin* 99:688–690.

McLaren, I. A. 1981. The incidence of vagrant landbirds on Nova Scotian Islands. *Auk* 98:243–257.

McLaughlin, R. L., and R. D. Montgomerie. 1985. Brood division by Lapland Longspurs. *Auk* 102:687–695.

McNally, R. C., and R. E. Lemon. 1985. Repeat and serial singing modes in American Redstarts *(Setophaga ruticilla):* a test of functional hypotheses. *Zeitschrift für Tierpsychologie* 69:191–202.

Meanley, B. 1966. Some observations on habitats of the Swainson's Warbler. *Living Bird* 5:151–165.

———— 1971. Natural history of the Swainson's Warbler. *North American Fauna* 69:1–90.

———— 1972. *Swamps, river bottoms, and canebrakes.* Barre, Mass.: Barre Publishing Company.

Meanley, B., and G. M. Bond. 1950. A new race of Swainson's Warbler from the Appalachian Mountains. *Proceedings of the Biological Society of Washington* 63:191–195.

Meinertzhagen, R. 1938. Winter in Arctic Lapland. *Ibis* 1938:754–759.

Mengel, R. M. 1964. The probable history of species formation in some northern wood warblers (Parulidae). *Living Bird* 3:9–43.

———— 1970. The North American Central Plains as an isolating agent in bird speciation. *Special Publication, Department of Geology, University of Kansas* 3:279–340.

Meyer de Schaunsee, R. 1966. *The species of birds of South America.* Wynnewood, Penn.: Livingston.

Miles, D. B., and R. E. Ricklefs. 1984. The correlation between ecology and morphology in deciduous forest passerine birds. *Ecology* 65:1629–1640.

Miller, A. H. 1963. Avifauna of an American equatorial cloud forest. *University of California Publications in Zoology* 66:1–72.

Miller, J. A. 1987. Ecology of a new disease. *BioScience* 37:11–15.

Miller, R. S. 1967. Pattern and process in competition. *Advances in Ecological Research* 4:1–74.

Minock, M. E. 1972. Interspecific aggression between Black-capped and Mountain chickadees at winter feeding stations. *Condor* 74:454–461.

Monroe, B. L., Jr. 1968. A distributional survey of the birds of Honduras. *Ornithological Monographs* 7:1–458.

———— 1970. Effects of habitat changes on population levels of the avifauna in Honduras. *Smithsonian Contributions to Zoology* 26:38–41.

Montgomerie, R. D., and B. E. Lyon. 1986. Does longevity influence the evolution of delayed plumage maturation in passerine birds? *American Naturalist* 128:930–936.

Moore, F., and P. Kerlinger. 1987. Stopover and fat deposition by North American wood-warblers (Parulinae) following spring migration over the Gulf of Mexico. *Oecologia* 74:45–54.

Moore, F. R., and P. A. Simm. 1986. Risk-sensitive foraging by a migratory bird (*Dendroica coronata*). *Experientia* 42:1054–1056.

Moore, M. C. 1980. Habitat structure in relation to population density and timing of breeding in Prairie Warblers. *Wilson Bulletin* 92:177–187.

Moore, N. W. 1967. A synopsis of the pesticide problem. *Advances in Ecological Research* 4:75–129.

Moreau, R. E., and J. Kikkawa. 1985. White-eye. In B. Campbell and E. Lack, eds., *A dictionary of birds*. Calton, England: Poyser.

Moreno, J. 1984. Parental care of fledged young, division of labor, and the development of foraging techniques in the Northern Wheatear (*Oenanthe oenanthe* L.). *Auk* 101:741–752.

Morris, R. F., W. F. Cheshire, C. A. Miller, and D. G. Mott. 1958. The numerical response of avian and mammalian predators during a gradation of the Spruce Budworm. *Ecology* 39:487–494.

Morrison, M. L. 1981. The structure of western warbler assemblages: analysis of foraging and habitat selection in Oregon. *Auk* 98:578–588.

Morrison, M. L., and J. W. Hardy. 1983. Hybridization between Hermit and Townsend's warblers. *Murrelet* 64:65–72.

Morse, D. H. 1966a. Notes on the Wren-thrush. *Condor* 68:520–521.

———— 1966b. The contexts of songs in the Yellow Warbler. *Wilson Bulletin* 78:444–455.

———— 1967a. Competitive relationships between Parula Warblers and other species during the breeding season. *Auk* 84:490–502.

———— 1967b. The contexts of songs in the Black-throated Green and Blackburnian warblers. *Wilson Bulletin* 79:62–72.

———— 1967c. Foraging relationships of Brown-headed Nuthatches and Pine Warblers. *Ecology* 48:94–103.

———— 1968. A quantitative study of foraging of male and female spruce-woods warblers. *Ecology* 49:779–784.

———— 1969. Distraction display of a pair of Black-throated Green Warblers. *Wilson Bulletin* 81:106–107.

———— 1970a. Ecological aspects of some mixed-species foraging flocks of birds. *Ecological Monographs* 40:119–168.

———— 1970b. Territorial and courtship songs of birds. *Nature* 226:659–661.

———— 1971a. The foraging of warblers isolated on small islands. *Ecology* 52:216–228.

———— 1971b. The insectivorous bird as an adaptive strategy. *Annual Review of Ecology and Systematics* 2:177–200.

———— 1973. The foraging of small populations of Yellow Warblers and American Redstarts. *Ecology* 54:346–355.

———— 1974a. Foraging of Pine Warblers allopatric and sympatric to Yellow-throated Warblers. *Wilson Bulletin* 86:474–477.

———— 1974b. Niche breadth as a function of social dominance. *American Naturalist* 108:818–830.

———— 1975. Ecological aspects of adaptive radiation in birds. *Biological Reviews* 50:167–214.

———— 1976a. Hostile encounters among spruce-woods warblers (*Dendroica,* Parulidae). *Animal Behaviour* 24:764–771.

———— 1976b. Variables determining the density and territory size of breeding spruce-woods warblers. *Ecology* 57:290–301.

———— 1977a. Feeding behavior and predator avoidance in heterospecific groups. *BioScience* 27:332–339.

———— 1977b. The occupation of small islands by passerine birds. *Condor* 79:399–412.

———— 1978a. Populations of Bay-breasted and Cape May warblers during an outbreak of the Spruce Budworm. *Wilson Bulletin* 90:404–413.

———— 1978b. Structure and foraging patterns of tit flocks in an English woodland. *Ibis* 120:298–312.

———— 1979. Habitat use by the Blackpoll Warbler. *Wilson Bulletin* 91:234–243.

———— 1980a. *Behavioral mechanisms in ecology.* Cambridge, Mass.: Harvard University Press.

———— 1980b. Foraging and coexistence of spruce-woods warblers. *Living Bird* 18:7–25.

———— 1980c. Population limitation: breeding or wintering grounds? In A. Keast and E. S. Morton, eds., *Migrant birds in the Neotropics.* Washington, D.C.: Smithsonian Institution Press, pp. 505–516.

———— 1981. Foraging speeds of warblers in large populations and in isolation. *Wilson Bulletin* 93:334–339.

———— 1985. Habitat selection in North American parulid warblers. In M. L. Cody, ed., *Habitat selection in birds.* New York: Academic Press, pp. 131–157.

———— 1989. Song patterns of warblers at dawn and dusk. *Wilson Bulletin* 101:26–35.

Morse, D. H. and R. S. Fritz. 1987. The consequences of foraging for reproductive success. In A. C. Kamil, J. R. Krebs, and H. R. Pulliam, eds., *Foraging behavior.* New York: Plenum.

Morse, D. H., and S. W. Kress. 1984. The effect of burrow loss on mate choice in the Leach's Storm-petrel. *Auk* 101:158–160.

Morton, E. S. 1970. Ecological sources of selection on avian sounds. Ph.D. diss., Yale University, New Haven, Conn.

——— 1975. Ecological sources of selection on avian sounds. *American Naturalist* 109:17–34.

——— 1977. On the occurrence and significance of motivation-structural rules in some bird and mammal sounds. *American Naturalist* 111:855–869.

——— 1980. Adaptations to seasonal changes by migrant land birds in the Panama Canal Zone. In A. Keast and E. S. Morton, eds., *Migrant birds in the Neotropics*. Washington, D.C.: Smithsonian Institution Press, pp. 437–453.

——— 1982. Grading, discreteness, redundancy, and motivation-structural rules. In D. E. Kroodsma and E. H. Miller, eds., *Acoustic communication in birds*, vol. 1. New York: Academic Press, pp. 183–212.

——— 1986. Predictions from the ranging hypothesis for the evolution of long distance signals in birds. *Behaviour* 99:65–86.

Morton, E. S., J. F. Lynch, K. Young, and P. Mehlhop. 1987. Do male Hooded Warblers exclude females from nonbreeding territories in tropical forest? *Auk* 104:133–135.

Morton, E. S., and K. Young. 1986. A previously undescribed method of song matching in a species with a single song "type," the Kentucky Warbler (*Oporornis formosus*). *Ethology* 73:334–342.

Mousely, H. M. 1924. A study of the home life of the Northern Parula and other warblers at Hatley, Stanstead County, Quebec. *Auk* 41:236–248.

——— 1926. A further study of the home life of the Northern Parula and of the Yellow Warbler and Ovenbird. *Auk* 43:184–197.

——— 1928. A further study of the home life of the Northern Parula Warbler (*Compsothlypis americana usneae*). *Auk* 45:475–479.

Moynihan, M. 1960. Some adaptations which help to promote gregariousness. *Proceedings of the International Ornithological Congress* 12:523–541.

——— 1962. The organization and probable evolution of some mixed species flocks of Neotropical birds. *Smithsonian Miscellaneous Collections* 143(7):1–140.

——— 1968. Social mimicry; character convergence versus character displacement. *Evolution* 22:315–331.

Mueller, H. C., N. S. Mueller, and P. G. Parker. 1981. Observation of a brood of Sharp-shinned Hawks in Ontario, with comments on the functions of sexual dimorphism. *Wilson Bulletin* 93:85–92.

Murray, B. G., Jr. 1965. On the autumn migration of the Blackpoll Warbler. *Wilson Bulletin* 77:122–133.

——— 1966. Migration of age and sex classes of passerines on the Atlantic coast in autumn. *Auk* 83:352–360.

——— 1971. The ecological consequences of interspecific territorial behavior in birds. *Ecology* 52:414–423.

——— 1976. The return to the mainland of some nocturnal passerine migrants over the sea. *Bird-banding* 47:345–358.

——— 1979. Fall migration of Blackpoll and Yellow-rumped warblers at Island Beach, New Jersey. *Bird-banding* 50:1–11.

Murray, B. G., Jr., and F. B. Gill. 1976. Behavioral interactions of Blue-winged and Golden-winged warblers. *Wilson Bulletin* 88:231–254.

Murton, R. K., A. J. Isaacson, and N. J. Westwood. 1971. The significance of gregarious feeding behavior and adrenal stress in a population of wood-pigeons *Columba palumbus. Journal of Zoology* (London) 165:53–84.

Nice, M. M. 1937. Studies in the life history of the Song Sparrow. I. *Transactions of the Linnaean Society of New York* 4:1–247.

———— 1943. Studies in the life history of the Song Sparrow. II. The behavior of the Song Sparrow and other passerines. *Transactions of the Linnaean Society of New York* 6:1–328.

Nisbet, I. C. T. 1963. American passerines in western Europe. *British Birds* 56:204–217.

———— 1970. Autumn migration of the Blackpoll Warbler: evidence for long flight provided by regional survey. *Bird-banding* 41:207–240.

Nisbet, I. C. T., W. H. Drury, Jr., and J. Baird. 1963. Weight-loss during migration. Part I. Deposition and consumption of fat by the Blackpoll Warbler *Dendroica striata. Bird-banding* 34:107–138.

Nisbet, I. C. T., and Lord Medway. 1972. Dispersion, population ecology and migration of Eastern Great Reed Warblers *Acrocephalus orientalis* wintering in Malaysia. *Ibis* 114:451–494.

Noble, G. K. 1939. The role of dominance on the social life of birds. *Auk* 56:263–273.

Nolan, V., Jr. 1955. Invertebrate nest associates of the Prairie Warbler. *Auk* 72:55–61.

———— 1958. Anticipatory food-bringing in the Prairie Warbler. *Auk* 75:263–278.

———— 1959. Additional invertebrate nest associates of the Prairie Warbler. *Auk* 76:352–357.

———— 1963. Reproductive success of birds in a deciduous scrub habitat. *Ecology* 44:305–313.

———— 1978. The ecology and behavior of the Prairie Warbler *Dendroica discolor. Ornithological Monographs* 26:1–595.

Nolan, V., Jr., and R. E. Mumford. 1965. An analysis of Prairie Warblers killed in Florida during nocturnal migration. *Condor* 67:322–338.

Noon, B. R., D. K. Dawson, D. B. Inkley, C. S. Robbins, and S. H. Anderson. 1980. Consistency in habitat preference of forest bird species. *Transactions of the North American Wildlife and Natural Resource Conference* 45:226–244.

Norris, R. A. 1961. A modification of the Miller method of aging live passerine birds. *Bird-banding* 32:55–57.

Odum, E. P. 1969. The strategy of ecosystem development. *Science* 164:262–270.

Odum, E. P., C. W. Connell, and H. L. Stoddard. 1961. Flight energy and estimated flight ranges of some migratory birds. *Auk* 78:515–527.

Odum, E. P., and E. W. Kuenzler. 1955. Measurement of territory and home range size in birds. *Auk* 72:128–137.

Olson, S. L. 1985. The fossil record of birds. *Avian Biology* 8:79–238.

Olson, S. L., and H. F. James. 1982. Prodromus of the fossil avifauna of the Hawaiian Islands. *Smithsonian Contributions to Zoology* 365:1–59.

Oniki, Y. 1979. Is nesting success of birds low in the tropics? *Biotropica* 11:60–69.

Orejuela, J. E., R. J. Raitt, and H. Alvarez. 1980. Differential use by North American migrants of three types of Colombian forests. In A. Keast and E. S. Morton, eds., *Migrant birds in the Neotropics*. Washington, D.C.: Smithsonian Institution Press, pp. 253–264.

Orians, G. H. 1980. Some adaptations of marsh-nesting blackbirds. *Monographs in Population Biology* 14:1–295.

Orians, G. H., and M. F. Willson. 1964. Interspecific territories of birds. *Ecology* 45:736–745.

Osterhaus, M. B. 1962. Adaptive modifications in the leg structure of some North American warblers. *American Midland Naturalist* 68:474–486.

Pachur, H.-J., and S. Kröpelin. 1987. Wadi Howar: paleoclimatic evidence from an extinct river system in the southeastern Sahara. *Science* 237:298–300.

Palmer, R. S. 1949. Maine birds. *Bulletin of the Museum of Comparative Zoology* 102:1–656.

Park, T. 1948. Experimental studies of interspecies competition. 1. Competition between populations of the flour beetles, *Tribolium confusum* Duval and *Tribolium castaneum* Herbst. *Ecological Monographs* 18:265–308.

Parkes, K. C. 1961. Intergeneric hybrids in the family Pipridae. *Condor* 63:345–350.

——— 1978. Still another parulid intergenic hybrid *(Mniotilta × Dendroica)* and its taxonomic and evolutionary implications. *Auk* 95:682–690.

——— 1985a. Audubon's mystery birds. *Natural History* 94(4):88–93.

——— 1985b. Warbler (2). In B. Campbell and E. Lack, eds., *A dictionary of birds*. Calton, England: Poyser, pp. 642–643.

Parnell, J. F. 1969. Habitat relations of the Parulidae during spring migration. *Auk* 86:505–521.

Payne, R. B. 1972. Mechanisms and control of molt. In D. S. Farner and J. R. King, eds., *Avian biology*, vol. 2. New York: Academic Press, pp. 103–155.

——— 1977. The ecology of brood parasitism in birds. *Annual Review of Ecology and Systematics* 8:1–28.

Payne, R. B., L. L. Payne, and S. M. Doehlert. 1984. Interspecific song learning in a wild Chestnut-sided Warbler. *Wilson Bulletin* 96:292–294.

Paynter, R. A. 1953. Autumnal migrants on the Campeche Bank. *Auk* 70:338–349.

Pearson, D. L. 1980. Bird migration in Amazonian Ecuador, Peru, and Bolivia. In A. Keast and E. S. Morton, eds., *Migrant birds in the Neotropics*. Washington, D.C.: Smithsonian Institution Press, pp. 273–283.

Pennycuick, C. J. 1975. Mechanics of flight. In D. S. Farner and J. R. King, eds., *Avian biology*, vol. 5. New York: Academic Press, pp. 1–75.

Perrins, C. M. 1979. *British tits*. London: Collins.

Petit, L. J., W. J. Fleming, K. E. Petit, and D. R. Petit. 1987. Nest-box use by Prothonotary Warblers *(Protonotaria citrea)* in riverine habitat. *Wilson Bulletin* 99:485–488.

Phillips, A. R. 1951. Complexities of migration: a review. *Wilson Bulletin* 63:129–136.

Pianka, E. R. 1966. Latitudinal gradients in species diversity: a review of concepts. *American Naturalist* 100:33–46.

Plattner, J., and E. Sutter. 1947. Ergebnisse der Meisen- und Kleiberberingung in der Schweiz. *Ornithologischer Beobachter* 44:1–35.

Popp, J. W., R. W. Ficken, and J. A. Reinartz. 1985. Short-term temporal avoidance of interspecific acoustic interference among forest birds. *Auk* 102:744–748.

Post, W. 1978. Social and foraging behavior of warblers wintering in Puerto Rican coastal scrub. *Wilson Bulletin* 90:197–214.

Powell, G. V. N. 1979. Structure and dynamics of interspecific flocks in a Neotropical mid-elevation forest. *Auk* 96:375–390.

——— 1980. Migrant participation in Neotropic mixed species flocks. In A. Keast and E. S. Morton, eds., *Migrant birds in the Neotropics*. Washington, D.C.: Smithsonian Institution Press, pp. 477–483.

——— 1985. Sociobiology and adaptive significance of interspecific foraging flocks in the Neotropics. *Ornithological Monographs* 36:713–732.

Power, D. M. 1972. Warbler ecology: diversity, similarity, and seasonal differences in habitat segregation. *Ecology* 52:434–443.

Probst, J. R. 1985. Summer records and management implications of Kirtland's Warbler in Michigan's Upper Peninsula. *Jack-Pine Warbler* 63:9–16.

——— 1986. A review of factors limiting the Kirtland's Warbler on its breeding grounds. *American Midland Naturalist* 116:87–100.

Probst, J. R., and J. P. Hayes. 1987. Pairing success of Kirtland's Warblers in marginal vs. suitable habitat. *Auk* 104:234–241.

Proctor-Gray, E., and R. T. Holmes. 1981. Adaptive significance of delayed attainment of plumage in male American Redstarts: tests of two hypotheses. *Evolution* 35:742–751.

Pulich, W. M. 1976. *The Golden-cheeked Warbler*. Austin, Tex.: Texas Parks and Wildlife Department.

Pumphrey, R. J. 1961. Sensory organs: vision. In A. J. Marshall, ed., *Biology and comparative physiology of birds,* vol. 2. New York: Academic Press, pp. 55–68.

Pyke, G. 1984. Optimal foraging theory: a critical review. *Annual Review of Ecology and Systematics* 15:523–575.

Pyke, G. H., H. R. Pulliam, and E. L. Charnov. 1977. Optimal foraging: a selective review of theory and tests. *Quarterly Review of Biology* 52:137–154.

Rabenold, K. N. 1978. Foraging strategies, diversity, and seasonality in bird communities of Appalachian spruce-fir forests. *Ecological Monographs* 48:397–424.

——— 1980. The Black-throated Green Warbler in Panama: geographic and seasonal comparison of foraging. In A. Keast and E. S. Morton, eds., *Migrant birds in the Neotropics*. Washington, D.C.: Smithsonian Institution Press, pp. 297–307.

Radabaugh, B. E. 1972. Polygamy in the Kirtland's Warbler. *Jack-Pine Warbler* 50:48–52.

——— 1974. Kirtland's Warbler and its Bahama wintering grounds. *Wilson Bulletin* 86:374–383.

Ralph, C. J. 1978. The disorientation and possible fate of young passerine coastal migrants. *Bird-banding* 49:237–247.

—————— 1981. Age ratios and their possible use in determining autumn routes of passerine migrants. *Wilson Bulletin* 93:164–188.

Ramos, M. A., and D. W. Warner. 1980. Analysis of North American subspecies of migrant birds wintering in Los Tuxtlas, Southern Veracruz, Mexico. In A. Keast and E. S. Morton, eds., *Migrant birds in the Neotropics.* Washington, D.C.: Smithsonian Institution Press, pp. 173–180.

Rappole, J. H. 1983. Analysis of plumage variation in the Canada Warbler. *Journal of Field Ornithology* 54:152–159.

Rappole, J. H., E. S. Morton, T. E. Lovejoy, III, and J. L. Ruos. 1983. *Nearctic avian migrants in the Neotropics.* Washington, D.C.: U.S. Fish and Wildlife Service.

Rappole, J. H., and D. W. Warner. 1976. Relationships between behavior, physiology and weather in avian transients at migration stopover sites. *Oecologia* 26:193–212.

—————— 1980. Ecological aspects of migrant bird behavior in Veracruz, Mexico. In A. Keast and E. S. Morton, eds., *Migrant birds in the Neotropics.* Washington, D.C.: Smithsonian Institution Press, pp. 353–393.

Rappole, J. H., D. W. Warner, and M. Ramos. 1977. Territoriality and population structure in a small passerine community. *American Midland Naturalist* 97:110–119.

Ratcliffe, D. A. 1970. Changes attributable to pesticides in egg breakage frequency and eggshell thickness in some British birds. *Journal of Applied Ecology* 7:67–115.

Rausch, R. L. 1983. The biology of avian parasites: helminths. In D. S. Farner, J. R. King, and K. C. Parkes, eds., *Avian biology,* vol. 7. New York: Academic Press, pp. 367–442.

Raveling, D. G., and D. W. Warner. 1978. Geographic variation of the Yellow Warblers killed at a TV tower. *Auk* 95:73–79.

Remington, C. L. 1968. Suture zones of hybrid interaction between recently joined biotas. *Evolutionary Biology* 3:321–428.

Remsen, J. V., Jr. 1984. High incidence of "leapfrog" pattern of geographic variation in Andean birds: implications for the speciation process. *Science* 224:171–173.

—————— 1986. Was Bachman's Warbler a bamboo specialist? *Auk* 103:216–219.

Remsen, J. V., Jr. and T. A. Parker III. 1984. Arboreal dead-leaf-searching birds of the Neotropics. *Condor* 86:36–41.

Rhijn, J. G. van. 1973. Behavioural dimorphism in male Ruffs, *Philomachus pugnax* (L.). *Behaviour* 47:153–229.

Rice, J. 1978. Ecological relationships of two interspecifically territorial vireos. *Ecology* 59:526–538.

Richards, D. G. 1981. Estimation of distance of singing conspecifics by the Carolina Wren. *Auk* 98:127–133.

Richards, D. G., and R. H. Wiley. 1980. Reverberations and amplitude fluctuations in the propagation of sound in a forest: implications for animal communication. *American Naturalist* 115:381–399.

Richardson, W. J. 1974. Spring migration over Puerto Rico and the western Atlantic, a radar study. *Ibis* 116:172–193.

———— 1978. Reorientation of nocturnal landbird migrants over the Atlantic Ocean near Nova Scotia in autumn. *Auk* 95:717–732.

Ricklefs, R.E. 1969. An analysis of nesting mortality in birds. *Smithsonian Contributions to Zoology* 9:1–48.

Ricklefs, R. E., and G. W. Cox. 1972. Taxon cycles in the West Indian avifauna. *American Naturalist* 106:195–219.

Ricklefs, R. E., and J. Travis. 1980. A morphological approach to the study of avian community organization. *Auk* 97:321–338.

Ridgway, R. 1902. The birds of North and Middle America. *Bulletin of the United States National Museum* 50(2):1–834.

Riper, C. van, III, S. G. van Riper, M. L. Goff, and M. Laird. 1986. The epizootiology and ecological significance of malaria in Hawaiian land birds. *Ecological Monographs* 56:327–344.

Ritchie, J. C. 1976. The late-Quaternary vegetational history of the western interior of Canada. *Canadian Journal of Botany* 54:1793–1818.

Ritchie, J. C., and C. V. Haynes. 1987. Holocene vegetation zonation in the eastern Sahara. *Nature* 330:645–647.

Robbins, C. S. 1980a. Effects of forest fragmentation on breeding bird populations in the Piedmont of the mid-Atlantic region. *Atlantic Naturalist* 33:31–36.

———— 1980b. Predictions of future Nearctic landbird vagrants to Europe. *British Birds* 73:448–457.

Robbins, C. S., D. Bystrak, and P. H. Geissler. 1986. The breeding bird survey: its first fifteen years, 1965–1979. *U.S. Fish and Wildlife Service Resource Publication* 157:1–196.

Roberts, J. O. L. 1971. Survival among some North American wood warblers. *Bird-banding* 42:165–184.

Robertson, R. J., and R. F. Norman. 1976. Behavioral defenses to brood parasitism by potential hosts of the Brown-headed Cowbird. *Condor* 78:166–173.

———— 1977. The function and evolution of aggressive host behavior towards the Brown-headed Cowbird *(Molothrus ater)*. *Canadian Journal of Zoology* 55:508–518.

Robinson, S. K. 1981. Ecological relations and social interactions of Philadelphia and Red-eyed vireos. *Condor* 83:16–26.

Robinson, S. K., and R. T. Holmes. 1982. Foraging behavior of forest birds: the relationships among search tactics, diet, and habitat structure. *Ecology* 63:1918–1931.

———— 1984. Effects of plant species and foliage structure on the foraging behavior of forest birds. *Auk* 101:672–684.

Rogers, D. T., Jr., D. L. Hicks, E. W. Wischusen, and J. R. Parrish. 1982. Repeats, returns and estimated flight ranges of some North American migrants in Guatemala. *Journal of Field Ornithology* 53:133–138.

Rogers, D. T., Jr., and E. P. Odum. 1966. A study of autumnal postmigrant weights and vernal fattening of North American migrants in the tropics. *Wilson Bulletin* 78:415–433.

Rohwer, S. 1975. The social significance of avian winter plumage variability. *Evolution* 29:593–610.

Rohwer, S., S. D. Fretwell, and D. M. Niles. 1980. Delayed maturation in passerine plumages and the deceptive acquisition of resources. *American Naturalist* 115:400–437.

Root, R. B. 1967. The niche exploitation pattern of the Blue-gray Gnatcatcher. *Ecological Monographs* 37:317–350.

Rosenzweig, M. L. 1973. Habitat selection experiments with a pair of coexisting heteromyid rodent species. *Ecology* 54:111–117.

Rosenzweig, M. L., and J. Winakur. 1969. Population ecology of desert rodent communities: habitats and environmental complexity. *Ecology* 50:558–572.

Rothstein, S. I. 1975a. Evolutionary rates and host defenses against avian brood parasitism. *American Naturalist* 109:161–176.

––––––– 1975b. An experimental and teleonomic investigation of avian brood parasitism. *Condor* 77:250–271.

Rothstein, S. I., J. Verner, and E. Stevens. 1984. Radio-tracking confirms a unique diurnal pattern of spatial occurrence in the parasitic Brown-headed Cowbird. *Ecology* 65:77–88.

Royama, T. 1966. A re-interpretation of courtship feeding. *Bird Study* 13:116–129.

––––––– 1970. Factors governing the hunting behaviour and selection of food by the Great Tit (*Parus major* L.). *Journal of Animal Ecology* 39:619–668.

––––––– 1984. Population dynamics of the Spruce Budworm *Choristoneura fumiferana*. *Ecological Monographs* 54:429–462.

Rue, D. J. 1987. Early agricultural and early postclassic Maya occupation in western Honduras. *Nature* 326:285–286.

Ruiter, C. J. S. 1941. Waarnemingen omtrent de levenswijze van de Gekraagde Roodstaart, *Phoenicurus ph. phoenicurus* (L.). *Ardea* 30:175–214.

Russell, E. W. B. 1983. Indian-set fires in the forests of the northeastern United States. *Ecology* 64:78–88.

Russell, S. M. 1980. Distribution and abundance of North American migrants in lowlands of Northern Columbia. In A. Keast and E. S. Morton, eds., *Migrant birds in the Neotropics*. Washington, D.C.: Smithsonian Institution Press, pp. 249–252.

Ryder, J. P. 1970. A possible factor in the evolution of clutch size in Ross' Goose. *Wilson Bulletin* 82:5–13.

Ryel, L. A. 1979. On the population dynamics of Kirtland's Warbler. *Jack-Pine Warbler* 57:76–83.

––––––– 1981. Population change in the Kirtland's Warbler. *Jack-Pine Warbler* 59:77–90.

––––––– 1984. Situation report, Kirtland's Warbler, 1984. *Michigan Department of Natural Resources, Wildlife Division Report* 2983:1–10.

Sabo, S. R. 1980. Niche and habitat relations in subalpine bird communities of the White Mountains of New Hampshire. *Ecological Monographs* 50:241–259.

Sabo, S. R., and R. T. Holmes. 1983. Foraging niches and the structure of forest bird communities in contrasting montane habitats. *Condor* 85:121–138.

Salomonsen, F. 1955. The evolutionary significance of bird migration. *Det Kongelige Danske Videnskabernes Selskab Biologiske Meddelelser* 22:1–62.

Sanders, W. T. 1971a. Cultural ecology and settlement patterns of the Gulf Coast. In R. Wauchope, G. F. Ekholm, and I. Bernal, eds., *Handbook of Middle American Indians*, vol. 11. Austin: University of Texas Press, pp. 543–557.

———— 1971b. Settlement patterns in Central Mexico. In R. Wauchope, G. F. Ekholm, and I. Bernal, eds., *Handbook of Middle American Indians*, vol. 10. Austin: University of Texas press, pp. 3–44.

Schoener, T. W. 1965. The evolution of bill size differences among sympatric congeneric species of birds. *Evolution* 19:189–213.

———— 1967. The ecological significance of sexual dimorphism to size in the lizard *Anolis conspersus*. *Science* 155:474–477.

———— 1974. Resource partitioning in ecological communities. *Science* 185:27–39.

———— 1983. Field experiments on interspecific competition. *American Naturalist* 122:240–285.

Scholander, S. I. 1955. Land birds over the western North Atlantic. *Auk* 72:225–239.

Schroeder, D. J., and R. H. Wiley. 1983. Communication with shared song themes in Tufted Titmice. *Auk* 100:414–424.

Schwartz, P. 1964. The Northern Waterthrush in Venezuela. *Living Bird* 3:169–184.

———— 1980. Some considerations on migratory birds. In A. Keast and E. S. Morton, eds., *Migrant birds in the Neotropics*. Washington, D.C.: Smithsonian Institution Press, pp. 31–34.

Sealy, S. G. 1979. Extralimital nesting of Bay-breasted Warblers: response to Forest Tent Caterpillars? *Auk* 96:600–603.

Searcy, W. A., P. Marler, and S. S. Peters. 1981. Species song discrimination in adult female Song and Swamp sparrows. *Animal Behaviour* 29:997–1003.

Selander, R. K. 1966. Sexual dimorphism and differntial niche utilization in birds. *Condor* 68:113–151.

Sheppard, D. H., P. H. Klopfer, and H. Oelke. 1968. Habitat selection: differences in stereotypy between insular and continental birds. *Wilson Bulletin* 80:452–457.

Sherry, T. W. 1979. Competitive interactions and adaptive strategies of American Redstarts and Least Flycatchers in a northern hardwoods forest. *Auk* 96:265–283.

Sherry, T. W., and R. T. Holmes. 1985. Dispersion patterns and habitat responses of birds in northern hardwoods forests. In M. L. Cody, ed., *Habitat selection in birds*. New York: Academic Press, pp. 283–309.

———— 1988. Habitat selection by breeding American Redstarts in response to a dominant competitor, the Least Flycatcher. *Auk* 105:350–364.

Short, L. L. 1963. Hybridization in the wood warblers *Vermivora pinus* and *V. chrysoptera*. *Proceedings of the International Ornithological Congress* 13:147–160.

———— 1969. Isolating mechanisms in the Blue-winged–Golden-winged warbler complex. *Evolution* 23:355–356.

Short, L. L., and A. R. Phillips. 1966. More hybrid hummingbirds from the United States. *Auk* 83:253–265.

Short, L. L., Jr. and C. S. Robbins. 1967. An intergeneric hybrid wood warbler (*Seiurus* × *Dendroica*). *Auk* 84:534–543.

Shuler, J. 1977. Bachman's Warbler habitat. *Chat* 41(2):19–23.

Shy, E., and E. S. Morton. 1986. The role of distance, familiarity, and time of day in Carolina Wrens' responses to conspecific songs. *Behavioral Ecology and Sociobiology* 19:393–400.

Sibley, C. G. 1968. The relationships of the "Wren-Thrush," *Zeledonia coronata* Ridgway. *Postilla* 125:1–12.

———— 1970. A comparative study of the egg-white proteins of passerine birds. *Peabody Museum of Natural History, Bulletin* 32:1–131.

Sibley, C. G., and J. E. Ahlquist. 1982. The relationships of the Yellow-breasted Chat *(Icteria virens)* and the alleged slowdown in the rate of macromolecular evolution in birds. *Postilla* 187:1–19.

———— 1983. The phylogeny and classification of birds, based on the data of DNA-DNA hybridization. *Current Ornithology* 1:245–292.

Sick, H. 1971. Blackpoll Warbler on winter quarters in Rio de Janeiro, Brazil. *Wilson Bulletin* 83:198–200.

Simms, E. 1985. *British Warblers*. London: Collins.

Skutch, A. F. 1949. Do tropical birds rear as many young as they can nourish? *Ibis* 91:430–458.

———— 1954. Life histories of Central American birds. *Pacific Coast Avifauna* 31:1–448.

———— 1957. The resident wood warblers of Central America. In L. Griscom and A. Sprunt, Jr., eds., *The warblers of America*. New York: Devin-Adair, pp. 275–285.

———— 1966. A breeding bird census and nesting success in Central America. *Ibis* 108:1–16.

———— 1967a. Adaptive limitations of the reproductive rate of birds. *Ibis* 109:579–599.

———— 1967b. Life histories of Central American highland birds. *Publications of the Nuttall Ornithological Club* 7:1–213.

———— 1976. *Parent birds and their young*. Austin: University of Texas Press.

———— 1985. Clutch size, nesting success, and predation on nests of Neotropical birds, reviewed. *Ornithological Monographs* 36:575–594.

Slagsvold, T. 1976. Arrival of birds from spring migration in relation to vegetational development. *Norwegian Journal of Zoology* 24:161–173.

Sloan, N. F., and G. A. Simmons. 1973. Foraging behavior of the Chipping Sparrow in response to high populations of the Jack Pine Budworm. *American Midland Naturalist* 90:210–215.

Slud, P. 1960. The birds of Finca "La Selva," Costa Rica: a tropical wet forest locality. *Bulletin of the American Museum of Natural History* 121:49–148.

———— 1964. The birds of Costa Rica. *Bulletin of the American Museum of Natural History* 128:1–430.

Smith, J. M. N. 1978. Division of labour by Song Sparrows feeding fledged young. *Canadian Journal of Zoology* 56:187–191.

Smith, J. M. N., and H. P. A. Sweatman. 1974. Food-searching behavior of titmice in patchy environments. *Ecology* 55:1216–1232.

Smith, K. G. 1977. Distribution of summer birds along a forest moisture gradient in the Ozark watershed. *Ecology* 58:810–819.

Smith, S. M. 1978. The "underworld" in a territorial sparrow: adaptive strategy for floaters. *American Naturalist* 112:571–582.

—— 1980. Henpecked males: the general pattern in monogamy? *Journal of Field Ornithology* 51:55–64.

—— 1984. Flock switching in chickadees: why be a winter floater? *American Naturalist* 123:81–98.

Smith, T. M., and H. H. Shugart. 1987. Territory size variation in the Ovenbird: the role of habitat structure. *Ecology* 68:695–704.

Smythe, N. 1970. On the existence of "pursuit invitation" signals in mammals. *American Naturalist* 104:491–494.

Snow, D. W. 1949. Jämförände studier över våra mesarters näringssökande. *Vår Fågelvarld* 8:159–169.

—— 1966. Annual cycle of the Yellow Warbler in the Galápagos. *Bird-banding* 37:44–49.

Stanwood, C. J. 1910. The Black-throated Green Warbler. *Auk* 27:289–294.

Steadman, D. 1985. Holocene vertebrate fossils from Isla Floreana, Galápagos. *Smithsonian Contributions to Zoology* 413:1–104.

Stein, R. C. 1958. The behavioral, ecological and morphological characteristics of two populations of the Alder Flycatcher, *Empidonax traillii* (Audubon). *Bulletin of the New York State Museum and Science Service* 371:1–63.

—— 1962. A comparative study of songs recorded from five closely related warblers. *Living Bird* 1:61–74.

Stenger, J. 1958. Food habits and available food of Ovenbirds in relation to territory size. *Auk* 75:335–346.

Stevenson, H. M. 1972. The recent history of Bachman's Warbler. *Wilson Bulletin* 84:344–347.

Stewart, P. A., and H. A. Conner. 1980. Fixation of wintering Palm Warblers to a specific site. *Journal of Field Ornithology* 51:365–367.

Stewart, R. E. 1953. A life history study of the Yellow-throat. *Wilson Bulletin* 54:99–115.

Stewart, R. E., and J. W. Aldrich. 1951. Removal and repopulation of breeding birds in a spruce-fir forest community. *Auk* 68:471–482.

Stewart, R. M. 1973. Breeding behavior and life history of the Wilson's Warbler. *Wilson Bulletin* 85:21–30.

Stewart, R. M., R. P. Henderson, and K. Darling. 1977. Breeding ecology of the Wilson's Warbler in the High Sierra Nevada, California. *Living Bird* 16:83–102.

Stiles, F. G. 1980. Evolutionary implications of habitat relations between permanent and winter resident landbirds in Costa Rica. In A. Keast and E. S. Morton, eds., *Migrant birds in the Neotropics.* Washington, D.C.: Smithsonian Institution Press, pp. 421–435.

Stoll, N. R., R. P. Dollfus, J. Forest, N. D. Riley, C. W. Sabrowsky, C. W. Wright, and R. V. Melville. 1961. *International code of zoological nomenclature adopted by the XV International Congress of Zoology.* London: International Trust for Zoological Nomenclature.

Studd, M. V., and R. J. Roberston. 1985a. Evidence for reliable badges of status in territorial Yellow Warblers *(Dendroica petechia)*. *Animal Behaviour* 33:1102–1113.

———— 1985b. Life span, competition, and delayed plumage maturation in male passerines: the breeding threshold hypothesis. *American Naturalist* 126:101–115.

———— 1985c. Sexual selection and variation in reproductive strategy in male Yellow Warblers *(Dendroica petechia)*. *Behavioral Ecology and Sociobiology* 17:101–109.

Sutton, G. M. 1951. A new race of Yellow-throated Warbler from northwestern Florida. *Auk* 68:27–29.

Svärdson, G. 1949. Competition and habitat selection in birds. *Oikos* 1:157–174.

Szaro, R. C., and R. P. Balda. 1979. Bird community dynamics in a Ponderosa Pine forest. *Studies in Avian Biology* 3:1–66.

Terborgh, J. W. 1974. Preservation of natural diversity: the problem of extinction prone species. *BioScience* 24:715–722.

———— 1980. The conservation status of Neotropical migrants: present and future. In A. Keast and E. S. Morton, eds., *Migrant birds in the Neotropics*. Washington, D.C.: Smithsonian Institution Press, pp. 21–30.

Terborgh, J. W., and J. R. Faaborg. 1980. Factors affecting the distribution and abundance of North American migrants in the Eastern Caribbean Region. In A. Keast and E. S. Morton, eds., *Migrant birds in the Neotropics*. Washington, D.C.: Smithsonian Institution Press, pp. 145–155.

Terborgh, J. W., and J. S. Weske. 1969. Colonization of secondary habitats by Peruvian birds. *Ecology* 50:765–782.

Terrill, S. B., and R. D. Ohmart. 1984. Facultative extension of fall migration by Yellow-rumped Warblers *(Dendroica coronata)*. *Auk* 101:427–438.

Thompson, C. F., and B. M. Gottfried. 1976. How do cowbirds find and select nests to parasitize? *Wilson Bulletin* 88:673–675.

Tinbergen, L. 1960. The natural control of insects in pinewoods. I. Factors influencing the intensity of predation by songbirds. *Archives Néerlandaises de Zoologie* 13:265–336.

Titterington, R. W., H. S. Crawford, and B. N. Burgason. 1979. Songbird responses to commercial clear-cutting in Maine spruce-fir forests. *Journal of Wildlife Management* 43:602–609.

Todd, W. E. C. 1963. *Birds of the Labrador Peninsula*. Toronto, Ontario: University of Toronto Press.

Tramer, E. J. 1974. On latitudinal gradients in avian diversity. *Condor* 76:123–130.

Tramer, E. J., and T. R. Kemp. 1980. Foraging ecology of migrant and resident warblers and vireos in the highlands of Costa Rica. In A. Keast and E. S. Morton, eds., *Migrant birds in the Neotropics*. Washington, D.C.: Smithsonian Institution Press, pp. 285–296.

———— 1982. Notes on migrants wintering at Monteverde, Costa Rica. *Wilson Bulletin* 94:350–354.

Tramer, E. J., and F. E. Tramer. 1977. Feeding responses of fall migrants to prolonged inclement weather. *Wilson Bulletin* 89:166–167.

Trautman, M. B. 1979. Experiences and thoughts relative to the Kirtland's Warbler. *Jack-Pine Warbler* 57:135–140.

Trivers, R. L. 1974. Parent-offspring conflict. *American Zoologist* 14:249–264.

Ulfstrand, S. 1976. Feeding niches of some passerine birds in a South Swedish coniferous plantation in winter and summer. *Ornis Scandinavica* 7:21–27.

———— 1977. Foraging niche dynamics and overlap in a guild of passerine birds in a South Swedish coniferous woodland. *Oecologia* 27:23–45.

Ulrich, D., and S. Ulrich. 1981. Observations of the 1980 Indiana Sutton's Warbler. *Atlantic Naturalist* 34:12–13.

Varley, G. C., G. R. Gradwell, and M. P. Hassell. 1973. *Insect population ecology*. Oxford, England: Blackwell Scientific Publications.

Verner, J. 1976. Complex song repertoire of male Long-billed Marsh Wrens in eastern Washington. *Living Bird* 14:263–300.

Verner, J., and M. F. Willson. 1966. The influence of habitats on mating systems of North American passerine birds. *Ecology* 47:143–147.

———— 1969. Mating systems, sexual dimorphism, and the role of the male North American passerine birds in the nesting cycle. *Ornithological Monographs* 9:1–76.

Voous, K. H. 1985. Classification. In B. Campbell and E. Lack, eds., *A dictionary of birds*. Calton, England: Poyser, pp. 89–90.

Wahlenberg, W. G. 1946. *Longleaf Pine: its use, ecology, regeneration, protection, growth, and managment*. Washington, D.C.: Charles Lathrop Pack Forestry Foundation, and Forest Service, U.S. Department of Agriculture.

Waide, R. B. 1980. Resource partitioning between migrant and resident birds: the use of irregular resources. In A. Keast and E. S. Morton, eds., *Migrant birds in the Neotropics*. Washington, D.C.: Smithsonian Institution Press, pp. 337–352.

———— 1981. Interactions between resident and migrant birds in Campeche, Mexico. *Tropical Ecology* 22:134–154.

Walkinshaw, L. H. 1938. Nesting studies of the Prothonotary Warbler. *Bird-banding* 9:32–46.

———— 1941. The Prothonotary Warbler, a comparison of nesting conditions in Tennessee and Michigan. *Wilson Bulletin* 53:3–21.

———— 1959. The Prairie Warbler in Michigan. *Jack-Pine Warbler* 37:54–63.

———— 1983. Kirtland's Warbler. *Bulletin of the Cranbrook Institute of Science* 58:1–207.

Walkinshaw, L. H., and W. Faust. 1975. 1974 Kirtland's Warbler nesting success in northern Crawford County, Michigan. *Jack-Pine Warbler* 53:54–58.

Walter, H. 1979. *Eleonora's Falcon*. Chicago: University of Chicago Press.

Ward, P., and A. Zahavi. 1973. The importance of certain assemblages of birds as "information-centres" for food-finding. *Ibis* 115:517–534.

Warner, R. E. 1967. The role of introduced diseases in the extinction of the endemic Hawaiian avifauna. *Condor* 70:101–120.

Wasserman, F. E. 1977. Intraspecific acoustical interference in the White-throated Sparrow *(Zonotrichia albicollis)*. *Animal Behaviour* 25:949–952.

Webster, J. D. 1958. Systematic notes on the Olive Warbler. *Auk* 75:469–473.

Weeden, J. S., and J. B. Falls. 1959. Differential responses of male Ovenbirds to recorded songs of neighboring and more distant individuals. *Auk* 76:343–351.

Werner, E. E. 1977. Species packing and niche complementarity in three sunfishes. *American Naturalist* 111:553–578.

Western, D., and J. Ssemakula. 1982. Life history patterns in birds and mammals and their evolutionary interpretation. *Oecologia* 54:281–290.

Weydemeyer, W. 1973. The spring migration pattern at Fortine, Montana. *Condor* 75:400–413.

Wheelwright, N. T. 1986. Review of "Neotropical Ornithology." *Condor* 88:404–405.

Whitcomb, R. F., C. S. Robbins, J. F. Lynch, B. L. Whitcomb, M. K. Klimkiewitz, and D. Bystrak. 1981. Effects of forest fragmentation on avifauna of the eastern deciduous forest. In R. L. Burgess and D. M. Sharpe, eds., *Forest island dynamics in man-dominated landscapes*. New York: Springer-Verlag, pp. 125–205.

Whitcomb, B. L., R. F. Whitcomb, and D. Bystrak. 1977. Island biogeography and "habitat islands" of eastern forest. III. Long-term turnover and effects of selective logging on the avifauna of forest fragments. *American Birds* 31:17–23.

White, T. C. R. 1969. An index to measure weather-induced stress of trees associated with outbreaks of psyllids in Australia. *Ecology* 50:905–909.

Whitmore, R. C. 1975. Habitat ordination of passerine birds of the Virgin River Valley, southwestern Utah. *Wilson Bulletin* 87:65–74.

——— 1977. Habitat partitioning in a community of passerine birds. *Wilson Bulletin* 89:253–265.

Whittle, C. L. 1922. A Myrtle Warbler invasion. *Auk* 39:23–31.

Wiens, J. A. 1975. Avian communities, energetics, and functions in coniferous forest habitats. In D. R. Smith, ed., *Proceedings of the symposium on management of forest and range habitats for nongame birds*. General Technical Report WO-1. Washington, D.C.: Forest Service, U.S. Department of Agriculture, pp. 226–265.

——— 1977. On competition and variable environments. *American Scientist* 65:590–597.

Wilcove, D. S. 1985. Nest predation in forest tracts and the decline of migratory songbirds. *Ecology* 66:1211–1214.

——— 1988. Changes in the avifauna of the Great Smoky Mountains: 1947–1983. *Wilson Bulletin* 100:256–271.

Wiley, R. H., and D. G. Richards. 1978. Physical constraints on acoustic communication in the atmosphere: implications for the evolution of animal vocalizations. *Behavioral Ecology and Sociobiology* 3:69–94.

——— 1982. Adaptations for acoustic communication in birds: sound transmission and signal detection. In D. E. Kroodsma and E. H. Miller, eds., *Acoustic communication in birds*, vol. 1. New York: Academic Press, pp. 131–181.

Williams, E. E. 1969. The ecology of colonization as seen in the zoogeography of anoline lizards on small islands. *Quarterly Review of Biology* 44:345–389.

Williams, G. C. 1966. Natural selection, the costs of reproduction and a refinement of Lack's Principle. *American Naturalist* 100:687–690.

Williams, T. C. 1985. Autumnal bird migration over the Windward Caribbean islands. *Auk* 102:163–167.

Williams, T. C., and J. M. Williams. 1978. An oceanic mass migration of land birds. *Scientific American* 239(4):166–176.

Williamson, P. 1971. Feeding ecology of the Red-eyed Vireo *(Vireo olivaceus)* and associated foliage-gleaning birds. *Ecological Monographs* 41:129–152.

Willis, E. O. 1966. The role of migrant birds at swarms of army ants. *Living Bird* 5:187–231.

––––––– 1973. Survival rates for visited and unvisited nests of Bicolored Antbirds. *Auk* 90:263–267.

––––––– 1974. Populations and local extinctions of birds on Barro Colorado Island, Panama. *Ecological Monographs* 44:153–169.

––––––– 1980. Ecological roles of migratory and resident birds on Barro Colorado Island, Panama. In A. Keast and E. S. Morton, eds., *Migrant birds in the Neotropics*. Washington, D.C.: Smithsonian Institution Press, pp. 205–225.

Willis, E. O., and Y. Oniki. 1978. Birds and army ants. *Annual Review of Ecology and Systematics* 9:243–263.

Willson, M. F., and G. H. Orians. 1964. Interspecific territories of birds. *Ecology* 45:736–745.

Wilson, D. S. 1975. The adequacy of body size as a niche difference. *American Naturalist* 109:769–784.

Wilz, K. J. and V. Giampa. 1978. Habitat use by Yellow-rumped Warblers at the northern extremities of their winter range. *Wilson Bulletin* 90:566–574.

Winkler, H., and B. Leisler. 1985. Morphological aspects of habitat selection in birds. In M. L. Cody, ed., *Habitat selection in birds*. Orlando, Fla.: Academic Press, pp. 415–434.

Winstanley, D., R. Spencer, and K. Williamson. 1974. Where have all the Whitethroats gone? *Bird Study* 21:1–14.

Witkin, S. R. 1977. The importance of directional sound radiation in avian vocalization. *Condor* 79:490–493.

Wolda, H. 1978. Fluctuations in abundance of tropical insects. *American Naturalist* 112:1017–1045.

Wolf, L. L., F. R. Hainsworth, and F. B. Gill. 1975. Foraging efficiencies and time budgets in nectar-feeding birds. *Ecology* 56:117–128.

Wood, N. A., and E. H. Frothingham. 1905. Notes on the birds of the Au Sable Valley, Michigan. *Auk* 22:39–54.

Woodward, P. W. 1983. Behavioral ecology of fledgling Brown-headed Cowbirds and their hosts. *Condor* 85:151–163.

Woolfenden, G. E., and J. W. Fitzpatrick. 1984. The Florida Scrub Jay. *Monographs in Population Biology* 20:1–406.

World Resources Institute and the International Institute for Environment and Development. 1986. *World Resources 1986*. New York: Basic Books.

Wright, H. E., Jr. 1971. Late Quaternary vegetational history of North America. In K. K. Turekian, ed., *The late Cenozoic glacial ages*. New Haven, Conn.: Yale University Press, pp. 425–464.

Wright, S. J. 1979. Competition between insectivorous lizards and birds in central Panama. *American Zoologist* 19:1145–1156.

——— 1981. Extinction-mediated competition: the *Anolis* lizards and insectivorous birds of the West Indies. *American Naturalist* 117:181–192.

——— 1982. Character change, speciation, and the higher taxa. *Evolution* 36:427–443.

Wurster, D. H., C. F. Wurster, Jr., and W. N. Strickland. 1965. Bird mortality following DDT spray for Dutch Elm Disease. *Ecology* 46:488–499.

Yarbrough, C. G. 1970. Summer lipid levels of some subarctic birds. *Auk* 87:100–110.

——— 1971. The influence of distribution and ecology on the thermoregulation of small birds. *Comparative Biochemistry and Physiology* 39A:235–266.

Yurtsev, B. A. 1985. Beringia and its biota in the late Cenozoic: a synthesis. In V. L. Kontrimachivus, ed., *Beringia in the Cenozoic era*. Rotterdam: Balkema, pp. 261–275. [Translated from the Russian (1976).]

Zach, R., and J. B. Falls. 1975. Response of the Ovenbird (Aves: Parulidae) to an outbreak of Spruce Budworm. *Canadian Journal of Zoology* 53:1669–1672.

——— 1976a. Do Ovenbirds (Aves: Parulidae) hunt by expectation? *Canadian Journal of Zoology* 54:1894–1903.

——— 1976b. Foraging behavior, learning, and exploration by captive Ovenbirds (Aves: Parulidae). *Canadian Journal of Zoology* 54:1880–1893.

——— 1976c. Ovenbird (Aves: Parulidae) hunting behavior in a patchy environment: an experimental study. *Canadian Journal of Zoology* 54:1863–1879.

——— 1977. Influence of capturing a prey on subsequent search in the Ovenbird (Aves: Parulidae). *Canadian Journal of Zoology* 55:1958–1969.

——— 1978. Prey selection by captive Ovenbirds (Aves: Parulidae). *Journal of Animal Ecology* 47:929–943.

——— 1979. Foraging and territoriality of male Ovenbirds (Aves: Parulidae) in a heterogeneous habitat. *Journal of Animal Ecology* 48:33–52.

Zumeta, D. C., and R. T. Holmes, 1978. Habitat shift and roadside mortality of Scarlet Tanagers during a cold wet New England spring. *Wilson Bulletin* 90:575–586.

Index